T0134085

Advanced Geoinformation Science

Advanced Geoinformation Science

Edited by
Chaowei Yang • David Wong
Qianjun Miao • Ruixin Yang

CRC Press
Taylor & Francis Group
Boca Raton   London   New York

CRC Press is an imprint of the
Taylor & Francis Group, an **informa** business

CRC Press
Taylor & Francis Group
6000 Broken Sound Parkway NW, Suite 300
Boca Raton, FL 33487-2742

Printed in the United States of America on acid-free paper
10 9 8 7 6 5 4 3 2 1

International Standard Book Number: 978-1-4398-1060-6 (Hardback)

Library of Congress Cataloging-in-Publication Data

Advanced geoinformation science / editors, Chaowei Yang ... [et al.].
    p. cm.
  "A CRC title."
  Includes bibliographical references and index.
  ISBN 978-1-4398-1060-6 (alk. paper)
  1. Geographic information systems. 2. Geographic information
systems--Technological innovations. 3. Geospatial data. I. Yang, Chaowei. II. Title.

G70.212.A385 2010
910.285--dc22
                               2010018969

Visit the Taylor & Francis Web site at
http://www.taylorandfrancis.com

and the CRC Press Web site at
http://www.crcpress.com

Contents

Preface

Why This Book?

Two motivations lead to the compilation of this edited volume. At the turn of the century, a variety of global and regional issues became increasingly challenging to global citizens and could only be best tackled by collaborations across countries and organizations. These challenges have physical dimensions, such as tsunamis, hurricanes, and climate change, and human dimensions, such as economic crises, epidemics, and emergency responses. This edited book is an effort to reflect how certain technical aspects of geoinformation have been used and could be used to address such global issues. Geoinformation describes characteristics of the Earth's surface and spheres above the surface. Its scope covers the range from global, to regional, and local scales. Editors of this volume and chapter authors have been involved in various global initiatives and research problems, such as Global Earth Observation System of Systems (GEOSS), Digital Earth, Air Quality, Public Health, and Cloud Computing. The aim of this book is to provide an overview of geoinformation science based on our experiences.

Another motivation is related to an international training program on advanced geoinformation science that was first conducted in 2006 at the Center for Intelligent Spatial Computing (CISC), George Mason University (GMU). Heilongjiang Bureau of Surveying and Mapping (HLJBSM) in China and the International Association of Chinese Professionals in Geographic Information Science (CPGIS) provided organizational support to solicit both student participants and top-notch researchers and scientists as instructors. These instructors came from academia, industries, and governments to share their knowledge on the latest developments in geoinformation science.

Based on the challenges facing our planet Earth and our successful 4-year training program, we selected topics, commissioned some program instructors to write about the selected topics, and compiled this volume to benefit a broader audience, including (1) senior undergraduate and graduate students to extend their knowledge from information technology, geography, geology, atmospheric sciences, and other geosciences disciplines to advanced geoinformation science; (2) managers using geoinformation to deal with global, regional, and local issues; and (3) others who are interested in the physical aspects of global issues.

How Did We Edit This Book?

In 2008, the third year of the training program, CRC Press/Taylor & Francis (Irma Shagla) and the book editors agreed to collect lecture materials of the training program and to publish them as a book. During the past 2 years, we followed nine steps to ensure a well-integrated book for our audience: (1) solicit contributors from our guest speakers; (2) select relevant topics according to the book design; (3) assign editors to be responsible for different chapters; (4) collect the contributions and have the editors review each section; (5) review each chapter by authors of the same chapter for coordination and fine-tuning; (6) send out chapters for external review; (7) address the external reviews among responsible editors and section authors; (8) finalize the chapters by proofreading the language among authors, editors, and language editors; and (9) finalize with the Taylor & Francis editors.

What Is This Book About?

This book provides a systematic overview of geoinformation science in 10 chapters. Chapter 1 is an introduction, stating the problems and reasons behind the needs of geoinformation science. Chapter 2 introduces different methods for collecting spatial data as the initial feeds to geoinformation science. Chapter 3 introduces geoinformation computing platforms. Chapter 4 discusses data systems for data management, and Chapter 5 discusses data analysis and information extraction. Chapter 6 is dedicated to the geoinformation infrastructure. Chapter 7 is dedicated to knowledge capturing, formatting, and utilization. Chapters 8 and 9 focus on a variety of geoinformation applications, with a focus on the environment in Chapter 8 and other focuses in Chapter 9. Chapter 10 provides visions and education requirements for geoinformation science for the next 10 to 15 years.

Acknowledgments

We would like to thank the students and chapter authors–instructors for participating in the summer training programs. Without the practice, lessons learned, and improvements in the past several years, we could not have completed this book. Support from the leadership of the three organizations, HLJBSM, GMU, and CPGIS, is greatly appreciated, and their indispensible collaborative support allowed us to make the program an international success. Thanks go to all the authors, who worked tirelessly with the book editors, meeting all types of demands and requests, as well as finally completing all the required paperwork. Their collaborative attitude made the lives of the editors much easier.

Thanks also go to Taylor & Francis acquiring editor, Irma Shagla and production coordinator David Fausel, who helped to ensure that the manuscript was formatted according to standards adopted by Taylor & Francis. They provided insightful comments and were patient when working with us. Students at CISC/GMU also assisted in the process, including proofreading, collecting material, and formatting manuscripts according to the publisher's specifications. Students also contributed to many sections in the book with other authors.

Unche A. Saydahmat and Stephen C. McClure at GMU provided tremendous editorial support. Finally, we would like to thank our family members for their tolerance and for bearing with us as we stole valuable family time to finish the book.

Editors

Chaowei Yang is the founder and director of the Center for Intelligent Spatial Computing (CISC) at George Mason University (GMU). He is an expert on geoinformation science in (1) utilizing computing techniques to solve geospatial problems, (2) model and data interoperability for integrating geospatial information and simulations, (3) spatial data integration and scheduling in the parallel computing environment, and (4) spatial Web portals and GEOSS applications. He has served as principal investigator for over 10 geoinformation science projects funded by NASA, FGDC, UCAR or NSF, EPA, ESIP, and NPS. He also participated as co-I in several large scale geoinformation science projects of over $10M in the past seven years. He cofounded the Cyberinfrastructure Specialty Group in AAG. He served as the president of CPGIS (2005) and Advisory BOD members for many national and international organizations. Dr. Yang has organized many national and international conference sessions, workshops, and training programs, such as the Advanced Geoinformation Science program, which has brought more than 20 geoinformation professionals from China to be trained at CISC each summer since 2006. He has published over 40 peer-reviewed journal papers and guest edited (1) the Distributed Geographic Information Processing special issue for *International Journal of Geographic Information Science*, (2) the Cyberinfrastructure special issue for the journal of *Computers, Environment, and Urban Systems*, (3) the Distributed GIS special issue for the *Annals of Geographic Information Science*, and (4) the Geospatial Cyberinfrastructure and Polar Research for the journal *Computer & Geosciences.*

David Wong, professor in the Department of Geography and Geoinformation Science at George Mason University, has research and teaching interests in population geography and the broadly defined geoinformation science. Besides publishing more than 40 refereed journal papers, about half of which are on measuring population distributions and racial–ethnic segregation. Professor Wong also has coauthored two books on spatial statistics and GIS. Other research interests include spatial analytical techniques, visualization, and spatial epidemiology or public health. He has served as the guest editor on special issues of several journals, such as the *Journal of Geographical Systems and Urban Geography*. Currently, he is a member of the editorial boards of several geoinformation science–related journals, and he has served on proposal review panels for NSF, NIH, and NASA. Research funding support for Dr. Wong has come from FGDC, NIH, and HUD.

Qianjun Miao is a post-doctoral researcher at Tsinghua University and an expert on digital city, ecocity, and informatization of surveying and mapping. He serves as the deputy director of the Heilongjiang Bureau of Surveying and Mapping (HLJBSM). He was a young academic leader of the State Bureau of Surveying and Mapping of China. He served as vice director of the Heilongjiang Provincial Ecology Plan Compilation Group and executive editor for the *Journal of Geomatics and Spatial Information Technology*. He is the chief scientist for Development Strategy Research of Chinese Surveying and Mapping and the vice chair of the Expert Committee of the China Society of Image and Graphics. He serves on the review panel of Chinese 863 projects on observation and navigation and the international cooperation project of Ministry of Science and Technology (MOST) and the vice chair of the Urban Specialty Group of the China GIS Society. He is a member of the China Standardization Committee for Surveying and Mapping and the chair of the Science and Technology Committee of the Heilongjiang Bureau of Surveying and Mapping and the vice chair of the Heilongjiang Provincial Society for Geodesy Photogrammetry and Cartography. He has served as principal investigator for over 10 scientific research projects supported by the "863 Projects" of MOST, science data sharing of MOST, and key laboratory foundation of State Geomatics Engineering. He also served as the chief editor for several edited volumes of proceedings, such as *Spatial Information Infrastructure and Interoperability* and *Development Strategy Research on the China Surveying and Mapping Industry*. He also organized many national and international conferences, such as the International Symposium on Digital City Sustainable Development, the International High-level Forum on Outsourcing Services and the Geomatics Industry Development, and the International Symposium on Eco-city Construction.

Ruixin Yang is associate professor in the College of Science at George Mason University. Ruixin Yang received his BS in fluid mechanics from Tsinghua University (Beijing, China) and a PhD in aerospace engineering from University of Southern California (Los Angeles). Dr. Yang has worked in several fields including accretion flows onto black holes in astrophysics, turbulence and water waves in fluid dynamics, weather and dispersion model evaluations in numerical weather prediction, geoscience data information systems, Earth systems science, and scientific data mining. He also has conducted research in remote sensing applications, information technology for Earth science data, and led the system development of several data information systems for Earth science data dissemination. His current research interests are in geosciences, data analysis, scientific data mining, hurricane intensity changes, climate data analysis techniques, and data information systems.

Contributors

Peggy Agouris
Department of Geography and
  Geoinformation Science
George Mason University
Fairfax, Virginia

Joan L. Aron
U.S. Environmental Protection
  Agency
Office of Research and
  Development
Washington, DC

Myra Bambacus
Chief Information Office
NASA GSFC
Greenbelt, Maryland

Ying Cao
Center for Intelligent Spatial
  Computing
George Mason University
Fairfax, Virginia

Long S. Chiu
Department of Geography and
  Geoinformation Science
George Mason University
Fairfax, Virginia

and

Institute of Space and Earth
  Information Science
Chinese University of
  Hong Kong, Shatin, NT
Harbin, People's Republic
  of China

Arie Croitoru
Department of Earth and
  Atmospheric Sciences
University of Alberta
Edmonton, Alberta, Canada

Meixia Deng
Center for Spatial Information
  Science and Systems
George Mason University
Fairfax, Virginia

Liping Di
Center for Spatial Information
  Science and Systems
George Mason University
Fairfax, Virginia

Kennith Foote
Department of Geography
University of Colorado at
  Boulder
Boulder, Colorado

Diansheng Guo
Department of Geography
University of South Carolina
Columbia, South Carolina

Xianjun Hao
Environmental Science and
  Technology Center
George Mason University
Fairfax, Virginia

Qunying Huang
Center for Intelligent Spatial
  Computing
George Mason University
Fairfax, Virginia

Laura E. Jackson
U.S. Environmental Protection
  Agency
Office of Research and
  Development
Washington, DC

Steven Kempler
NASA Goddard Earth
 Sciences (GES) Data and
 Information Services
 Center (DISC)
NASA/Goddard Space Flight
 Center
Greenbelt, Maryland

Jing Li
Center for Intelligent Spatial
 Computing
George Mason University
Fairfax, Virginia

Wenwen Li
Center for Intelligent Spatial
 Computing
George Mason University
Fairfax, Virginia

Zhenlong Li
Center for Intelligent Spatial
 Computing
George Mason University
Fairfax, Virginia

Zhong Liu
Center for Earth Observing and
 Space Research
George Mason University
Fairfax, Virginia

Sheryl Luzzadder-Beach
Department of Geography and
 Geoinformation Science
George Mason University
Fairfax, Virginia

Carol B. Meyer
Foundation for Earth Science
Raleigh, North Carolina

Qianjun Miao
Heilongjiang Bureau of Surveying
 and Mapping
Harbin, People's Republic
 of China

Lenard Milich
Arid Lands Resource Sciences
University of Arizona
Tucson, Arizona

Doug Nebert
Federal Geographic Data
 Committee
United States Geological Survey
Reston, Virginia

Qisheng Pan
Department of Urban Planning
 and Environmental Policy
Texas Southern University
Houston, Texas

John J. Qu
Department of Geography and
 Geoinformation Science
George Mason University
Fairfax, Virginia

Hampapuram K. Ramapriyan
NASA Goddard Space Flight
 Center
Greenbelt, Maryland

Rob Raskin
NASA JPL
Pasadena, California

Ronald G. Resmini
Department of Geography and
 Geoinformation Science
George Mason University
Fairfax, Virginia

Hualan Rui
ADNET Systems
Rockville, Maryland

Xuan Shi
Center for Geographic Information
    Systems
Georgia Institute of Technology
Atlanta, Georgia

Anthony Stefanidis
Department of Geography and
    Geoinformation Science
George Mason University
Fairfax, Virginia

Donglian Sun
Department of Geography and
    Geoinformation Science
George Mason University
Fairfax, Virginia

Min Sun
Center for Intelligent Spatial
    Computing
George Mason University
Fairfax, Virginia

Greg Susanke
U.S. Environmental Protection
    Agency
Office of Research and
    Development
Washington, DC

Jiqiang Tan
State Key Laboratory of
    Information Engineering in
    Surveying, Mapping and
    Remote Sensing
Wuhan University
Wuhan, People's Republic of China

Yanli Tang
Heilongjiang Bureau of Surveying
Harbin, People's Republic of China

William Teng
Wyle Information Systems
McLean, Virginia

Fahui Wang
Department of Geography and
    Anthropology
Louisiana State University
Baton Rouge, Louisiana

Edward Washburn
U.S. Environmental Protection
    Agency
Office of Research and
    Development
Washington, DC

John White
EPA OAQPS
Research Triangle Park,
    North Carolina

David Wong
Department of Geography
    and Geoinformation
    Science
George Mason University
Fairfax, Virginia

Huayi Wu
Center for Intelligent Spatial
    Computing
George Mason University
Fairfax, Virginia

and

State Key Laboratory of
    Information Engineering
    in Surveying, Mapping
    and Remote Sensing
Wuhan University
Wuhan, People's Republic
    of China

Jibo Xie
Center for Intelligent Spatial
    Computing
George Mason University
Fairfax, Virginia

Chaowei Yang
Center for Intelligent Spatial
    Computing
George Mason University
Fairfax, Virginia

Haihong Yang
Communications Resource, Inc.
McLean, Virginia

Ruixin Yang
Department of Geography and
    Geoinformation Science
George Mason University
Fairfax, Virginia

Stephen Young
U.S. Environmental Protection
    Agency
Office of Environmental Information
Washington, DC

Genong Yu
Center for Spatial Information
    Science and Systems (CSISS)
George Mason University
Fairfax, Virginia

Hongwei Yu
Heilongjiang Bureau of Surveying
    and Mapping
Harbin, People's Republic of
    China

Chuanrong Zhang
Department of Geography and
    Center for Environmental
    Sciences and Engineering
University of Connecticut
Storrs, Connecticut

Ming Zhang
School of Architecture
The University of Texas at Austin
Austin, Texas

Peisheng Zhao
Center for Spatial Information
    Science and Systems
George Mason University
Fairfax, Virginia

Naijun Zhou
Department of Geography
University of Maryland
College Park, Maryland

1

Advanced Geoinformation Science: An Overview

Chaowei Yang, David Wong, and Ruixin Yang

CONTENTS

1.1 Introduction

If one regards the Canadian Geographic Information System (CGIS) as the first operational Geographic Information System (GIS), then GIS has more than five decades of history. At about the same time, GIS pioneers at Harvard Graphics Lab developed other general mapping systems (e.g., SYMAP) as well as systems with a strong flavor of modern GIS (e.g., ODYSSEY). Although the names of these systems can only be found in historical archives or history of GIS literature today, current GIS users may be convinced that today's GIS technology is way more advanced than those "dinosaur" systems, and the past is no better than the present—there are no "good old days" in GIS. On the other hand, the importance of GIS has grown beyond the point that GIS is just one of many types of information systems. Many disciplines support the development of GIS: geography, geodesy, computer science, psychology, statistics, and so on. Goodchild (1992) first advocates the interdisciplinary subject of Geographic Information Science (GIScience), arguing that we need a concerted effort from supporting disciplines to advance GIS. Therefore, GIS has evolved from a system to an interdisciplinary science (GIScience).

Over the decades, GIS operating environments have migrated from mainframes (CGIS on IBM) to minicomputers (Arc-Info on VMS/VAX), workstations (e.g., Intergraph Microstation), personal computers (e.g., AtlasGIS, MapInfo, ArcView 3.X), and now hand-held and location-aware devices (e.g., ArcPad on Palm), the Internet (e.g., MapQuest, Google Earth), and virtual computing networks. Since GISs are now operating in new computing environments with new devices (hardware), new technologies are needed to support the functioning of various spatial computing processes and procedures in such settings. The rapid evolution of information science and technologies (1) not only enhance our ability to collect, archive, and process spatial data more efficiently, but also to generate and utilize geographic information and knowledge more efficiently than what we can do in the past; (2) enable us to integrate remote sensing, global position systems, and GIS with other computing technologies in various domain application areas; and (3) provide us the capabilities to tackle complex scientific and application problems that were intractable in the past. The term *Geoinformation Science* captures this trend and consists of components such as Earth observations, network computing platforms, information infrastructure, distributed data collection and distribution systems, and the applications of these components to support the studies of various geospatial phenomena.

However, certain concepts, functionalities, and methodological issues in Geoinformation Science and supporting sciences in today's world are not different than what they were several decades ago. For instance, we are still trying to improve our sensor and remote sensing technologies so that we can more precisely capture different aspects of the Earth systems, by increasing spatial and spectral resolutions. We are leveraging advancements in information technology to develop geosensor webs so that data in multiple formats and from dispersed locations and sources can be coalesced to provide a more comprehensive picture and, possibly, more dimensions of the environment or events. We have developed many new tools to analyze and visualize spatial data, but we are still struggling to extract meaningful information from data. In addition, we are contending with the new challenge of dealing with massive amount of data. We are still trying to make GIS and its related tools more user-friendly, with more intuitive procedures and more understandable interfaces, partly through research in cognitive psychology, ontology, semantics, and knowledge. However, the current standards of user-friendliness or usability are not set by GIS professionals but by the general public.

On the one hand, we continue our effort to remove impediments in using GIS and related technologies. On the other, we have to deal with new challenges, most of which, by nature, are reactions or responses to the advancements in information technology. The development of the Internet and the advancements in network technology push the computing environment from a monolithic to a distributed GIS architecture (Yang et al., 2005). Components of GIS and Geoinformation processing procedures can reside over dispersed locations, and many new developments and challenges in

Geoinformation Science are directed toward overcoming impediments that are brought along by the distributed computing infrastructure. Further, the demand for real- or near-real-time analysis has increased tremendously due to various social needs. Shortening the cycle and processing time between various spatial data processes and procedures has been one of the emphases in various research activities.

Various GIS and spatial technologies are being used by academics and practitioners across many fields and have intruded into the daily life of most people in one way or another. Some years ago in an ESRI conference, Jack Dangermond, the president of ESRI, expressed the goal to put GIS on every desktop. We have not yet seen GIS installed on every computer (although many people have downloaded and installed Google Earth), but we are definitely in the era in which almost every computer can access some aspects of GIS and Geoinformation technology through the Internet.

1.2 Spatial Data Collection

Capturing or inputting spatial data is the first step of Geoinformation processing. Technological advancement has substantially changed the way we capture spatial data. However, traditional methods of capturing or deriving spatial data are still widely applicable in many parts of the world, especially in countries in which spatial infrastructure has just begun to develop. Before we discuss new ways of capturing spatial data, it will be necessary to briefly discuss the definition of spatial data and the traditional methods of capturing spatial data.

1.2.1 What Are Spatial Data?

Many definitions of spatial data exist. A very general one simply defines spatial data as "data that contain spatial information." A slightly different, but somewhat restrictive, definition is, "data that are georeferenced." However, the latter definition often constrains our understanding of spatial data to that which includes latitude–longitude readings, coordinate information based on some geodetic systems or grid layouts, or addresses such that the data can easily pinpoint locations on a map. This understanding of spatial data is a "retrospective" approach in which spatial data are first captured in maps, and the definition of spatial data is derived from maps. This narrower definition also reflects the traditional data capturing methods of converting map documents into digital spatial data or capturing the location information of features on the ground to be stored in GIS.

On the contrary, spatial ontology literature reveals the fact that spatial data are present almost everywhere in our daily life (e.g., Smith, 2002; Crane, 2004; Scharl and Tochtermann, 2007). Spatial information can be extracted from conversations and writings. Sometimes, Geoinformation extracted from these sources is highly personal and is of no value to others, but at other times, some of this information may have public relevance. A wide variety of spatial data is found in between the two extreme types of information sources of traditional map documents and personal conversations. Some are well structured, such as images from satellites in formats similar to a typical map. Others may be unstructured, or even buried, such as place descriptions in historical texts and in pictures of geographical features. These types of information can be spatialized if we identify their geographical locations or link spatial tags with these information items. Therefore, data originally without intrinsic spatial referencing information can become part of spatial data if we relate the features or events to space. Thus, the definition of spatial data becomes quite fluid, partly due to our better understanding of the nature of Geoinformation and partly due to our increasing capability to extract spatial data from various sources.

1.2.2 Traditional Methods for Spatial Data Capture and Input

Early methods for spatial data capture and input are tied to the historical development of GIS. Actually, GIS was used to convert map documents into digital formats. The goal was to transfer information from hardcopy documents (most likely maps) to digital formats, which can be in the formats of images or files recording geographical features. As a result, scanning was a major process to capture data as images. Complementing scanning techniques is digitizing, in which features or objects on hardcopy maps are traced by a cursor on a digitizing table or tablet to record the coordinates representing those features. Documents can also be scanned into images for rapid digitizing. With the advancements of image processing and analysis algorithms, features on images may be traced and extracted semiautomatically or fully automatically.

Although many early GIS projects commenced with the process of transferring spatial data from hardcopy documents to digital formats, the roles of remote sensing in capturing and providing spatial data have always been indispensable. Remotely sensed data are probably the most logical sources of spatial data for organizations to start building their spatial data infrastructure. Remote sensing provides data sources from which we can update geographical features, such as changes in coast lines or expansion of urban areas. In addition, these data also offer very rich descriptions of geographical features due to the high sensitivity of optical and nonoptical sensors in capturing the electromagnetic and other characteristics of the features. Recent trends in remote sensing technology have expanded our ability in several dimensions: increasing spatial resolution (submeter level images); increasing spectral resolution (hyperspectral imaging); augmenting sensor capabilities to observe all aspects of the Earth systems (atmosphere, hydrosphere, lithosphere, and biosphere/anthrosphere); and increasing coverage frequency for monitoring. As a result, remote sensing data are not only available in massive volumes but also being created at a rate faster than scientists can analyze them.

Parallel to the advancements of remote sensing technology is the increasing popularity of global positioning system (GPS). GPS technology complements remote sensing for surveying and mapping tasks and makes field collection of spatial data easier. A GPS unit with appropriate software not only replaces, to some degree, the traditional survey equipment but can also capture geographical features more efficiently; for instance, by carrying the unit along features or adopting the GPS van technology to collect data. In fact, additional information can also be tagged onto the features to expand our domain in collecting spatial data (Coetsee et al., 1994).

The traditional methods of gathering spatial data are most likely used to gather or update spatial data that fall under one of the seven themes of the U.S. National Spatial Data Infrastructure (NSDI) framework data: geodetic control, orthoimagery, elevation, transportation, hydrography, governmental units, and cadastral information (http://www.fgdc.gov/framework/). These framework data are the core or the most commonly used base data and, therefore, they serve as the foundations of many mapping projects.

1.2.3 Nonconventional Methods to Capture Spatial Data

Since Geoinformation is no longer confined to traditional maps, the challenges become identifying where (sources) the data can be captured and determining how (methods) spatial data can be extracted. Data extracted from nonconventional sources are not likely part of the framework data, and these data sources can be highly heterogeneous in formats and widely dispersed physically. Captured data are probably more suitable for specific domain applications rather than for general use across various applications. Although there are many possible and creative methods of capturing data from nontraditional data sources, this section discusses just a few common examples, such as from texts, GPS, and sensor technologies.

As mentioned earlier, semantics of language has trapped rich Geoinformation. Texts and documents provide many spatial descriptions. Names of feature objects are often mentioned with no explicit spatial information. However, if they are cross-referenced to a geographic name gazetteer, then the spatial context can be extracted. Apart from the descriptions of features or spatial phenomena, texts and documents may include spatial relationship descriptions, which are often imprecise (Mark and Frank, 1991). Utilizing texts and documents as spatial data is challenging. This approach of extracting Geoinformation has been used in the intelligence and surveillance communities. In addition to texts and documents, data submitted by users online to systems [volunteered geographic information (VGI)], such as the OpensStreetMap (http://www.openstreetmap.org/), used by the public offer new sources of spatial data (Goodchild, 2007; Sui, 2008). To a certain extent, this general approach of extracting Geoinformation combines data mining techniques for texts and geocoding methods.

The advancement in GPS technology not only enhances our capability to capture location information more precisely and expeditiously to create framework data and maps, but it also has the "side-effect" of tracking individuals' locations. Different forms of GPS are used in our daily life. Obviously, navigation systems heavily rely on GPS. However, GPS is embedded in many personal devices we use daily, including various types of mobile phones. These location-aware devices have the capability to capture our locations at different times. Data generated from these devices contain very rich Geoinformation that can shed light on human spatial and social behavior. Unfortunately, much of this type of data is proprietary in nature and is highly sensitive, as personal and confidential information is involved.

Parallel to the advancement of GPS technology is the development of sensor technology. In the context of Earth observing, we have more powerful sensors enhancing our ability to monitor the Earth surface. However, sensor technology has also proliferated in daily life. Surveillance cameras and sensors have been widely used, particularly in the context of public safety and public work. Sensors have the capability to capture both still and motion images. With fixed sensor installations, spatial tags can be linked to images or videos captured by these sensors to spatialize these data. Further, the capability of these sensors in capturing spatial data can be expanded along two dimensions, spatial and temporal, to provide a more comprehensive coverage of the Earth surface. An obvious limitation of the stationary sensor is the fixed and limited ground coverage constrained by the viewing angle. However, with multiple installations in an "optimal" pattern, large areas can be monitored by a network of sensors. When these sensors are linked with live networks, the sensor network serves as a powerful tool to capture not just real-time data but also spatiotemporal data depicting movements of objects and events (Nittel et al., 2008). The information can be corroborated with other information sources in the intelligence network. This topic will be addressed in greater detail in Chapter 2.

Regardless of the nature and sources of these nontraditional spatial data, these data are usually not well structured and Geoinformation is often not explicit. These data have to be processed and analyzed by special tools or procedures in order to derive Geoinformation. Spatial data mining techniques are needed to extract Geoinformation from these unstructured data sources.

1.3 Spatial Data Management and Integration

Petabytes (PB) of optical and nonoptical remote sensing data are collected by numerous Earth observation satellites on a daily basis. For example, the

Earth Observing System (EOS) satellite Terra of the National Aeronautics and Space Administration (NASA) collects integrated measurements of Earth's processes, including land surface, biosphere, solid Earth, atmosphere, and oceans. The Earth Observing System Data and Information System (EOSDIS), one of the largest scientific data systems, gathers spatial data from more than 30 satellites, such as the NASA's Terra, Aqua, and Aura. As part of the EOSDIS, NASA Goddard Earth Sciences Data and Information Services Center (GES DISC) archived and distributed remote sensing data from the EOS at the rate of 1 TB per day (Leptoukh, 2005). Another example is that the National Oceanic and Atmospheric Administration (NOAA) is expected to archive 100 PB of data in 2015 and 162 PB of data in 2020 (NOAA, 2004). Together with other spatial data collecting efforts, such as data gathered from airborne platforms, spatial data are collected at an exploding rate.

Accessing spatial data is the first step in using Geoinformation and the process involves subsetting, processing, and visualization: (1) Subsetting refers to the process of extracting datasets based on data requests. (2) Processing refers to any intentional manipulation of datasets for specific applications. The data processing module within a spatial data management and integration system should provide the functionality to generate meta-data for different data resources, including both spatial attributes (e.g., location, resolution, and projection) and nonspatial attributes (e.g., providers, time, and data formats), so that the data sets can be searched and accessed. Further, the data processing module should provide basic spatial processing capabilities, such as format transformation and reprojection, to facilitate the access and sharing of data sets. (3) Visualization is the graphical representation of requested results, and it masks the complexity of spatial data. A typical visualization process involves the conversion of numerical or non-numerical data into visual representations such as maps, imagery, graphics, 3D models, and other visual products.

For example, an Earth scientist who is interested in El Nino may want to examine the Sea Surface Temperature (SST) over the past 30 years. The SST data are hosted by NOAA, recording the variations of SST from the 1970s with different spatial and temporal granularity (TRITON Data Delivery, http://www.pmel.noaa.gov/tao/data_deliv/deliv.html). Thus, NOAA provides a public accessible portal for users to customize data subsetting requests. Both spatial and temporal criteria can be used to select and extract data. After subsetting, data are processed, including format conversion, on the server side. Meanwhile, a general sketch is generated to verify the data requests. Finally, the visualization module in the portal will generate a 2D image of the requested data for end users. Chapter 4 will discuss some data processing issues.

This process also involves several issues in data management and integration: (1) Spatial data are multidimensional (e.g., 3D, 4D, and 4D+) and are collected from multiple sources at multiple scales. They can be in large quantities and may be geographically dispersed . How to subset and identify the interested data from the distributed data volumes is a daunting task for efficient Geoinformation management. It is relatively easy to select the requested datasets from one database, but subsetting datasets from distributed databases could be overwhelming. (2) Data processing allows access to heterogeneous spatial datasets in customized formats and contents. (3) Visualization in the distributed environment should also be adjusted. Online visualization partially addresses the adjustment by separating the process to a sequence of modules. Chapter 6 discusses how these challenging issues could be resolved.

Multiple data sources that describe the same phenomenon may exist. In such case, data fusion becomes important to generate a more comprehensive understanding of the scientific phenomenon for end users. Data fusion is recognized as a valuable tool for integrating spatial data. The most common type of data fusion is remote sensing image fusion. The actual fusion process can take place at different stages of data processing, from raw data preprocessing to final data product generation. Fusion methods include linear superposition, nonlinear methods, optimization approaches, artificial neural networks, image pyramids, wavelet transform, and generic multiresolution fusion schemes (Shum et al., 2007). Normally, data fusion is time consuming and involves complex algorithms and methodologies.

Another practical issue for data management and integration is data heterogeneity. Spatial datasets are collected by different vendors for different applications in different communities. The structural and schematic heterogeneity makes the exchange of data almost impossible without a data conversion tool (Bernard et al., 2003). Data conversion is a very time-consuming process and has obvious limitations to satisfy the requirements of real-time applications, such as traffic routing and emergency response. Meanwhile, spatial datasets with different schema often require vendor-specific invoking or accessing methodologies. In addition, metadata describing the spatial datasets or services are not standardized, and manual operations are often required to assist the access of data instead of relying completely on a machine-enabled process. Therefore, communicating and exchanging resources across different Geoinformation systems is very difficult. The heterogeneous system environment creates another problem—data duplication. Early efforts in spatial data interoperability and integration relied on data translation and offline synchronization. However, this strategy still cannot correct massive data redundancy and updating problems. To tackle these issues, the current approach is to ensure interoperability through standardizations of spatial data contents, spatial data formats, processing module interfaces, model interfaces, and semantics. Relevant topics are discussed in Chapter 3 (for computing), Chapter 6 (for infrastructure), and Chapter 7 (for semantics and knowledge).

1.4 Spatial Data Mining

The volume of spatial data is increasing at accelerated speeds through business activities, scientific observations, and numerical simulations. Traditional data analysis methods that are developed for sparse data would encounter limitations for efficiently handling large volume of data. Data mining, especially mining for Geoinformation, will become a prominent technology for extracting information and knowledge from the ever-increasing volume of spatial data.

In traditional data analysis, researchers form hypotheses based on experience and data in hand and use data to verify the proposed hypotheses. In contrast, data mining techniques are used as exploratory methods to generate hypotheses, and statistical analysis is used to confirm hypotheses, if necessary (e.g., Hand et al., 2001). One important category of data mining is to identify patterns hidden in large volume of diverse data. As a discipline, data mining was originally developed mainly for handling transaction data in business; the methodologies are well developed and documented (e.g., Han and Kamber, 2006).

Spatial data are different from transaction data because they are associated with both location and time. Spatiotemporal data pose new challenges for data mining, and much effort has been paid to tackle the difficulties. For example, Ertöz et al. (2002, 2003) discuss the density-based clustering problem and developed the shared nearest-neighbor (SNN) algorithm to cluster spatial data. This method was applied to large-scale climatic data for the discovery of climate indices (Steinbach et al., 2003). These indices can be considered as patterns extracted from the vast data, but the process does not require the knowledge of the underlying physics connecting the patterns. In other words, pattern extractions from the climatic data are data mining for Geoinformation. In parallel, scientists can use data mining tools to identify the hidden physical relationships among a large number of geophysical measurements in geoscience research and exploratory analysis. One successful example of mining geosciences data is using the association rules to discover the combinations of conditions that affected the rapid intensification of Atlantic hurricanes (Yang, R. et al., 2007, 2008).

The difficulties of spatial data mining or Geographic Knowledge Discovery (GKD) can be traced back to the presence of spatial dependency and heterogeneity of spatial data. Locations and other associated properties, such as size, shape, orientation, and proximity, may all contribute to the complexity of the mining rules and patterns. By taking Geoinformation into account, traditional data mining methodologies are extended to spatial classification, spatial clustering, spatial associations, and so on. Recent developments in sensor Web technology have the capabilities to monitor the dynamic environment and movements of objects. Databases created using such technology capture both spatial and temporal information. Mining this type of data has an additional challenge, as temporal relationship needs to be taken into account in addition to spatial relationship. Recent developments in spatial data mining and GKD are summarized in the edited volume by Miller and Han (2009). Chapters 4 and 5 in this book address selected issues related to mining data for Geoinformation.

1.5 Geoinformation Applications

Geoinformation is valuable in different disciplines, such as geography and demography. By providing spatial data and techniques, Geoinformation Science helps implement theories and concepts in these domains to solve problems. Applications of Geoinformation Science may be broadly divided into two categories: physical science and social science. Among the Geoinformation applications in physical science, applications in environmental science probably have the largest share. Developing a thorough understanding of the environment is critical to preserve our planet. For example, atmospheric scientists calibrate simulation models with field survey data to examine atmospheric dynamics. Remote sensing data are frequently used to detect various signs leading to the possible occurrence of natural hazards. Vegetation indices are used to monitor changes in forests, grasslands, and wetlands, or land cover in general. The potential influences of such changes on animals under the threat of extinction could also be evaluated. Spatial datasets are frequently used in numerous environmental applications. The scope of these studies can be at local, regional, and global scales, and spatial data in all these scales are critical for the studies.

Geoinformation is also of great importance in various social science disciplines. Since most human activities are spatial in nature, having accurate, timely, and reliable Geoinformation is highly desirable. Geoinformation applications in social science may cover issues in population, public health, crime study, and market planning. The roles played by spatial data in social science research have been ever increasing. Boundary data of various political, administrative, and statistical units support mapping of social phenomena. Demographic data are not only used frequently but also provide the foundation of investigation of almost all issues. Spatial analysis performed in social science may be simple or complex, whereas the accuracy and quality of spatial data adds another dimension of complexity to social science inquiries. Another challenge is to facilitate social scientists to fully exploit the potentials of Geoinformation Science in enhancing social science research, part of the mission of the Center for Spatially Integrated Social Science (CSISS).

No doubt that Geoinformation facilitates studies in many domains. Generic issues in using Geoinformation in these applications include (1) defining and analyzing complex applications, such as global Earth observation societal benefit areas (geoss.org), so that heterogeneous spatial data from different sources can be integrated to enhance our understanding of complex systems such as the Earth system; (2) consolidating the understanding of different Geoinformation databases at different scales, such as reach consensus on global warming among different countries; (3) facilitating domains to share knowledge for scientists to collaboratively solve problems across disciplines; and (4) harnessing the social implications and social networking complexities within a global environment to solve multiregion problems, such as water crisis. These issues add to the challenges of Geoinformation Science and will be discussed and demonstrated through examples in Chapters 8 and 9.

1.6 Challenges for Geoinformation Science

Geoinformation Science faces many grand challenges. Many spatial analytical processes and models create computational challenges in at least three aspects: (1) The PB of data are collected by and archived at different data processing centers. Using these data in a seamless manner will require significant effort in data integration and processing. (2) Computationally intensive analysis methods and models, which extend across a broad spectrum of spatial and temporal scales, have gained wide acceptance (Armstrong et al., 2005). These models and methods require the support of advanced computing methods to solve complex scientific problems (Yang and Raskin, 2009). (3) The development of information science, technology, and the Internet-popularized Web-based applications, which require fast access and response for concurrent users (Yang et al., 2005). All these challenges are involved when spatial data, Geoinformation services, and domain knowledge are all brought together to build an infrastructure. A variety of computing paradigms have been utilized to tackle these challenges. Some of these paradigms are detailed in Chapter 3, including Geoinformation Distributed Computing (GDC) (Yang, C. et al., 2008), High Performance Computing (HPC) (Clematis et al., 2003), Grid Computing (GC) (Armstrong et al., 2005), and, most recently, Cloud Computing (Buyya et al., 2008; Armbrust et al., 2009).

Geoinformation Science is a domain that involves sophisticated and complex ontology, semantics, and knowledge. The Geoinformation knowledge obtained is of great importance to help us automate the Geoinformation processing and application–building process. However, it is difficult to (1) precisely capture the ontologies that are required for the automation process; (2) present and organize Geoinformation knowledge in computable models to support Geoinformation processing automation; and (3) develop sophisticated middleware that can leverage the Geoinformation knowledge in a fast and learning fashion. For example, a spatial database manager may prefer to classify datasets according to well-organized semantics within an organization to facilitate an automated data discovery task. The database may work perfect locally and can provide sufficient information for the user community (Bishr et al., 1999). However, when the database is integrated with other domains' databases, spatial semantics captured in the first database may not be easily integrated with other databases. Many aspects of the databases may not be understood by all the parties involved. Integration may have to start from the ground by laying out a schema for the semantics according to knowledge gained through previous efforts in integrating and utilizing similar databases. This example reflects the challenge of *Semantic Interoperability* for the Geoinformation community (Sheth, 1999). Chapter 7 discusses this challenge.

Data analysis and fusion is another challenge for effectively integrating geospatial data from different sources to augment Geoinformation. The increasing number of sensors, geospatial data at different levels of spatial and temporal resolution, at different levels of accuracy and measurement scale add to the complexity of this problem (Goodchild, 2009). Fusing datasets of dramatically different properties or characteristics, such as data at extreme levels of resolution or accuracy, may not be possible (or in fact, should not be considered). However, data integration and analysis, in general, can enhance information content, and this line of research has attracted a great deal of interest and attention, as reflected in Chapter 5.

Spatial data are valuable resources that can be shared within the research and practitioner's communities. In addition to data, other resources supporting Geoinformation Science and applications, such as spatial Web services and spatial models, also can be shared among users. To facilitate the sharing and integration of these resources, the development of Geoinformation Infrastructure is warranted. Such infrastructure leverages the existing standards to deal with issues associated with heterogeneous data and systems. Geoinformation infrastructure of different scopes have been proposed and developed over the past decade or so. An example is the Spatial Data Infrastructure (SDI), which grows rapidly and covers different geographical scales, from Global SDI (GSDI) to National SDI (NSDI) and Local SDI (LSDI) (Yang et al., 2006). Spatial Web Portals (SWP), SDI, Web 2.0, Mashup, and other technologies bring along opportunities to build Geoinformation infrastructure and advantages for addressing the challenges, as detailed in Chapter 6 (Yang, C. et al., 2007).

This edited volume consists of a diverse set of topics ranging from data collection, observations, and sensing to advanced computing platforms, data system and data analysis, infrastructure, knowledge management, applications, and the future of Geoinformation Science. Through this volume, we intend to provide overviews of state-of-the-art of selected topics or issues for scientists, researchers, developers, educators, and students. Chapter 2 focuses on acquiring spatial data from different spheres of the Earth systems and the use of geosensor Web technology. Chapter 3 reviews various types of computing technology or platforms that have been used or are of great potential for Geoinformation processing. Since data serve as the foundation of all information systems, we devote Chapter 4 to discuss spatial data systems. Developing more effective methods to extract information from spatial data has been a long-term effort. Chapter 5 reviews selected issues on statistical and visualization approaches and the development of extracting information from images and videos. Chapter 6 reviews some pertinent developments and issues on Geoinformation infrastructure. Chapter 7 discusses some of the issues on ontology and Geoinformation knowledge management. Chapters 8 and 9 focus on applications of Geoinformation technology. Selected applications in various fields, from physical science to social science, are briefly demonstrated. Chapter 10 looks forward to some future directions and developments, capturing the future visions of Geoinformation Science.

References

Armbrust, M., A. Fox, R. Griffith et al. 2009. *Above the Clouds: A Berkeley View of Cloud Computing*. University of California, Berkeley, Berkeley, CA.

Armstrong, M. P., M. Cowles, and S. Wang. 2005. Using a computational grid for geographic information analysis. *Professional Geographer*, 57(3): 365–375.

Bernard, L., U. Einspanier, S. Haubrock et al. 2003. Ontologies for intelligent search and semantic translation in spatial data infrastructures. *Photogrammetrie—Fernerkundung—GeoInformation (PFG)*, 6: 451–462.

Bishr, Y., H. Pundt, W. Kuhn, and M. Radwan. 1999. Probing the concept of information communities—a first step toward semantic interoperability. In *Interoperating Geographic Information Systems*, eds. M. F. Goodchild, M. Egenhofer, R. Fegeas, C. Kottman, pp. 55–71. Kluwer Academic, Boston, MA.

Buyya, R., C. S. Yeo, and S.Venugopal. 2008. Market-oriented cloud computing: Vision, hype, and reality for delivering IT services as computing utilities. *CoRR*, vol. abs/0808.3558.

Clematis, A., M. Mineter, and R. Marciano. 2003. High performance computing with geographical data. *Parallel Computing*, 29(10): 1275–1279.

Coetsee, J., A. Brown, and J. Bossler. 1994. GIS data collection using the GPSVan supported by a GPS/Inertial mapping systems. In *Proceedings of GPS-94*. The Institution of Navigation (ION), Salt Lake City, UT.

Crane, G. 2004. Georeferencing in Historical Collections. *D-Lib Magazine*, 10(5). http://www.dlib.org/dlib/may04/crane/05crane.html

Ertöz, L., M. Steinbach, and V. Kumar. 2002. A new shared nearest neighbor clustering algorithm and its applications. In *Workshop on Clustering High Dimensional Data and Its Applications*, SIAM Data Mining 2002, Arlington, VA.

Ertöz, L., M. Steinbach, and V. Kumar. 2003. Finding clusters of different sizes, shapes, and densities in noisy, high dimensional data. In *Proceedings of Second SIAM International Conference on Data Mining*, San Francisco, CA.

Goodchild, M. F. 1992. Geographical information science. *International Journal of Geographical Information Systems*, 6(1): 31–45.

Goodchild, M. F. 2007. Citizens as seniors: the world of volunteered geography. *GeoJournal*, 69(4): 211–221.

Goodchild, M. F. 2009. Geographic information systems and science: Today and tomorrow. *Procedia Earth and Planetary Science*, 1(1): 1037–1043.

Han, J. and M. Kamber. 2006. *Data Mining: Concepts and Techniques*. Morgan Kaufmann Publishers, San Francisco, CA.

Hand, D., H. Mannila, and P. Smyth. 2001. *Principles of Data Mining*. M.I.T Press, Cambridge, MA.

Leptoukh, G. 2005. NASA Remote Sensing data in earth sciences: Processing, archiving, distribution, applications at the GES DISC. In *Proceedings of the 31st International Symposium on Remote Sensing of Environment*, Saint Petersburg, June 20–34. http://www.isprs.org/publications/related/ISRSE/html/papers/217.pdf

Mark, D. M. and A. U. Frank. 1991. *Cognitive and Linguistic Aspects of Geographic Space*. Kluwer Academic Publishers, Dordrecht.

Miller, H. J. and J. Han. 2009. *Geographic Data Mining and Knowledge Discovery*. CRC Press, Boca Raton, FL.

Nittel, S., A. Labrinidis, and A. Stefanidis. 2008. *Advances in GeoSensor Networks*. Lecture Notes in Computer Science, Springer, Berlin/Heidelberg, Vol. 4540.

NOAA. 2004. *Strategic Direction for NOAA's Integrated Global Environmental Observation and Data Management System*. National Oceanic and Atmospheric Administration, Washington, DC.

Scharl, A. and K. Tochtermann. 2007. *The Geospatial Web: How Geobrowsers, Social Software and the Web 2.0 are Shaping the Network Society*. Springer, Boston.

Sheth, A. P. 1999. Changing focus on interoperability in information systems: From system, syntax, structure to semantics. In *Interoperating Geographic Information Systems*, eds. M.F. Goodchild, M. Egenhofer, R. Fegeas, C. Kottman, pp. 5–30. Kluwer Academic, Boston, MA.

Shum, H. Y., S. C. Chan, and S. B. Kang. 2007. *Image-based Rendering*. Springer.

Smith, D. A. 2002. Detecting and browsing events in unstructured text. In *Proceedings of the 25th Annual ACM SIGIR Conference*. ACM, Tampere, Finland.

Steinbach, M., P. N. Tan, V. Kumar, C. Potter, and S. Klooster. 2003. Discovery of climate indices using clustering. In *Proceedings of the 9th ACM SIGKDD International Conference on Knowledge Discovery and Data Mining*. Washington, DC.

Sui, D. Z. 2008. The wikification of GIS and its consequences: Or Angelina Jolie's new tattoo and the future of GIS. *Computers, Environment and Urban Systems*, 32(1): 1–5.

Yang, C., Y. Cao, M. Kafatos, J. Evans, and M. Bambacus. 2006. Interoperable web portal supporting spatial data infrastructure. *Geographic Information Sciences*, 12(1): 38–43.

Yang, C., J. Evans, M. Cole, N. Alameh, S. Marley, and M. Bambacus. 2007. The emerging concepts and applications of the spatial web portal. *PE&RS*, 73(6): 691–698.

Yang, C., W. Li, J. Xie, and B. Zhou. 2008. Distributed geospatial information processing: Sharing earth science information to support digital earth, *International Journal of Digital Earth*, 1(3): 259–278.

Yang, C. and R. Raskin. 2009. Introduction to distributed geographic information processing research. *International Journal of Geographic Information Science*, 23(5): 553–560.

Yang, C., D. W. S. Wong, R. Yang, M. Kafatos, and Q. Li. 2005. Performance improving techniques in web-based GIS. *International Journal of Geographical Information Science*, 19(3): 319–342.

Yang, R., D. Sun, and J. Tang. 2008. A "sufficient" condition combination for rapid intensifications of tropical cyclones. *Geophysical Research Letters*, 35, L20802, doi:10.1029/2008GL035222.

Yang, R., J. Tang, and M. Kafatos. 2007. Improved associated conditions in rapid intensifications of tropical cyclones. *Geophysical Research Letters*, 34, L20807, doi:10.1029/2007GL031241.

2

Earth Observations

Long S. Chiu, Xianjun Hao, Ronald G. Resmini, Donglian Sun,
Anthony Stefanidis, John J. Qu, and Ruixin Yang

CONTENTS

Earth observation (EO) data, especially the remotely sensed data, are one of the important sources of spatial data. These data may come from remote sensing instruments deployed on orbital satellites, by aircraft, or just in ground-based stations. This chapter reviews remote sensing from satellites, the air, and on ground. The chapter covers remote sensing for the atmosphere, land, and ocean by starting with EO history and remote sensing applications for land cover or land use. Next, atmospheric remote sensing is discussed, emphasizing the meteorological satellite missions, NASA EOS missions with atmospheric components, including precipitation retrievals in detail. Then, land remote sensing through hyperspectral remote sensing, mainly from airborne platforms, is presented. After the discussion on land and atmospheric remote sensing, the oceanic remote sensing retrieval algorithms for two selected typical parameters, SST and ocean color (chlorophyll-a), are examined. Finally, the ground-based sensors are reviewed with emphases on GeoSensor Networks (GSN) and SensorWeb.

2.1 EO History and Global Land Cover/Land Use*

2.1.1 Introduction

Space-based EOs have become critical for human beings in various applications, including environmental monitoring, natural hazards detection, weather forecasting, land management, and coastal management, because they monitor where we live. Our survival can sometimes depend on them. This section provides a brief introduction to space-based EO and discusses applications of space-based measurements for global land cover or land use change.

2.1.2 History of Space-Based Earth Observation

The history of space-based EO can be traced back to the nineteenth century (Joseph 2005; Nicholas 2009), when photographs from elevated platforms were proposed to monitor the Earth's surface. During World War II, EO technology advanced remarkably with the capability to expand from the aerial visible

---

* This section is contributed by Xianjun Hao and John J. Qu.

photographs to infrared (IR) and radar systems. However, before the launch of the first man-made satellite in 1957, EO platforms had limited capabilities in temporal and spatial coverage. The first satellite-based picture of the Earth was taken by the United States' Explorer 6 in 1959. The launch of the Television Infrared Observation Satellite (TIROS-1) in 1960 marked a new milestone for EO. Space-based multispectral channel photography began with the Apollo 9 mission in 1968, with four cameras mounted to take pictures in the green, red, photo IR, and color IR bands. Major milestones for systematic EO were the launch of the Earth Resources Technology (ERTS-1) satellite in 1972 (the name of the series was changed to Landsat in 1975) and the TIROS-N satellite and the Nimbus-7 satellite in 1978. The satellites of the Landsat series (Landsat 1-7) have been providing continuous and consistent measurements of the Earth from the early 1970s to the present day. The Multispectral Scanner System (MSS) was the primary sensor of Landsats 1-3 platforms and was carried on Landsats 4-5 platforms for continuity. The MSS had four spectral bands covering the visible and IR regions, except that MSS on Landsat 3 had an additional band in the thermal IR region. The spatial resolution of MSS is 68 m in cross-track direction and 83 m in along-track direction. The Thematic Mapper (TM) on board Landsats 4-5 platforms is an advanced multispectral sensor with higher spatial resolution, more spectral channels, and greater radiometric accuracy than MSS. The TM has seven spectral channels with spatial resolution at 30 m for solar reflective bands (bands 1–5, band 7) and spatial resolution at 120 m for the thermal emissive band (band 6). The Enhanced Thematic Mapper Plus (ETM+) on board Landsat 7 has a panchromatic band with spatial resolution at 15 m and a thermal IR channel with a fourfold improvement in spatial resolution. The revisit period of Landsat 1-3 is 18 days, and the revisit period of Landsat 4-5 is 16 days. Landsat MSS/TM/ETM+ data have been widely used for land applications, especially for land cover and land use monitoring at a regional scale. The Advanced Very High Resolution Radiometer (AVHRR) is the primary sensor on board the NOAA polar-orbiting (PO) satellites. The AVHRR is a radiometer with four to six channels and daily global coverage at the spatial resolution of 1.1 km (AVHRR/1 had four channels, AVHRR/2 had five channels, and the latest version of AVHRR, that is, AVHRR/3, has six channels). The AVHRR data have been collected and continuously archived since 1981 and have become one of the most important data sets for global change study.

Since the 1980s, with the rapid advance of electro-optical technology and information technology, many remote sensing systems for EO have been developed and operated by governments and the private industry, such as SPOT, ERS, Envisat, RESURS, OKEAN, IRS, ResourceSat, JERS, ALOS, and CBERS (Kramer et al. 2002; Krige et al. 2000). In the United States, the NOAA launched a series of PO and geostationary satellites for operational weather and environmental monitoring, and NASA launched an EOS series of satellites for global observation. The Moderate Resolution Imaging Spectroradiometer (MODIS) is a key instrument aboard the Terra and Aqua satellites of NASA EOS missions. Terra MODIS and Aqua MODIS can cover the entire Earth's surface every 1–2 days, acquiring data in 36 spectral bands ranging in wavelength from 0.4 μm to 14.4 μm (http://modis.gsfc.nasa.gov/). The MODIS has three spatial resolutions, with two bands at 250 m, five bands at 500 m, and the remaining 29 bands at 1 km. The MODIS provides the capability to improve our understanding of global dynamics and processes occurring on the land, ocean, and atmosphere (Justice et al. 1998; Salomonson et al. 2006). The MODIS satellite data products have been widely used for global and regional applications for environment monitoring, natural hazards detection, weather forecasting, and climate change studies (Salomonson et al. 2006).

For space-based EO, considering satellite orbits, there are two major types of platforms: geostationary and PO. A geostationary satellite appears to be fixed at a specified position around 35,786 km above the equator. Geostationary satellites have the advantage of continuous observation at a high temporal resolution, but they usually have a low spatial resolution. Polar-orbiting satellites circle around the Earth at a near-polar inclination and usually operate in a sun-synchronous orbit, that is, passing overhead at essentially the same solar time throughout all the seasons in a year. This characteristic is important for long-term comparisons and global change monitoring. From the aspect of energy sources, EO sensors can be classified as passive or active. Passive sensors, such as imaging spectrometer and spectroradiometer, detect reflected or scattered or emitted electromagnetic radiation from natural sources, whereas active sensors, such as Radar and Lidar, detect reflected responses of an artificially generated source (Herbert 2002; Joseph 2005).

EO systems also have diverse spectral, spatial, and temporal characterizations. There are various sensors for different ranges of the electromagnetic spectrum, such as ultraviolet (UV), visible, IR, microwave (MW), and radio regions. From the aspects of spectral resolution, sensors can be classified as broad band sensors and hyperspectral sensors. Broad band sensors, such as MSS, AVHRR, and MODIS, usually have a limited number of channels, and each channel covers a relatively wide range of the electromagnetic spectrum. Hyperspectral sensors can have hundreds or thousands of channels over a specified region, and each channel covers a very narrow region of the electromagnetic spectrum. From the aspects of spatial resolution, sensors can be classified as high spatial resolution (<4 m), medium spatial resolution (4–30 m), and low spatial resolution (>30 m).

The diversity of current EO techniques and systems is related to the broad applications about various aspects of the Earth system at different spatial and temporal scales. High spatial resolution systems are usually used for local and regional applications. The PO systems with medium and low spatial resolution are usually used for global studies, including environmental monitoring (such as pollution, aerosol, and ozone), global change analysis (such as land cover, land use, vegetation phenology, climate, surface temperature, etc.), and natural hazards detection (such as wildland fire, hurricane, flooding, drought, etc.).

2.1.3 Remote Sensing of Global Land Cover and Land Use Change

Land cover is the physical material of the Earth's surface, including vegetation, water, bare soil, snow, ice, and so on. It is one of the most important properties of the Earth's system and is closely related to the ecosystem and climate. Land cover changes can be caused by human activities and natural disturbances (Asner et al. 2005; Boose et al. 1994; DeFries 2002; Fu 2003). Human activities, such as forest logging and urbanization, are the primary causes of land cover changes. Natural disturbances, such as severe wildland fires, drought, hurricanes, volcanoes, Earthquakes, and so on, can also change land cover. For global climate change analysis, ecosystem change monitoring, global carbon cycle, and surface radiation budget monitoring, it is crucial to observe land cover and land use changes and investigate the impacts of land cover and land use changes on the Earth's environment, ecosystem, and climate (Houghton et al. 1999, 2000; IPCC 2000). In the United States, extensive related work has been conducted with support from NASA, USGS, NOAA, and EPA (Janetos et al. 2000). In this section, we briefly discuss some of the essential topics of global land cover and land use change monitoring, including land cover mapping; anthropogenic and natural disturbances; and their impacts of land cover and land use change on the ecosystem, environment, and climate.

Land cover mapping and change analysis. Most previous work on land cover and land use mainly focused on mapping and the change detection of land cover types with satellite remote sensing measurements, among which Landsat TM/ETM+ data, AVHRR data, and MODIS data were mostly used (DeFries et al. 2000; DiGregorio et al. 2000; Friedl et al. 2002; Fry et al. 2009; Hansen et al. 1998, 2000a, 2000b; Homer et al. 2004; Townshend et al. 1988; Turner et al. 1994; Vogelmann et al. 2001). Global and regional land cover data products have been developed at various resolutions. At a regional scale in the United States, the Multi-Resolution Land Characteristics (MRLC) Consortium (http://www.mrlc.gov/), a group of federal agencies, developed national land cover products at 30 m resolution based on Landsat data, that is, the National Land Cover Dataset (NLCD 1992) and the National Land Cover Database (NLCD 2001) (Fry et al. 2009; Homer et al. 2004; Vogelmann et al. 2001). These 30 m land cover data products provided details about the land cover and land use change in the United States during the 1990s.

On the global scale, AVHRR and MODIS are the two primary sensors for land cover and land use change monitoring. The Global Land Cover Facility (GLCF, http://glcf.umiacs.umd.edu/) at the University of Maryland at College Park generated global land cover classification data products in 1998, using AVHRR imagery at 1°, 8 km, and 1 km spatial resolutions (Hansen et al. 1998). In comparison to AVHRR, MODIS have more bands for remote sensing of land; especially band 1 (red) and band 2 near-infrared (NIR) have a nominal resolution of 250 m at nadir, whereas bands 3–7 have a nominal resolution of 500 m at nadir. The operational MODIS land cover data products provide global classifications of the land surface for every 96 days (Friedl et al. 2002). Time series of MODIS land cover data provide the capability for understanding the location, rates, and patterns of land cover change.

For a more detailed analysis of dynamics of land surface, vegetation greenness indices have been used, including the Normalized Difference Vegetation Index (NDVI) (Tucker 1979; Tucker et al. 2004, 2005) and the MODIS Enhanced Vegetation Index (EVI) (Huete et al. 2002). The GIMMS (Global Inventory Modeling and Mapping Studies) NDVI data product is derived from AVHRR measurements to monitor monthly changes in terrestrial vegetation (Tucker et al. 2004, 2005). Since GIMMS NDVI is available for time-series analysis of the surface biophysical parameters with a long period from 1981, it can be used in climate models for global change study. The MODIS 16-day composite NDVI and EVI data products at 250 m and 1 km resolutions (Huete et al. 2002) from the year 2000 are also important for global land cover study, especially vegetation phenology analysis (Zhang et al. 2003, 2004, 2006). The MODIS global land cover dynamics data product provides estimates of the timing of vegetation phenology at global scales (Friedl et al. 2002; Zhang et al. 2003, 2004, 2006).

Anthropogenic and natural disturbance. Anthropogenic and natural disturbance are the primary factors for land cover changes (Asner et al. 2005; Boose et al. 1994; DeFries et al. 2002; Fu et al. 2003; IPCC 2000; Justice et al. 2002; Mildrexler et al. 2009; Potter et al. 2005; Wang et al. 2009). Human activities such as commercial forest logging and urbanization can cause remarkable and permanent land cover changes, whereas other land management activities may cause either temporary or permanent land cover changes (Asner et al. 2005). Each year, there are a huge number of natural disasters, including wildfires, hurricanes, flooding, drought, volcanoes, and earthquakes. These natural disturbances also contribute to land cover changes, either directly or indirectly (Boose et al. 1994; DeFries et al. 2002; Justice et al. 2002; Mildrexler et al. 2009; Wang et al. 2010). Wildland fires can burn forest areas into sparsely vegetated areas or even bare soil. Hurricanes may change forest live biomass significantly, produce large amount of dead fuels, and hence increase the potential for fire danger (Wang et al. 2010), and indirectly affect land cover. Currently, it is possible to identify the locations of land cover disturbances and monitor the recovery process of temporary changes. For example, fire locations, fire burned areas, and large logging areas can be effectively detected by remote sensing data, and vegetation time-series indices can reflect the recovery process of burned forests. However, it is still a challenging problem to quantify the extent of a disturbance, such as burn severity and biomass change (Mildrexler et al. 2009; Wang et al. 2010), which are important to more accurately characterize land cover changes caused by anthropogenic and natural disturbances.

Impacts of land cover and land use change on ecosystem, environment, and climate. Large-scale land cover and land use changes have significant impacts on environment, ecosystem, and even climate. Due to urbanization over recent decades, fast expansions of the wildland–urban interface (WUI) have introduced the vulnerability of ecosystems and significantly increased the risk of wildfires. Carbon dioxide and other greenhouse gases (GHGs) released from forest fires can affect the climate system as well as industrial development may cause environmental problems. For example, recently the Taihu Lake in China experienced a water crisis due to a severe bloom of blue algae. This event is believed to have a close relationship with the rapid industrialization of the regions around the Taihu Lake since the 1980s. Due to the complicated interactions among human activities, ecosystem, and climate, the characterization of the impacts of land cover and land use change at regional and global scales is a very challenging issue in EO and relies on synergistic integration of measurements from multiple EO missions and more reliable models of the Earth system.

2.1.4 Summary and Discussions

Space-based EO has become the primary approach for global change monitoring. Land cover is a crucial component of the Earth system. Satellite remote sensing is critical for identifying the drivers of land cover change, detecting global land cover change, and understanding potential impacts. Despite many efforts toward land cover mapping and land cover or land use change monitoring using remote sensing measurements over the past few decades, characterization of anthropogenic and natural disturbances and understanding of the impacts of land cover and land use change on ecosystem, environment, and climate are still complex problems vexing the science communities. These problems are interdisciplinary in nature, requiring investigation of the Earth system in multiple dimensions at regional and global scales over a relatively long period. Each EO sensor has a limited life cycle and collects specific measurements for certain aspects of the Earth. The sensors may have different spectral and spatial characterizations, so construction of long-term consistent measurements, that is, climate data records (CDRs) with multiple sensors, is critical for further characterization and understanding of global land cover and land use change.

2.2 Atmospheric Remote Sensing: A Brief Overview*

2.2.1 Introduction

Since its first introduction, atmospheric remote sensing has advanced from only basic science uses to every day applications. Satellite observation is now

---

* This section is contributed by Long S. Chiu.

an important tool for monitoring the Earth and its components. The need for environmental, climate, and hazard monitoring and for national security prompts investment in satellite remote sensing technologies in both developed and developing countries. The high cost of satellite and sensor development and deployment and the benefits of sharing data to tackle global problems foster collaborations among the nations.

Atmospheric remote sensing can be classified into passive and active sensing, and the approach involves physical and statistical retrievals and pattern analyses. Sensors have advanced from a single channel to multichannel and hyperspectral observations, and retrieval algorithms have utilized the combination of sensors and channel information or merging different sensors on different platforms. Atmospheric remote sensors are carried on board by airborne, space-borne, or ground-based platforms. Orbital mechanics provides a description of the motion of space-borne platforms for sampling; the theory of radiative transfer enables a description of the propagation of electromagnetic waves through the atmosphere and its constituents. Image analysis provides a tool for recognizing targets or patterns for analyzing imagery. An important component of remote sensing is in-situ measurements, which validate retrieval. Theories and techniques for atmospheric remote sensing have been discussed in a number of texts (Acker et al. 2002; Campbell 2005; Elachi and Van Zyl 2006; Kidder and Vondar Haar 1995; Qu et al. 2006a, 2006b; Ulaby et al. 1991). The Web-based material provided by NASA (Short, N., Remote Sensing tutorial, http://rst.gsfc.nasa.gov/, retrieved January 15, 2010) provides a good overview of the current remote sensing techniques and applications.

2.2.2 Overview of Sensors and Missions

Meteorological satellites are broadly classified into geosynchronous (GEO) and low Earth orbiting (LEO) satellites. The LEOs include PO satellites, such as the NOAA series, and non-PO satellites, such as the Tropical Rainfall Measuring Mission (TRMM) satellite. Figure 2.1 shows examples of the GEOs and POs and their nominal orbital characteristics.

Atmospheric remote sensing provides information on the atmospheric constituents, trace gases, radiation, temperature and humidity, air motion, aerosol, cloud and precipitation, and lightning. Table 2.1 shows the Earth orbiting remote sensing systems primarily designed for atmospheric remote sensing. Kramer (2002) provides a complete list and description of legacy, current, and planned sensors and missions up to the date of publication.

After the success of NASA's NIMBUS series of satellites, the EOS) is an international program that provides long-term observations of the Earth's atmosphere, hydrosphere, lithosphere, cryosphere, biosphere, and anthrosphere (http://en.wikipedia.org/wiki/Earth_Observing_System, retrieved January 25, 2010; Gurney et al. 1993). The satellite component of the EOS started with the launch of the Sea-viewing Wide Field-of-view Sensor (SeaWiFS) and TRMM in 1997.

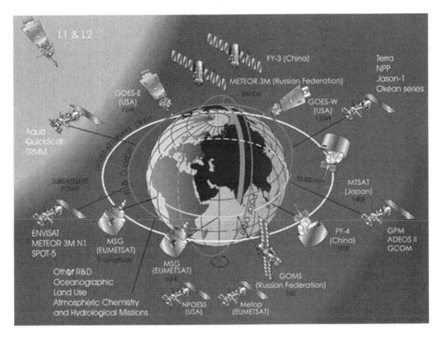

FIGURE 2.1
(**See color insert following page 144.**) Schematic showing nominal orbital characteristics of Geosynchronous (GEO), low Earth orbit (LEO), and polar-orbiting (PO) satellites. (Adapted from http://www.eohandbook.com/eohb05/images/fig_03_(weather).jpg)

TABLE 2.1

Meteorological Remote Sensing Systems in Earth's Orbit

Earth Observing System (EOS)	SeaWiFS · TRMM · Terra · ACRIMSAT · Meteor 3M-1/Sage III · Aqua · ICESat · SORCE · Aura · CloudSat · CALIPSO · NPOESS · GPM[a] · GCOM
A-train satellites	Aqua · Aura · PARASOL[b], CALIPSO · CloudSat · GCOM-W[c]
Geostationary Satellites	GOES · Fengyun-2 · Meteor · MTSAT (GMS) · Meteosat · INSAT · COMS[d]
Polar satellites	· COSMIC · DMSP · Envisat · Fengyun · MetOp · NOAA-N' (POES) · RADARSAT-1 · RADARSAT-2 ·
Completed	Nimbus · SEASAT · TIROS · Vanguard · OCO[e]

[a] The GPM is scheduled for launch in 2013.
[b] PARASOL was moved off the A-Train orbit on December 2, 2009.
[c] It is decided to include the GCOM-W in the A-Train.
[d] As the first Korean meteorological GEO, COMS is scheduled to be launched in 2010.
[e] The Orbiting Carbon Observatory (OCO) was lost during launch on February 24, 2009.

Since it is the major energy source of the Earth's system, monitoring the sun's total output is of importance to climate change. Total solar irradiance has been monitored by the Active Cavity Radiometer Irradiance Monitor (ACRIM) on board the Solar Maximum Mission (SMM) and the Upper Atmosphere Research Satellite (UARS) (Wilson 1997). The ACRIMSAT and the Solar Radiation and Climate Experiment (SORCE) are the most advanced instruments to continue measuring solar output. The Clouds and the Earth's Radiant Energy System (CERES), and its predecessor, the Earth Radiation Budget Experiment (ERBE), measures the reflected shortwave and outgoing radiative fluxes at the top of the atmosphere (TOA) (Wielicki et al. 1998). The CERES combines with other instruments such as MODIS and Multi-angle Imaging SpectroRadiometer (MISR) to give detailed TOA radiative fluxes (Loeb et al. 2006).

The SeaWiFS flown on board the Seastar satellite is designed to measure global ocean color (Craknell et al. 2000). In retrieving ocean optical properties, the atmospheric aerosol contributions must be accurately determined and removed; hence, aerosols over the oceans are also retrieved (Wang et al. 1999).

The Terra (EOS AM-1) satellite, launched in December 2001, has been collecting data since February 24, 2002. It carries the Advanced Spaceborne Thermal Emission and Reflection Radiometer (ASTER), a CERES, the MISR, the MODIS, and the Measurements of Pollution in the Troposphere (MOPITT) sensor. The ASTER is a high-resolution (15–90 m) radiometer measuring radiation in the visible to thermal IR range to provide surface temperature, reflectance, emission, and elevation. The MISR is a push-broom radiometer pointing at nine angles and operating at four frequencies to provide bidirectional reflectance measurements of clouds and aerosols. In such a configuration, the spherical asymmetry of aerosol particulates is observed. The MOPITT is a nadir sounding instrument that measures upwelling IR radiation at 4.7 μm and 2.2–2.4 μm. It measures total columnar and profiles of carbon monoxide and methane in the lower atmosphere (Bowman 2006; Drummond and Mand 2006). The MODIS, as the centerpiece carried on board the NASA's Terra and Aqua satellites, is a multispectral radiometer with 36 frequency channels. It provides atmospheric, oceanic, and land parameters (Salomonson 2002). Qu et al. (2006a, 2006b) provided a good description of the algorithms and products.

The Aqua (EOS PM-1) satellite, launched in May 2002, is aimed at providing a detailed description of the water and energy cycles. It carries an Advanced Microwave Scanning Radiometer-EOS (AMSR-E), which measures cloud properties, SST, near-surface wind speed, radiative energy flux, surface water, ice, and snow. A MODIS and a CERES are also on board the satellite. The centerpiece is the Atmospheric Infrared Sounder (AIRS). When combined with the Advanced Microwave Sounding Unit (AMSU-A) and the Humidity Sounder for Brazil (HSB), which stopped functioning in 2003, these sensors provide temperature and humidity profiles with high vertical resolution and accuracy unmatched by their predecessors of MW sounders (see Section 2.1.3).

Aerosol parameters have been estimated from many sensors, including AVHRR, Total Ozone Mapping Spectrometer (TOMS), SeaWiFS, MODIS, MERIS, AIRS, and Ozone Monitoring Instrument (OMI). The Aura (EOS Ch-1) satellite carries four instruments that aim at studying the atmospheric compositions of the stratosphere and upper troposphere (Schoeberl et al. 2006). The Microwave Limb Sounder (MLS) measures natural MW thermal emissions from the edge of the atmosphere to provide vertical profiles of ozone, chlorine and other trace gases, and cloud ice. The OMI uses UV and visible radiation to produce high-resolution ozone maps. The Tropospheric Emission Spectrometer (TES) measures tropospheric ozone in the IR portion of the spectrum and also provides trace gas measurements, such as carbon monoxide, methane, and nitrogen oxides by examining the spectral signature of these gases. The High Resolution Dynamics Limb Sounder (HIRLDS) measures IR radiation from ozone, water vapor, CFCs, methane, and nitrogen compounds.

The CloudSat satellite, launched on April 28, 2006, carries a Cloud Profiling Radar (CPR) operating at 94 GHz. It is flown next to the Cloud-Aerosol Lidar and Infrared Pathfinder Satellite Observation (Calipso). The Calipso carries (1) a Cloud-Aerosol Lidar with Orthogonal Polarization (CALIOP), which provides high-resolution vertical profiles of aerosols and clouds; (2) a Wide Field Camera with channels matching the MODIS channels on the Aqua satellite; and (3) an Imaging Infrared Radiometer (IIR) for detecting cirrus emissivity and particle size. The CALIOP is aligned with the center of the IIR image to optimize joint CALIOP/IIR observations.

PARASOL stands for "Polarization and Anisotropy of Reflectances for Atmospheric Sciences coupled with Observations from a Lidar." It carries a POLDER (POLarization and Directionality of the Earth's Reflectances), a multispectral imaging radiometer providing unique measurements of the anisotropy and polarization of the solar radiation reflected by the Earth–atmosphere system.

The first European Remote Sensing Satellite (ERS-1) carried five instruments that provided information on the Earth system components. The instruments included (1) a Radar Altimeter, which was a nadir-pointing radar altimeter operating in the Ku band; (2) an Along-Track Scanning Radiometer (ATSR-1), which was a four-channel IR radiometer and an MW sounder for measuring temperatures at the sea surface and the top of clouds; (3) a Synthetic Aperture Radar (SAR) operating in C band, which can detect changes in surface heights with sub-millimeter precision; (4) a Wind Scatterometer (WS) for measuring wind speed and direction; and (5) a Microwave Radiometer (MWR), used in measuring atmospheric water. The ERS-1 failed on March 10, 2000. Second in the series was ERS-2, which was launched on April 21, 1995. It carries a Global Ozone Monitoring Experiment (GOME), which is a nadir-scanning UV and visible spectrometer, and an ATSR-2 measuring visible radiation in three spectral bands specialized for chlorophyll and vegetation measurements.

The ENVISAT satellite, as a successor to ERS-2, carries the following instruments:

- ASAR (Advanced Synthetic Aperture Radar) operates in C band and is capable of detecting changes in surface heights with sub-millimeter precision.
- MERIS (MEdium Resolution Imaging Spectrometer) measures the reflectance of the Earth (surface and atmosphere) in the solar spectral range (390–1040 nm) and transmits 15 spectral bands.
- AATSR (Advanced Along Track Scanning Radiometer) can measure the temperature of the sea surface.
- RA-2 (Radar Altimeter 2) is a dual-frequency nadir-pointing radar operating in the Ku and S bands. It is used to define ocean topography, map or monitor sea ice, and measure land heights.
- MWR is used to measure water vapor in the atmosphere and estimate the tropospheric delay for the altimeter.
- DORIS (Doppler Orbitography and Radiopositioning Integrated by Satellite) is for orbit determination within 10 cm, or less, accuracy.
- GOMOS (Global Ozone Monitoring by Occultation of Stars) looks toward the stars as they descend through the Earth's atmosphere and change color. This instrument also indicates the presence of gases such as O_3 (ozone) and allows for the first time a space-based measurement of the vertical distribution of trace gases.
- MIPAS (Michelson Interferometer for Passive Atmospheric Sounding) is a spectrometer for detection of limb emission spectra in the middle and upper atmosphere. It provides a better understanding of atmospheric chemistry.
- SCIAMACHY (SCanning Imaging Absorption spectroMeter for Atmospheric CHartographY) compares light coming from the sun with light reflected by the Earth, which provides information of the atmosphere through which the Earth-reflected light has passed.

To continue the series of TIROS for weather monitoring, MetOp (Meteorological Operational) is a series of PO meteorological satellites operated by the European Organization for the Exploitation of Meteorological Satellites (EUMETSAT). The satellites are all part of the EUMETSAT Polar System (EPS). The satellites, the first of which was launched on October 19, 2006, are equipped with the same equipment as the TIROS satellites plus extra atmospheric measuring instruments:

- IASI (Infrared Atmospheric Sounding Interferometer) (3.7–15 µm, 8461 channels)
- MHS (Microwave Humidity Sounder)

- GRAS (Global Navigation Satellite System Receiver for Atmospheric Sounding)
- ASCAT (Advanced Scatterometer)
- GOME-2 (Global Ozone Monitoring Experiment-2)
- AMSU-A1/AMSU-A2 (Advanced Microwave Sounding Units)
- HIRS/4 (High-resolution Infrared Radiation Sounder)
- AVHRR/3

The Japanese Global Climate Observing Mission follows the mission objectives of the Advanced Earth Observation Satellite (ADEOS) and consists of two components: the Water Cycle component (GCOM-W) and the climate component (GCOM-C).

Several satellites are pulled together to form a constellation called the "A-Train." With the A-train constellation of satellites and sensors, the same earth location can be viewed from the visible, IR, and MW within a 15-minute envelope. The A-Train series includes the Aqua, Aura, PARASOL, CALIPSO, CloudSat, and GCOM-W. The PARASOL has been moved to a lower orbit to allow for the insertion of GLORY. The GLORY satellite carries an Aerosol Polarimetry Sensor to collect visible, near infrared, and shortwave infrared scattered radiation from aerosols and clouds and a Total Irradiance Monitor (TIM) to provide solar irradiance measurements for climate studies.

The series of GEOs consist of the Geosynchronous Operational Earth Satellite (USA, GOES), MTSAT (Japan, continuation of GMS), Fengyun (Chinese, FY) series, Geosynchronous Operational Meteorological Satellite (Russia), Meteosat (European Union), Indian Satellite (India, INSAT), and Communication, Ocean, and Meteorology Satellite (Korea, COMS). The GOES project science Web site developed by Dr. Dennis Chester of NASA/GSFC and NOAA contains detailed information about the GOES satellites (http://goes.gsfc.nasa.gov/) and other GEOs (http://goes.gsfc.nasa.gov/text/geonews.html).

The Polar Orbiting Environmental Satellites (POES) are a series of environmental satellites for monitoring the environment and long-range weather forecasting (Rao et al. 1990). In this mission, NASA is responsible for sensor development, integration, and launching, and NOAA is responsible for satellite management after launching. Current POES consist of two PO satellites and the Advanced TV Infrared Observation Satellites (TIROS-N). The NOAA-KLM series science instruments include AVHRR; HIRS/3; and AMSU-A1, AMSU-A2, and AMSU-B. The NOAA-NN's instruments include AVHRR; HIRS/4; AMSU-A1 and -A2; MHS; and the Solar Backscatter Ultraviolet Radiometer (SBUV/2).

The major meteorological sensors on board the Defense Meteorological Satellite Program (DMSP) satellites are the Special Sensor Microwave Imager (SSM/I) and the Optical Line Scanner (OLS). The SSMIS (SSMI Sounder) will be carried on board the next generation of DMSP satellites (http://nsidc.org/data/docs/daac/ssmi_instrument.gd.html, retrieved January 15, 2010).

The National Polar-orbiting Operational Environmental Satellite System (NPOESS) replaces the DMSP and POES satellites. The NPOESS Preparatory Project (NPP) program aims at bridging the gap between old and new systems by flying new instruments on a satellite originally to be launched in 2005 (Murphy 2006). The five instruments include the Advanced Technology Microwave Sounder (ATMS), the Cross-track Infrared Sounder (CrIS), the CERES, the Visible Infrared Imager Radiometer Suite (VIIRS), and the Ozone Mapping and Profiler Suite (OMPS). The NPP project is under review, and the launch date has been postponed until September 2011.

The FengYun 1 primary payload consists of two Very High Resolution Scanning Radiometers (VHRSR). The five spectral bands used were 0.58–0.68 µm, 0.725–1.1 µm, 0.48–0.53 µm, 0.53–0.58 µm, and 10.5–12.5 µm. The system swath was 2860 km with a 1.08-km resolution in the High Resolution Picture Transmission (HRPT) mode and a 4-km resolution in the Automatic Picture Transmission (APT) mode. Chinese FY-3s are the second generation of China's PO meteorological satellite series. This satellite series, which includes seven satellites, will be operational during the period 2005–2020. The first two satellites, FY-3A and FY-3B, as well as the onboard instruments are being designed and manufactured. On board the satellites of FY-3A and FY-3B, there are three passive microwave (PMW) payloads: Microwave Temperature Sounder (MWTS), MHS, and Microwave Radiation Imager (MWRI) (Zhang et al. 2006). Active radar observations such as those provided by the RADARSAT-1 and RADARSAT-2 can provide information about the atmospheric boundary layer and rain cell identification over ocean surfaces (Jackson and Apel 2004).

Constellation Observing System for Meteorology, Ionosphere, and Climate (COSMIC, or FORMOSAT-3) is a program designed to provide advances in meteorology, ionospheric research, climatology, and space weather by using GPS satellites in conjunction with LEO satellites. It is the third meteorological satellite launched by Taiwan.

2.2.3 Atmospheric Sounding

An atmospheric sounding is a measurement of vertical distribution of physical properties of the atmospheric column, such as pressure, temperature, liquid water content, ozone concentration, pollution, and other properties. Retrievals can be performed using nadir (downward) or limb (side) scanning. Based on IR and/or MW channels, the sensor signal at a particular frequency is a weighted average of the atmospheric property at the atmospheric level. The forward problem of computing sensor radiance is based on the radiative transfer theory with atmospheric parameters as inputs. The sounding retrievals, or the inverse problem, involve the solution to an integral equation to estimate the atmospheric parameters. The equations are nonlinear and under-constrained and hence prior knowledge or additional constraints have to be imposed (Milman 1999; Rodgers 2000). Commonly used IR and MW frequencies include the 11 μm and 3.7 μm, which are highly sensitive to surface parameters; the 15 μm and 4.3 μm CO_2 absorption bands and the MW 56 GHz O_2 bands, sensitive to atmospheric temperature in different layers of the atmosphere; the 6.7 μm for water vapor and temperature determination; and the 9.6 μm O_3 band, sensitive to ozone distributions. The moisture sensitive channels, such as 50, 183 GHz, are used to estimate moisture profiles. The vertical resolution depends on the number of channels. The retrievals of temperature and moisture profiles are coupled, as temperature and humidity have a strong correlation.

Scanning mode includes limb and nadir scanning. Examples of limb sounders are the Limb Infrared Monitor of the Stratosphere (LIMS) on board the NIMBUS-7 satellite and the Stratospheric Aerosol and Gas Experiment II (SAGE II). Examples of nadir sounders are HIRS and HIRS2. The IR sounding instruments are accompanied by MW sounders, such as the Microwave Sounding Unit (MSU), to provide better clear area coverage for profile retrieval under cloudy conditions (Susskind 1993). Atmospheric sounders evolve from multispectral to hyperspectral sensors. The AIRS, working with its companions, the AMSU and HSB, retrieves temperature at approximately 1 km vertical resolution with accuracy of approximately 1 K and water vapor at 2 km layers with accuracy of approximately 15% globally (Chachine et al. 2006; http://airs.jpl.nasa.gov). The retrieval accuracies are dependent on the cloud conditions. For the clearest fraction, a retrieval accuracy of 0.6-K rms for temperature and 10% for water near the surface can be achieved. For 55% cloud-cleared retrievals, AIRS achieves about 1-K rms accuracy over the ocean and about 1.7 K over land. The latest hyperspectral sounder is the IASI flown on board the EUMETSAT MetOp satellite. The IASI has 8461 channels; the AIRS has 2378 channels; and the CrIS scheduled to be flown on board the NPOESS has 1305 channels. Weather forecasts have improved, particularly in the Southern Hemisphere, due to the incorporation of satellite retrieved soundings (Le Marshal et al. 2006).

2.2.4 Aerosol and Cloud

Clouds are aggregates of minute suspended particles of water or ice, or both, which are in enough concentration to be visible. Clouds are classified into high, middle, and low clouds. The International Satellite Cloud Climatology Project (ISCCP, http://isccp.giss.nasa.gov/) uses cloud optical thickness and cloud top pressure to classify clouds. Figure 2.2 schematically shows the ISCCP cloud classification scheme.

Aerosols are fine solid or liquid particulates that are suspended in air. Aerosol sources include sea spray, volcanic ashes, dust, and pollen. Anthropogenic sources include black carbon, fossil fuel generation from power plants, vehicular emission, and others. Aerosols are removed by dry

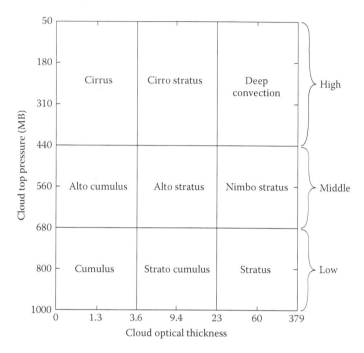

FIGURE 2.2
Classification scheme of cloud types for ISCCP. (Adapted from http://isccp.giss.nasa.gov/newalg. html#rev, retrieved January 15, 2010.)

(fall out due to gravity) and wet (scavenging and cloud conversion) depositions. Aerosols form cloud condensation nuclei (CCNs), which are necessary for cloud formation, growth, and, ultimately, precipitation (Potter et al. 2003).

Aerosol retrieval algorithms rely on the selection of clear pixels. Over land, the surface reflectance is highly variable, and radiometrically "dark" pixels have to be selected for aerosol determination. Cloud- and surface-contaminated pixels are first screened out to extract the clear aerosol pixels. The spectral radiance (0.55–2.1 µm) is fitted to calculated radiance from aerosol models. The retrieved parameters include aerosol optical depth, Angstrom exponent, fine-mode fraction, and aerosol effective radius. Secondary parameters such as CCNs are obtained from look-up-tables (Chu and Remer 2006). The spectral properties in the *Deep Blue* region of the spectrum are utilized over high reflectance semiarid or arid regions, in which atmospheric dust originates. The Deep Blue algorithm is capable of distinguishing between atmospheric dust and pollution aerosols. For the Deep Blue algorithm, a surface reflectance database is first developed, and aerosol reflectance in the blue region (412-, 490-, and 670-nm) of the spectrum is matched to that calculated from radiative transfer modeling (Hsu et al. 2006).

The MISR, which consists of nine cameras, measures directional radiation. It provides information on the radiative properties of aerosols, cloud types,

Earth's surface, and surface reflectance (Abdou et al. 2005; Diner 1999; Diner et al. 2007; MISR (http://www-misr.jpl.nasa.gov/, accessed January 15, 2010); IEEE 2002).

Active sensing of aerosols and clouds are provided by the Calipso satellite (http://www-calipso.larc.nasa.gov/). The CALIPSO combines an active lidar instrument with passive IR and visible imagers to construct the vertical structure and properties of thin clouds and aerosols. Combining with the active cloud radar on board the CloudSat satellite, they provide a three-dimensional picture of clouds and aerosols over the globe. CloudSat carries a CPR operating at 94 GHz (http://cloudsat.atmos.colostate.edu/, retrieved January 15, 2010). The synergy between CloudSat, CALIPSO, MODIS, and MISR offers cloud and aerosol information at much more detailed levels than previous missions did (Hu et al. 2009; Hunt et al. 2009; Young et al. 2009).

2.2.4 Precipitation

Barrett and Martin (1985) provided a thorough review of satellite rainfall estimation techniques at the date of their publication. Remote sensing of precipitation began with the first meteorological satellite, the TIROS. The indirect approaches are based on relations between cloud morphology and rain. The technique used by NOAA (NESDIS) is based on AVHRR imagery. Cloud types are manually identified by a trained meteorologist. The rain rates associated with each cloud type are empirically obtained. The areas enclosed by the cloud types are estimated, and rain rates are assigned to each cloud type. Initially, four types of clouds were used. It was later determined that the most important cloud type is the cumulonimbus. This is known as the cloud indexing technique.

Arkin (1979) found an empirical relation between the area of cold clouds and surface rain rate from radar observation during the Global Atmospheric Research Program Atlantic Tropical Experiment (GATE). He assigned a rain rate of 3 mm/day to the area of clouds with cloud top temperature colder than 235 K. The technique was initially based on the GOES satellite and was termed the "GOES Precipitation Index" (GPI). This technique works remarkably well over the tropics, except in regions where the presence of high cirrus clouds introduces a high bias (Chiu et al. 1993). This technique was extended to the global tropics to include all operational geostationary satellites.

Rain area and total rainfall have a strong relation. The Area Time Integral (ATI), which is the time integral of the rain area, is correlated with the total rainfall of each storm (Doneaud et al. 1984). Over a large area, the area average rainfall is strongly related to the areal rainfall (Chiu 1988; Kedem et al. 1990). This is due to the fact that rain rates follow certain distributions as they go through their life cycles. Tsonis and Issac (1985) used both visible and IR measurements to identify the cloud area. Areas identified from radar are used for training. This technique, termed RAINSAT, was used operationally by the Canadian Meteorological Service for tracking intense storms.

Microwave sensors are well suited to estimate rainfall over the oceans. The MW emissivity of sea water decreases with temperature, hence the brightness temperature of the ocean is remarkably constant, providing a constant background against which the bright hydrometeors (high emissivity) can be clearly distinguished. The launch of the Electrically Scanning Microwave Imager (ESMR-5) on board the Nimbus 5 satellite allows for the production of the first estimated rain maps based on MW data (Theon et al. 1975). The varying spatial resolution of the cross-track scanning ESMR-5 produces rain maps that are difficult to interpret. Nonetheless, the gross features of the global rain pattern are available over the oceans.

Microwaves interact directly with hydrometeors. The MW techniques usually involve estimating the cloud and environmental parameters. The spectral information from MW radiometers is insufficient to account for all atmospheric variables in the precipitation system. A Bayesian approach has been used in estimating rain profiles. The approach computes the received sensor measurements from a database of precipitation profiles and picks a profile that best fits the observed sensor measurements. An example of the Bayesian approach is the Goddard Profiling algorithm (GPROF) (Kummerow et al. 2001). The coupling between the inhomogeneity within the sensor's field of view and the nonlinear relation between the observed MW brightness temperature and rain rate introduces a bias in the estimation of the rain rate from the coarse resolution of the MW sensors (Chiu et al. 1991).

A major challenge to space or time estimates of rainfall is the intermittence and local scale of rainfall. Rainfall patterns can span from the frontal scale (tens of thousands of kilometers) to the cumulus scale (kilometers) and, temporally, in as short a time period as a few minutes. To mitigate poor space or time sampling, techniques that combine the superior temporal sampling by the visible and IR sensors on board geosynchronous satellites and the direct estimates of MW sensors have been developed. The adjusted GPI (or AGPI) uses MW rain estimates to calibrate IR estimates to increase temporal sampling. The Microwave Emission Brightness Temperature Histogram technique (METH; Chiu and Chokngamwong 2010; Wilheit et al. 1991) uses a combination of MW channels to mitigate the effect of water vapor on the MW rain signal and a rain–rate–brightness temperature relation based on an atmospheric radiative transfer model to fit the brightness temperature histogram to a mixed lognormal rain rate distribution. An empirical freezing height-dependent beamfilling correction is applied. The advantage of this approach is its robustness. The nonrain portion of the brightness temperature histogram is determined for each space or time cube and hence it can be considered self-calibrating. This technique has been applied to the TRMM Microwave Imager (TMI) and the Special Sensor on board the DMSP satellites. The SSM/I rain products are inputs to the oceanic components of the WCRP/GEWEX Global Precipitation Climatology Project (GPCP) rain maps.

The TRMM satellite (Figure 2.3), originally at a height of 350 km, was boosted into a higher orbit of 402 km in 2002 to conserve fuel. The robustness of the METH algorithm in maintaining a consistent ocean rainfall data set was demonstrated by calibrating the algorithm to the pre- and post-boost configuration, eliminating the jump in the rain rate products introduced by the boost (Chiu et al. 2010).

Based on the need of the hydrological community, higher spatial and temporal scale rainfall data are desirable. The TRMM Multi-satellite Precipitation Analysis (TMPA) uses the TMI MW rain estimates to calibrate the IR, and the 3-hourly amounts are constrained by the surface gauge analysis (Adler et al. 2003; Huffman et al. 1997). The Precipitation Estimation from Remotely Sensed Information using Artificial Neural Networks (PERSIANN) technique

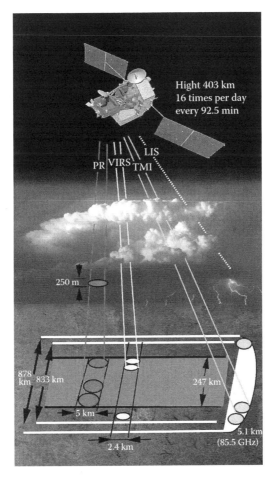

FIGURE 2.3
(**See color insert following page 144.**) Schematic showing the Tropical Rainfall Measuring Mission (TRMM) satellite with its payloads—the TRMM Precipitation Radar (PR), the TRMM Microwave Imager (TMI), the Visible IR Spectrometer (VIRS), and the Lightning Imaging Sensor (LIS). (Adapted from http://trmm.gsfc.nasa.gov/overview_dir/trmm_instrument_large.jpg)

is based on IR and visible properties and NN training of the rain field (Sorooshian et al. 2000). The Microwave calibrated Infrared Split-window Technique (MIST) makes use of the split channels at 10- and 12 μm to eliminate nonraining high clouds (Chokngamwong and Chiu 2008) to provide 3-hourly rainfall estimates at 4 km resolution. Lu et al. (2007) used topographic data coupled with meteorological wind direction, constrained by rainfall estimates at the satellite MW sensor resolution to provide rainfall data compatible with topographic data resolution. The high-resolution rainfall data developed by this technique has found applications in landslide triggering forecasts.

As a follow-on mission to TRMM, the Global Precipitation Mission (GPM; http://gpm.gsfc.nasa.gov/) is scheduled to launch in 2013. The satellite component consists of a series of satellites. The mother satellite carries a Core: a Dual-frequency Precipitation Radar (DPR) and a high-resolution, multichannel PMW rain radiometer known as the GPM Microwave Imager (GMI). The Core will also serve as the calibration reference system for a constellation of supporting satellites. In addition to the Core, a constellation of up to eight satellites will comprise the GPM sensor Web. The radiometer carried on board the satellite constellation will be similar to the GMI on the Core. Other vehicles in the constellation will be satellites of opportunity contributed by domestic agency partners, such as NOAA and the Department of Defense (DOD), and GPM international partners. One specific example of a potential satellite of opportunity is the proposed French or Indian mission Megha-Tropiques.

2.2.5 Trace Gases

One purpose of the Aura (EOS Ch-1) mission is to study air quality and trace gases. The sensors are designed to measure the spectra of atmospheric trace gases at various heights, as shown in Figure 2.4. As a hyperspectral sensor with over 3000 spectral lines in the visible to IR range containing the spectra of various trace atmospheric gases, AIRS is also capable of retrieving these trace gases. With the identification of the spectral signatures, a number of GHGs, such as carbon monoxide, carbon dioxide, and methane, other trace gases such as sulfur dioxide, and dust can also be identified (Chachine et al. 2006).

2.3 Land Remote Sensing: Hyperspectral Techniques*

2.3.1 Introduction to Hyperspectral Imagery

Hyperspectral imagery (HSI) remote sensing is the simultaneous acquisition of hundreds of coregistered images of a scene over a range of wavelengths in

---

* This section is contributed by Ronald Resmini.

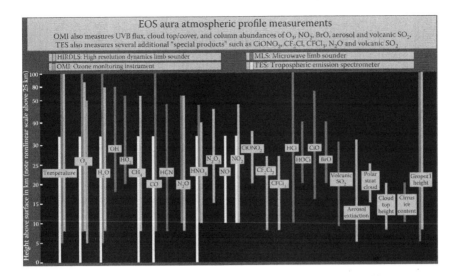

FIGURE 2.4

(See color insert following page 144.) Table summarizing the atmospheric parameters measured by HIRDLS, MLS, OMI, and TES. The altitude range in which these parameters are measured is shown as the vertical scale. In several cases, the measurements overlap, which provides independent perspectives and cross calibration of the measurements. (Adapted from http://aura.gsfc.nasa.gov/instruments/index.html, retrieved January 15, 2010.)

the visible (~0.40 micrometers or μm) to longwave infrared (~14.0 μm) region of the electromagnetic spectrum. Each image or band is a sample over a small wavelength interval. For example, Figure 2.1 shows that a black-and-white image is formed by one sampling function covering the visible range of the spectrum (~0.40 μm to ~0.70 μm—essentially the range of human vision), thus giving one image. A normal color composite image may be formed by acquiring three images centered on the red (~0.67 μm), green (~0.54 μm), and blue (~0.45 μm) regions and then displaying them simultaneously with the red, green, and blue guns, respectively, of a computer monitor. The HSI data are acquired with hundreds of sampling functions across the spectrum. The majority of HSI images sample regions of the spectrum that are beyond the range of human vision. Another key characteristic of HSI is that in addition to one caveat (to be described in the next paragraph) the spectrum is contiguously sampled across a broad range of wavelengths. Each HSI image is the result of the interaction of photons of light with matter within a small wavelength interval (in fact, all images are the result of the interaction of photons of light with matter). Materials reflect, absorb, or transmit electromagnetic radiation (see, e.g., Hapke 1993 and Solé et al. 2005 for detailed discussions of these foundational topics of HSI). Imaging sensors detect the reflected (or scattered) radiation.* Most HSI sensors are passive,

---

* As will be discussed in the next paragraph, some sensors measure self-emitted radiation; this is radiation reemitted at a longer wavelength than which was originally absorbed.

that is, they record reflected (or scattered) photons of sunlight or photons self-emitted by the materials in a scene that can likely trace their origin ultimately to solar energy. Active HSI sensors, not nearly as common as passive systems, supply photons for illumination (e.g., a laser). The HSI is an extension of multispectral imagery remote sensing (MSI; see, e.g., Landgrebe 2003; Richards and Jia 1999); The MSI is the acquisition of (on the order of) tens of broad bands of the electromagnetic spectrum that are not necessarily contiguous (Figure 2.5).

The Traditional Remote Sensing Regions. The *traditional* Earth remote sensing regions of the spectrum are, by convention, the visible to near-infrared or shortwave infrared (VNIR/SWIR, 0.40–2.5 µm); the midwave infrared (MWIR, 3.0–5.0 µm); and the longwave infrared (LWIR, 7.0–14.0 µm). The major gaps between the regions are due to atmospheric opacity, that is, the atmosphere absorbs the light at the gap wavelengths such that airborne or space-borne sensors will not detect a signal from the ground. Conversely, the traditional remote sensing regions are in the so-called atmospheric windows through which photons may pass from the ground to the sensor. Within each region (i.e., the VNIR/SWIR, MWIR, or LWIR), HSI sensors acquire on the order of hundreds of contiguous images covering the respective wavelength interval. Sensor design and engineering considerations, detector materials,

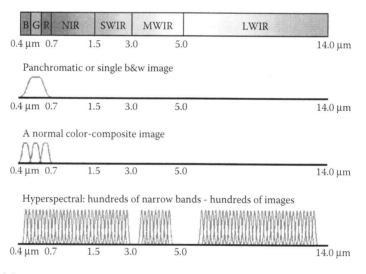

FIGURE 2.5
(See color insert following page 144.) Sampling the electromagnetic spectrum in the 0.4 µm to 14 µm range: B = blue; G = green; R = red; NIR = near infrared; SWIR = shortwave infrared; MWIR = midwave infrared; and LWIR = longwave (or thermal) infrared. A panchromatic image is formed with one sampling function covering the visible range of the spectrum. Acquiring three images centered on the R, G, and B regions forms a normal color-composite image. The HSI data are acquired with hundreds of narrow sampling functions across the spectrum.

and community practices have also influenced the convention of the traditional remote sensing regions. Generally speaking, current HSI sensors are either VNIR, VNIR or SWIR, MWIR and LWIR, or LWIR.

Three Factors Make HSI Possible. Three factors make HSI remote sensing possible: (1) there is an observable, measurable signal (reflected and emitted radiance); (2) the measurable signal varies in intensity as a function of the wavelength yielding a spectrum or signature; and (3) different materials have different spectral signatures, thus facilitating material identification and characterization. Spectral signatures are the fingerprints used to identify and characterize materials in a scene. The many bands of HSI data make evident, that is, resolve, the spectral signature features that characterize different substances. The HSI is imaging spectrometry, and the analysis methods of analytical spectroscopy (and other techniques as well) may be applied to the data.

2.3.2 Why HSI?

The HSI is the foremost remote material identification technique. This important capability distinguishes it from all other remote imaging technologies. The HSI remote sensing uses spectroscopy. The high-resolution sampling in the wavelength dimension generates signatures with resolved spectral features. The bands of an HSI data set and the spectral signatures they form provide numerous, distinct (though related) descriptors, each conveying a different token of information about the objects and materials captured in a scene. More descriptors allow greater separability between entities. For example, gender as a descriptor allows only the classification of a person as a man or a woman. Many additional descriptors are required to identify a specific individual or to facilitate a more detailed clustering of a set of people. In spectral remote sensing, additional bands facilitate the generation of more detailed signatures, which facilitate material identification and characterization (e.g., particle size effects in the LWIR; see Salisbury 1993).

The Utility of HSI. The HSI may be utilized for an extremely wide variety of land remote sensing applications, such as geological mapping, vegetation mapping and health assessment, littoral zone characterization (including optical bathymetry), and characterizing snow and ice, to name a few. Ultimately, each application has associated with its materials that may be mapped (and characterized) with HSI based on their spectral signatures and imprints on the signatures due to particle size effects, moisture, mixtures of materials within a single pixel, orientation of surfaces with respect to the illumination and viewing angles of the remote sensing scenario, and surface roughness, among many factors. The spectral resolution of HSI facilitates a detailed identification and examination of materials in a scene. The spectral signatures of HSI contain a wealth of information. The spatial context of hyperspectral images complements and extends the information content of the spectra.

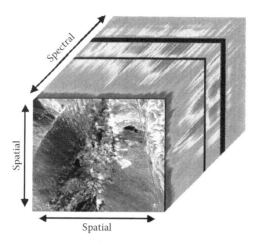

Spatial

Spatial

FIGURE 2.6
(See color insert following page 144.) The image or data cube: the fundamental data structure of HSI. The cube is a three-dimensional data structure with two spatial dimensions (samples and lines of the image space) and one spectral dimension (sampling in wavelength space). The image shown is a false-color composite of NASA AVIRIS data of Cuprite, NV (R = 2.1088 μm, G = 2.2086 μm, B = 2.3381 μm).

2.3.3 HSI Data and the Image Cube

The fundamental data structure of HSI is the image (or data) cube (Figure 2.6). The image cube is a stack of spatially coregistered images; each image is the result of the interaction of photons of light with matter within a small wavelength interval (Figure 2.5). The intensity or grayscale value of each picture element or pixel in an image is proportional to the reflection (or emission) of photons back to the sensor. It is thus a three-dimensional data structure with two spatial dimensions (samples and lines of the image space) and one spectral dimension (sampling in wavelength space). Each pixel of the image cube provides a spectrum—a plot of the reflected (and/or emitted) intensity received at the aperture of the sensor versus wavelength (Figure 2.7).

2.3.4 Radiance and Reflectance

In practice, HSI sensors measure radiance.* After a process called atmospheric compensation (or correction), reflectance spectra are obtained for data collected in the VNIR or SWIR or the solar-reflected region of the spectrum. For data collected in the LWIR or emissive region of the spectrum, atmospheric compensation followed by temperature or emissivity separation (TES) yields emissivity spectra, which are related to reflectance spectra

---

* Radiance has the units of microwatts per meter squared per steradian per micrometer (μW/ m² · sr · μm).

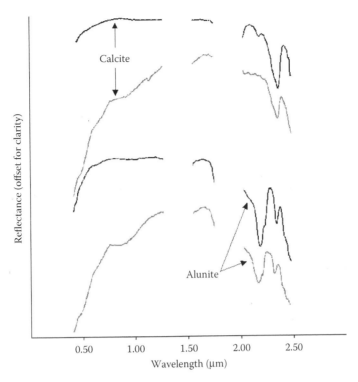

FIGURE 2.7
(See color insert following page 144.) Spectral signatures derived from the atmospherically corrected NASA AVIRIS Cuprite, NV, HSI data. The signatures are compared with those of similar (though not identical) materials acquired with a laboratory spectrometer (Clark et al., 1993). The green spectra are from the AVIRIS data; the black spectra are the laboratory measurements. The two upper spectra are of calcite (a carbonate); the lower are of the mineral alunite. The NIR portions of the AVIRIS spectra show the presence of iron in the minerals. Diagnostic spectral features for both minerals occur in the SWIR.

via Kirchhoff's Law. Data collected in the MWIR region are complicated by the combined reflected and emitted radiance received by the sensor. The development of techniques for the retrieval of reflectance (or emissivity) in the MWIR is a current area of research. A review of MWIR remote sensing (albeit MSI) is given in Boyd and Petitcolin (2004).

2.3.5 The Spectrum

Although the image cube is the fundamental data structure of HSI, the fundamental datum is the spectrum. The HSI data are indeed images, but the power and uniqueness of HSI reside in the information contained in the wavelength dimension. Figure 2.7 shows example spectral signatures derived from atmospherically corrected airborne HSI data. The signatures are compared with those of similar materials acquired with laboratory spectrometers (Clark et al. 1993). A spectrum is obtained by skewering the image cube along the spectral dimension and building a plot such as shown in Figure 2.7 with the values at each band.

2.3.6 HSI Sensors

The HSI remote sensing is based on the dispersion of electromagnetic radiation (e.g., light) emanating from a scene into its constituent frequency components. The familiar rainbow is the result of the dispersion of sunlight. The HSI sensors achieve dispersion by channeling light through a prism, onto a grating, or into an interferometer. The dispersed signal or interferogram is sampled by an array of detectors. Two-dimensional focal plane arrays (FPAs) are the most common detectors in current HSI systems; one-dimensional arrays are also employed. With the exception of interferometers, the need to disperse the input radiance requires collection modes that are different than point-and-shoot framing imagers. The HSI sensors record one line of imagery at a time through a slit aperture in the so-called push-broom mode. Alternatively, a point spectrometer is rapidly swept to form one line of imagery at a time; this is called the whisk-broom mode. Interferometers are generally framing imagers. Sensor type and mode of operation are closely tied to the motion of the platform. For example, the forward motion of an aircraft advances the slit aperture of a push-broom HSI sensor along the ground. Stationary, ground-based HSI sensors are also available as are field-portable point (nonimaging) spectrometers. Additional details on optics and sensor design are beyond the scope of this book but may be found in Wolfe (1997), Schott (1997), and Hecht (1987).

Example HSI sensors currently in operation are NASA AVIRIS,[*] SEBASS,[†] HyMap,[‡] Probe-1,[§] and ProSpecTIR[**]—all airborne. NASA Hyperion[††] and NASA AIRS[‡‡] are Earth-orbiting space-based HSI sensors. NASA Compact Reconnaissance Imaging Spectrometer for Mars (CRISM)[§§] sensor is in orbit around the planet Mars, and NASA's Moon Mineralogy Mapper (M³)[***] sensor is in orbit about the Moon.

2.3.7 Working with HSI Data

Processing and analysis techniques for HSI data are often different from those applied to multispectral (and other types of image) data. The rich

---

[*] See: http://aviris.jpl.nasa.gov/
[†] See: http://www.lpi.usra.edu/science/kirkland/Field/home_sebass.htm, Kirkland et al. (2002), and Hackwell et al. (1996).
[‡] See: http://www.hyvista.com/main.html *and* http://www.intspec.com/
[§] See: http://www.Earthsearch.com/index.php?sp=10
[**] See: http://www.spectir.com/DUAL.htm
[††] See: http://eo1.gsfc.nasa.gov/Technology/Hyperion.html
[‡‡] http://airs.jpl.nasa.gov/
[§§] See: http://crism.jhuapl.edu/
[***] See: http://moonmineralogymapper.jpl.nasa.gov/

information content of HSI has driven the development of new techniques of information extraction. One such family of techniques is the spectral matched filter. The matched filters are powerful and rapid techniques for mineral mapping and require only the spectrum of the mineral to be mapped. Algorithms such as spectral angle mapping (SAM), Euclidean distance, and filters based on second-order statistics (i.e., covariance) of the data (e.g., constrained energy minimization [CEM]) are routinely applied to HSI (Chang 2003; Stocker et al. 1990). Algorithm results highlight pixels containing the material of interest and render the pixels that do not contain the material a dark color; the resulting gray-level values are proportional to the amount of material present in a pixel.

Another technique particularly applicable to HSI (and MSI) is linear spectral mixture analysis (SMA; Adams et al. 1993; Adams and Gillespie 2006). In contrast (though related) to the matched filters, SMA requires the spectra of the several different materials present in a scene for material mapping. An endmember is a substance or material captured in an HSI data set with a spectral signature that is not a mathematical combination of other spectra from the same image cube. Endmembers may combine to yield other spectra observed in an image cube. The results of SMA, called fraction planes, may be viewed as grayscale images or can be combined to form color composites. The fraction planes are material maps.

Principal components analysis (PCA), a data transformation based on the second-order statistics of the data, may also be applied to HSI for the purpose of inspecting data quality and for general data exploration. The minimum noise fraction (MNF) transformation, a relative of PCA, is also commonly utilized.

Analysis strategies for HSI vary among practitioners and may often be set by the goal of the exploitation effort. Generally speaking, the following procedure may be followed: All analyses must begin with an examination of the radiance spectra for data quality. A PCA should be applied and the results inspected. Next, atmospheric compensation takes place, and the reflectance (or emissivity) spectra must be inspected. A PCA may again be applied. Reflectance spectra are examined and used in *reconnaissance* applications of SAM and/or Euclidean distance. Signatures may be selected by inspection of PCA results, examination of various color composite images, and/or by examining dark and bright pixels (or other "eye-catching" pixels). Endmember finder tools such as the Pixel Purity Index (PPI) may also be applied, and the resulting signatures used in SMA and/or various matched filter algorithms. Signatures extracted from the cube are matched against a spectral library; the resulting library matches may then be applied by algorithms to the image cube. Ultimately, specific material maps should be generated. The signatures used may be from a spectral library and from the scene. The material maps (i.e., algorithm results) are raster images that may be converted to vector layers or shapefiles and overlaid on context images and/or fused with other data and information.

2.3.8 Example: Cuprite, Nevada, USA

The canonical application area of HSI for land remote sensing is geology, notably the mapping of minerals and rocks. Mineral mapping at Cuprite, Nevada, will be described next.

The Cuprite mining district in the state of Nevada, The United States, is a well-characterized, well-studied region in which a diversity of remotely sensed and ground-based data have been collected. The region is particularly well known for NASA AVIRIS HSI data analyses that showcase the region's uncommon diversity of materials with prominent spectral features. In addition to AVIRIS, however, the region has also had data collected by the HYDICE (now defunct), SEBASS, Probe-1, and ProSpecTIR HSI sensors.

The Geological Setting of Cuprite, Nevada. A description of the geology and geomorphology of Cuprite is given in Resmini et al. (1997) as is a bibliography of early field and remote sensing studies. Only a brief description of the region is given here. The Cuprite mining district is located in the Great Basin or Basin and Range province in SW Nevada, The United States, and is an approximately 12 km² area of hydrothermally altered and unaltered exposed rocks. The region has been the subject of several remote sensing studies (e.g., Abrams et al. 1977a, 1977b; Clark et al. 2003; Goetz et al. 1985; Resmini et al. 1997; Swayze et al. 2003). The lithologies at Cuprite are comprised of rhyolitic ash-flow and air-fall tuffs; felsite dikes; basalt flows; interbedded chert and limestone; and orthoquartzite sandstone and siltstone. Three degrees of hydrothermal alteration are discernible at Cuprite. The altered rocks in order from the most altered to the least altered are referred to as *silicified*, *opalized*, and *argillized*. A suite of alteration or replacement minerals roughly defines each degree of alteration. Silicified regions contain predominantly quartz with minor alunite, kaolinite, and calcite; opalized areas—the most widespread—contain alunite, disseminated opal, some calcite replacing opal, and kaolinite; and the argillized areas primarily contain kaolinite with montmorillonite and minor amounts of opal.

Figure 2.8 shows AVIRIS, ProSpecTIR (VNIR/SWIR sensors), and SEBASS (LWIR) imagery of Cuprite.* The AVIRIS data have a ground sampling distance (GSD; i.e., "pixel size") of approximately 22 m, whereas the other sensors have GSDs in the order of approximately 3–4 m. The AVIRIS false color composite is the same as shown in Figure 2.6. The ProSpecTIR false color composite is derived from displaying the first, third, and fifth PCA images as RGB, respectively, of a PCA applied to a 70-band spectral subset from the SWIR (2.02–2.46 µm). The SEBASS image is an LWIR *panchromatic* grayscale formed by summing all of the bands in a cube comprised of two adjacent flight lines (the SEBASS MWIR data are not shown). The full-spectrum, multitemporal HSI data set is a unique and information-packed resource for

---

* The AVIRIS data were acquired in 1997; SEBASS and ProSpecTIR were acquired simultaneously in 2008.

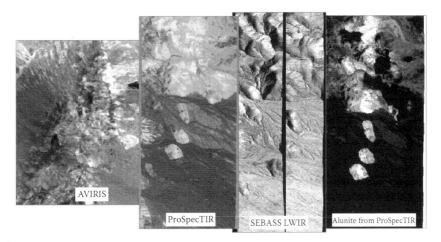

FIGURE 2.8
(**See color insert following page 144.**) AVIRIS, ProSpecTIR, and SEBASS imagery of Cuprite, NV. Note the one elongate and two semicircular features common to all images. The color composite of AVIRIS data is the same as shown in Figure 2.6; the ProSpecTIR image is a PCA-derived false color composite (see text); and the SEBASS image is an LWIR "panchromatic" grayscale formed by summing all of the bands in the cube. The right-most image is a map of alunite distribution using SAM and an in-scene spectrum with a 70-band SWIR spectral subset of the ProSpecTIR HSI data.

studying the geology and geomorphology at Cuprite. Here, we focus on the exploitation of the SWIR portion of the ProSpecTIR data.

Analyzing the Cuprite HSI Data. The Cuprite HSI data were processed and analyzed with the Environment for Visualizing Images (ENVI®)* software package. The ENVI® is a versatile, widely used package for the analysis of MSI and HSI. It is the *de facto* standard though other packages are available commercially and as freeware.

Results. The right-most frame in Figure 2.8 shows the results of SAM applied to a 70-band SWIR spectral subset of the ProSpecTIR HSI data with a scene-derived alunite spectrum. The display color table has been inverted so that the grayscale representation of SAM results shows bright areas corresponding to regions of alunite occurrence whereas darker regions contain less alunite. The natural variability of alunite (i.e., various cation substitutions in the crystal lattice; particle size effects; nonlinear spectral mixing effects with other minerals) dictates that alunite maps should ideally be constructed with several different alunite spectra and then be combined. Nonetheless, the result in Figure 2.8 captures the general distribution of this mineral. Alunite presence is proportional to grayscale intensity though a quantitative measure of alunite abundance may be obtained by applying a spectral mixture analysis. Figure 2.8 is a material map and is a typical

---

* http://www.ittvis.com/envi/

product of HSI analysis. The application of SAM and other matching algorithms is the process of material identification and mapping.

2.3.9 Summary

HSI remote sensing is the simultaneous acquisition of hundreds of coregistered images of a scene over a range of wavelengths in the visible to long-wave infrared (0.4–14.0 μm) region of the electromagnetic spectrum. The HSI is remote material identification and characterization, otherwise known as spectroscopy. The fundamental data structure is the image cube, but the fundamental datum is the spectrum; the power and uniqueness of HSI reside in the information contained in the wavelength dimension. The application of HSI to geological mapping was presented as was a discussion of processing approaches. Though not discussed, HSI data and derived maps should be part of a larger picture drawn by data and information fusion. The HSI is a powerful remote sensing technology with a vigorous community of practice crossing all sectors (academic, government/civil, and private industry).

2.4 Ocean Remote Sensing*

The ocean covers three-quarters of the Earth's surface. It is also important for transportation, recreation, and resources that include food and pharmaceuticals. Conventional or *in situ* measurements are very limited, and monitoring such a large resource is very difficult. Remote sensing will have to play an increasing role in all aspects of ocean science. Remote sensing is the technique or process of obtaining data or images from a distance, as from satellites or aircraft. The advantage of using remote sensing is that it covers a larger geographic area more frequently than field sampling. Satellites can be used to monitor SST and chlorophyll—a concentration on a daily basis. In this section, we will introduce some remote sensing algorithms for these two most widely used parameters.

2.4.1 Infrared Remote Sensing of Sea-Surface Temperature

The variability of SST has a significant socioeconomic impact. For example, slow variations of SST over the tropics and subtropics can substantially change the planetary atmospheric flow and are the dominant factors in global and regional climate changes. Variations of SST of less than 1 K can occur anywhere over the ocean. However, variations greater than 1 K over a

---

* This section is contributed by Donglian Sun.

large area usually occur only during El Niño events. In some El Niño events, the SST over the eastern equatorial Pacific may be 4–5 K higher than the climate mean. The SST is also a good indicator of global warming. However, ground truth data over the vast oceans are limited. Therefore, satellite-based SST measurements combined with ground truth information have been the major source of high-resolution SST data (Reynolds 1988; Reynolds and Smith 1994).

The accuracy of satellite SST determination has improved significantly due to the development of radiometers with two or more atmospheric window channels within mid-IR and far-IR bands (e.g., McClain et al. 1983). The fundamental basis of multichannel SST (MCSST) algorithms is the differential water vapor absorption in the various atmospheric window regions of the spectrum. The current satellite MCSST algorithm can allow global SST retrievals at spatial scales of 8 km with a root mean square error <0.7 K from the AVHRR currently flown on the NOAA polar orbiters (Barton et al. 1993; Legeckis and Zhu 1997; May et al. 1998; McClain et al. 1985). The MODIS measurement accuracy requirement for SST is 0.35 K (Brown and Monnett 1996). The MODIS SST retrieval is a follow-up to the AVHRR SST algorithm (Brown 1996). The operational AVHRR MCSST and the MODIS SST algorithms are statistical methods. These methods, which combine the satellite observation and *in situ* observation, have proved to be very successful in producing reliable global SST data sets. The current operational SST retrieval methods are based on two windows within the 10–13 μm interval in the daytime and an additional window within the 3.5–4.2 μm interval in the nighttime. Some research studies have also used water vapor information in the statistical method (e.g., Emery et al. 1993). Although physical retrievals have not been used for operational SST retrieval due to the large computational requirement and possible instability, they are promising methods for improving the retrieval precision.

Although satellite (IR sensors measure radiance from the skin of the ocean, oceanographers are more interested in SSTs for the upper several meters of the oceans, commonly referred to as the "bulk temperature" (Schluessel et al. 1990). This interest in the bulk temperature has led to the practice of calibrating satellite-derived SSTs with *in situ* bulk SSTs measured by ocean buoys. The difference between skin and bulk temperatures contributes an added level of uncertainty to the satellite SST retrieval. The relationship between skin and bulk SSTs has been investigated by a number of scientists (e.g., Schluessel et al. 1990). Currently, the AVHRR SST is calibrated to bulk SST, whereas the ATSR measures the skin temperature (Zavody et al. 1994). The next NPOESS will integrate the existing polar systems (NOAA) from the Department of Commerce (DOC) and the DMSP in the DOD. With the new IR channel (8.55 μm) of the VIIRS on board the NPOESS, some new SST algorithms for the future NPOESS/VIIRS SST retrieval were developed (Ji et al. 2000). The accuracy requirement for the NPOESS VIIRS SST was improved to 0.25 K.

2.4.2 Theoretical Description of SST Retrieval

2.4.2.1 Physics of the Problem

In clear sky conditions, the outgoing IR spectral radiance at the TOA can be represented by

$$L(\lambda,\mu) = \tau(\lambda,\mu)\varepsilon(\lambda,\mu)B(\lambda,T_s) + L_a(\lambda,\mu) + L_s(\lambda,\mu,\mu_0,\varphi_0)$$

$$+ L_d(\lambda,\mu,\mu_0,\varphi_0) + L_r(\lambda,\mu,\mu_0,\varphi_0) \tag{2.1}$$

where τ is the transmissivity, ε the surface spectral emissivity, B the Planck function, L_a the thermal path radiance, and L_s the path radiance resulting from scattering of solar radiation. L_d is the solar diffuse radiation and L_r the atmospheric thermal radiation reflected by the surface. λ is the wavelength. $\mu = \cos(\theta)$, and $\mu_0 = \cos(\psi)$, where θ is the satellite zenith angle and ψ is the solar zenith angle. φ_0 is the azimuth angle.

The wavelength is the wavelength center of a narrow interval, because there is no way to measure the exact monochromatic signal as a continuous function of wavelength by satellite sensors. We should choose window channels with little or no atmospheric contribution to infer the surface information. As shown in Figure 2.9, the wavelengths between 3.5 and 4.2 μm; between 8 and 9 μm; and between 10 and 13 μm are some typical atmospheric windows. For a perfect window, the total atmospheric transmittance $\tau (\lambda\ \mu)$ should be 1.0. However, as indicated in Figure 2.9, the transmittances at these

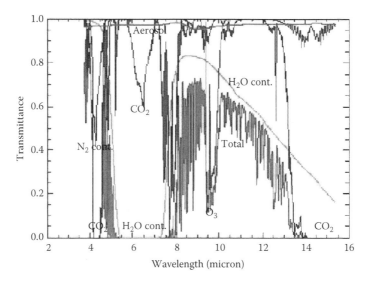

FIGURE 2.9
(See color insert following page 144.) Atmospheric transmittance versus wavelength.

windows are not 1.0. The main absorber for these windows is atmospheric water vapor.

Since the pioneering work of Anding and Kauth (1970) and Prabhakara et al. (1974), it has been known that atmospheric water vapor absorption effects in the IR can be corrected with high accuracy using linear combinations of multiple channel measurements.

$$T_s = a_0 + a_1 T_{11} + a_2 (T_{11} - T_{12}) T_b + a_3 (\sec \theta - 1) \tag{2.2}$$

where T_s is surface temperature, T is brightness temperature, and subscripts 11 and 12 refer to the two split window channels at 11 and 12 μm. a_0 to a_3 are regression coefficients. θ is the satellite zenith angle and T_b is the environmental temperature. The inclusion of an environmental temperature, T_b, as a multiplier for a brightness temperature difference between the two bands provides a different behavior at higher temperatures. A problem with implementing this version of the algorithm is the T_b term.

One must have an estimate of the temperature for the pixel within ±2σ before estimating its value. This is typically done using a climatology or an MCSST-type algorithm as a first guess.

With the improvements in satellite instruments and advanced algorithm developments, more advanced nonlinear and multichannel algorithms have been developed.

Nonlinear SST (NLSST) and MCSST algorithms have been used by AVHRR/MODIS/VIIRS.

Day (NLSST):
$$T_s = a_0 + a_1 T_{11} + a_2 T_{FG} (T_{11} - T_{12}) + a_3 (T_{11} - T_{12}) (\sec \theta - 1) \tag{2.3}$$

T_s retrieved SST; T_{11}, T_{12} brightness temperatures in channels 10.8 and 12 μm; θ view zenith angle; and T_{FG} first guess (*a priori*) SST (climate or analyses/forecast SST, e.g., Reynolds).

Night (MCSST):
$$T_s = a_0 + a_1 T_4 + a_2 T_{11} + a_3 T_{12} + a_4 (T_4 - T_{12}) (\sec \theta - 1) + a_5 (\sec \theta - 1) \tag{2.4}$$

T_4, T_{11}, T_{12} brightness temperatures in channels 3.8, 10.8, and 12 μm.; a_n, $n = 0$–5 Regression coefficients (derivation referred to as "SST Algorithm calibration"). Customarily calculated early in the satellite mission empirically against *in situ* SST (using 1–3 month matchups) and may be calculated using Radiative Tranfer Model (RTM) simulations or against a global reference SST field (e.g., Reynolds). A bias correction against *in situ* measurements is still needed.

The NLSST formulation was initially derived for AVHRR and it is currently used for MODIS and VIIRS. As a first-guess SST, the T_{FG} is used as a proxy for variations in atmospheric water vapor content. The band in 3.9 μm is not used during the daytime due to sunlight scattering and reflection. The MCSST formulation was initially derived for AVHRR and it is currently also used for MODIS. The band in 3.9 μm is only used during nighttime when the radiance received by satellite sensors is not contaminated by sunlight scattering and reflection.

The following regression methods are used in VIIRS SST retrieval testbed to derive the final algorithm and to drive the sensor design (Ji et al. 2001):

Daytime:

Split window (10.8 + 12 μm bands) nonlinear (modified from AVHRR operational, May et al. 1998):

$$SSM = a_0 + a_1 T_{11} + a_2(T_{11} - T_{12}) + a_3(\sec\theta - 1) + a_4(T_{11} - T_{12})^2 \qquad (2.5)$$

Triple window (10.8, 12, 8.55 μm bands) nonlinear:

$$SST = a_0 + a_1 T_{11} + a_2(T_{11} - T_{12}) + a_3(\sec\theta - 1)$$
$$+ a_4(T_{11} - T_{12})^2 + a_5(T_{11} - T_{8.55})^2 \qquad (2.6)$$

Quad (10.8, 12, 8.5, 4.05 μm bands) algorithm:

$$SST = a_0 + a_1 T_{11} + a_2 T_{12} + a_3(\sec\theta - 1) + a_4 T_{4.0} + a_5 T_{8.5}$$
$$+ a_6 T_{4.0} \cos(\theta s) + a_8(T_{11} - T_{12})^2 \qquad (2.7)$$

Water vapor algorithm (Emery et al. 1994):

$$SST = a_0 + a_1 T_{11} + a_2(T_{11} - T_{12}) + a_3(\sec\theta - 1) + a_4 * wat \qquad (2.8)$$

where θ is the satellite zenith angle at the Earth's surface, ranging from 0° to 53°; θs is the solar zenith angle (0° to 80° as daytime); and *wat* is the total column water. The temperatures T_{11}, T_{12}, $T_{4.0}$, and $T_{8.5}$ correspond to the brightness temperature at 10.8, 12, 4.05, and 8.55 μm bands, respectively.

Nighttime:

Split window (10.8 μm + 12 μm bands) nonlinear (modified from AVHRR operational, May et al. 1998):

$$SST = a_0 + a_1 T_{11} + a_2(T_{11} - T_{12}) + a_3(\sec\theta - 1) + a_4(T_{11} - T_{12})^2 \qquad (2.9)$$

Triple window (10.8 μm, 12 μm, and 3.7 μm bands) nonlinear (modified from AVHRR nighttime retrieval, May et al. 1998):

$$SST = a_0 + a_1 T_{11} + a_2(T_{3.7} - T_{12}) + a_3(\sec\theta - 1)$$

$$+ a_4(T_{11} - T_{12})^2 + a_5(T_{11} - T_{3.7})^2 \tag{2.10}$$

Triple window (10.8 μm, 12 μm, and 8.55 μm bands) nonlinear:

$$SST = a_0 + a_1 T_{11} + a_2(T_{8.55} - T_{12}) + a_3(\sec\theta - 1)$$

$$+ a_4(T_{11} - T_{12})^2 + a_5(T_{11} - T_{8.55})^2 \tag{2.11}$$

Quad (10.8, 12, 8.55, and 4.05 μm bands) algorithm:

$$SST = a_0 + a_1 T_{11} + a_2 T_{12} + a_3(\sec\theta - 1)$$

$$+ a_4 T_{8.5} + a_5 T_{4.0} + a_8(T_{11} - T_{12})^2 \tag{2.12}$$

Water vapor algorithm (Emery et al., 1994):

$$SST = a_0 + a_1 T_{11} + a_2(T_{11} - T_{12}) + a_3(\sec\theta - 1) + a_4 * wat \tag{2.13}$$

where θ is the satellite zenith angle with respect to the Earth's surface, ranging from $0°$ to $53°$, and *wat* is the total column water. The temperatures T_{11}, T_{12}, $T_{3.7}$, $T_{4.0}$, and $T_{8.5}$ correspond to the 10.8 μm, 12 μm, 3.7 μm, 4.05 μm, and 8.55 μm bands, respectively.

The triple window (2.6 and 2.11) and Quad (2.7 and 2.12) are the new algorithms developed for the NPOESS or VIIRS by Ji et al. (2000). Only a few of these equations will be used in the VIIRS SST retrieval. This will depend on the final design of the VIIRS instrument. The higher-order polynomial terms may not improve the results, therefore only second-order polynomial terms are used in the VIIRS algorithm. In order to improve uncertainty and accuracy, the SST field can be stratified into a few groups, and regression equations can be derived for each group.

Sun and Pinker (2007) proposed a four-channel or dual-split window algorithm for the EUMETSAT Meteosat Second Generation (MSG) imager SEVIRI (Spinning Enhanced Visible and InfraRed Imager).

Daytime:

$$T_s = a_0 + a_1 T_{11} + a_2(T_{11} - T_{12}) + a_3(T_{3.9} - T_{8.7})$$

$$+ a_4(T_{11} - T_{12})^2 + a_5(\sec\theta - 1) + a_6 T_{3.9} \cos\theta_s \tag{2.14}$$

Nighttime:

$$T_s = a_0 + a_1 T_{11} + a_2(T_{11} - T_{12}) + a_3(T_{3.9} - T_{8.7})$$

$$+ a_4(T_{11} - T_{12})^2 + a_5(\sec\theta - 1) \tag{2.15}$$

where the temperatures T_{11}, T_{12}, $T_{3.9}$, and $T_{8.7}$ correspond to the 11, 12, 3.9, and 8.7 μm bands, respectively. All other symbols are the same as mentioned in the SST formulas earlier.

Figure 2.10 compares the monthly mean SST in July 2005 as derived from the SEVIRI with Sun and Pinker's four-channel or dual-split window algorithm, SST from AVHRR at an 18 km resolution obtained from NASA JPL (http://podaac.jpl.nasa.gov) and AVHRR from Reynolds at 0.25° and 1° resolutions as obtained from the National Climate Data Center (NCDC). We can see that during the summer over the north tropical region, especially at the eastern Atlantic Ocean, monthly mean SST in the summer with diurnal

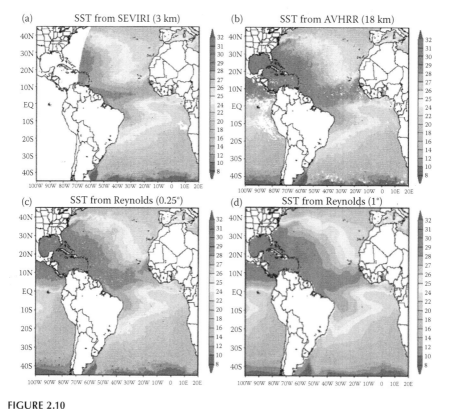

FIGURE 2.10
(See color insert following page 144.) Comparison of monthly mean SST from (a) SEVIRI, (b) AVHRR, (c) Reynolds (0.25°), and (d) Reynolds (1°) in July 2005.

sampling derived from the SEVIRI is usually cooler than the AVHRR-based SST with twice daily data (Figure 2.10).

Figure 2.11 shows the monthly mean, daily maximum and minimum, and the Diurnal Temperature Range (DTR = maximum – minimum) distribution in July 2005, as derived from the SEVIRI with the Sun and Pinker (2007) algorithm. The DTR is the most evident over the north tropical region. The major differences among the SST algorithms for ACGRR, MODIS, VIIRS, and SEVIRI are the different channels used, and, consequently, the regression coefficients will also be different. It is expected that with more IR window channels used, the accuracy can be improved.

2.4.3 Ocean Color Retrieval

The ideal ocean color measurement is made from a ship in which the seawater is collected and analyzed. The problem is that the oceans are vast and it is not practical in terms of time or cost. In fact, it would take a ship 10 years to collect an equivalent amount of data as a satellite can collect in 1 min.

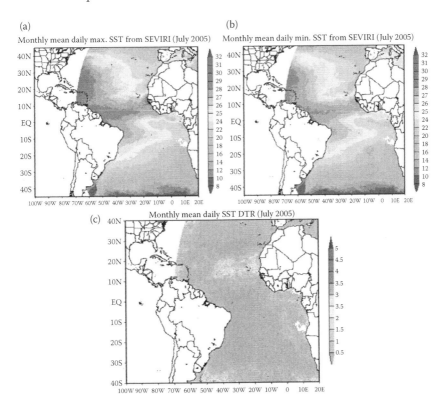

FIGURE 2.11
(See color insert following page 144.) Monthly mean daily (a) maximum, (b) minimum, and (c) DTR distribution in July 2005.

2.4.3.1 Ocean Color Sensors

There are many ocean color sensors that have been employed since 1978. Some of them have been listed later in the table. The design of these sensors varied considerably with a range of spectral bands and the swath and resolution. Some of the sensors described in this chapter are listed in bold in the table .

CZCS. The first satellite-based ocean color sensor was the Coastal Zone Color Scanner (CZCS). Other sensors had measured the reflected light of the ocean, but CZCS was the first to be specifically designed with spectral bands and a dynamic range appropriate for ocean color measurements. The four spectral bands selected for ocean color were 443, 520, 550, and 670 nm with 20 spectral band widths. One additional band was included at 750 nm for the land. There were also two thermal IR bands at 10.5 and 12.5 μm for measuring temperature. The sensor design was based on a scan mirror at 45° to nadir, which redirected the light into the cassegrain telescope and then through a polychromator. The CZCS sensor was on the Nimbus-7 space craft at an altitude of 940 km in a sun-synchronous orbit. The ground resolution was 825 m and the swath was 1600 km. The revisit time was 2 days but this was limited due to the sharing of power with other instruments. The CZCS was a proof-of-concept instrument and it led the way for later sensors for ocean color (Table 2.2).

Figure 2.12 shows an example of ocean color derived from the CZCS. The Gulf Stream can be clearly identified.

SeaWiFS. The SeaWiFS was the follow-on mission for ocean color after the CZCS. It was launched from an aircraft via a Pegasus launch vehicle in 1997. It has eight bands in the solar reflective region. It is at a sun-synchronous orbit at 705 km and has a 1-day revisit time. The "WiFS" in the name implies a wide field of view, which is 1500 km with a 1.1 km spatial resolution. The wide field of view significantly improves the spatial coverage and the temporal frequency. The radiometric resolution is 10 bit. The design of the sensor includes a half-angle mirror to minimize the polarization from the water, and it also has the ability to tilt 20°, which allows the avoidance of sun glint at the near equatorial latitudes. Figure 2.13 shows an example of global ocean color distribution as derived from the SeaWiFS.

MODIS. The MODIS includes the current ocean color satellite sensor. There are sensors in use aboard the Aqua and Terra satellite platforms launched in 1999 and 2002, respectively. The sensor has a total of 36 bands ranging from 0.4 to 14.4 μm with a 12-bit radiometric resolution. The spatial resolution varies from 250 to 1000 m depending on the spectral band. The swath is 2330 km and has a revisit time of every 1–2 days. The sensor has three different onboard calibration systems including a solar diffuser, a spectral calibrator, and a black body. The 36 bands provide a variety of data products in addition to ocean color. Some of these include clouds, aerosols, and surface temperatures. With an estimated lifespan of 6 years, the need for replacement hardware becomes more urgent. No schematic of the MODIS sensor can be found.

TABLE 2.2

Ocean Color Sensors on Satellites

Sensor	Agency	Satellite	Operating Dates	Swath (km)	Resolution (m)	Number of Bands	Spectral Coverage (nm)
CZCS	**NASA (USA)**	**Nimbus-7 (USA)**	**24/10/78–22/6/86**	**1556**	**825**	**6**	**433–12,500**
OCTS	NASDA (Japan)	ADEOS (Japan)	17/8/96–1/7/97	1400	700	12	402–12,500
POLDER	CNES (France)	ADEOS (Japan)	17/8/96–1/7/97	2400	6 km	9	443–910
MOS	DLR (Germany)	IRS P3 (India)	21/3/96	200	500	18	408–1600
MOS	DLR (Germany)	Priroda (Russia)	23/4/96	85	650	17	408–1010
SeaWiFS	**NASA (USA)**	**OrbView-2 (USA)**	**1/8/97**	**2806**	**1100**	**8**	**402–885**
OCI	NEC (Japan)	ROCSAT-1 (Taiwan)	Launched (January 1999)	690	825	6	433–12,500
MODIS	NASA (USA)	EOS AM-1 (USA)	Scheduled (August 1999)	2330	1000	36	405–14,385
MISR	NASA (USA)	EOS AM-1 (USA)	Scheduled (August 1999)	360	250	4	446–867
OCM	ISRO (India)	IRS-P4 (India)	Scheduled (May 1999)	1420	360	8	402–885
GLI	NASDA (Japan)	ADEOS-2 (Japan)	Scheduled (June 2000)	1600	250/1000	36	375–12,500
MERIS	ESA (Europe)	ENVISAT-1 (Europe)	Scheduled (end 2000)	1150	300/1200	15	412–1050
OSMI	KARI (Korea)	KOMPSAT (Korea)	Scheduled (August 1999)	800	850	6	400–900
POLDER-2	CNES (France)	ADEOS-2 (Japan)	Scheduled (June 2000)	2400	6000	9	443–910
S-GLI	NASDA (Japan)	ADEOS-3 (Japan)	Scheduled (March 2003)	1600	750	11	412–865

Source: Adapted from IOCCG 1999. Historical Ocean-Color Sensors, http://www.ioccg.org/sensors/sensors/historical.html, last accessed August 3, 2010.

Note: The table shows, in part, the number and ranges of sensors from countries other than the United States.

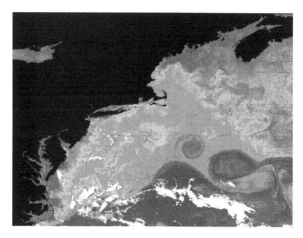

FIGURE 2.12
(**See color insert following page 144.**) The CZCS ocean color scene of the Gulf Stream.

Figure 2.14 shows an example of global ocean color distribution as derived from the MODIS.

VIIRS. The VIIRS is likely to be the first successor of the MODIS. It will provide higher spatial and radiometric resolution than prior sensors. The VIIRS is not a dedicated ocean color sensor. The data products produced will range from ocean color to forest fires and aerosols. The VIIRS will provide almost two dozen spectral bands with a 14-bit radiometric resolution. The

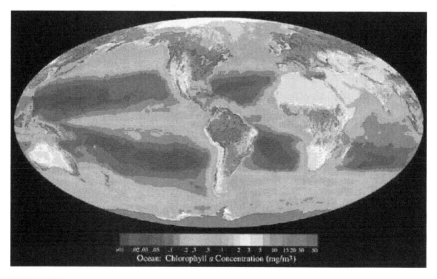

FIGURE 2.13
(**See color insert following page 144.**) The SeaWiFS ocean color. (Adapted from http://Earthobservatory.nasa.gov/images/imagerecords)

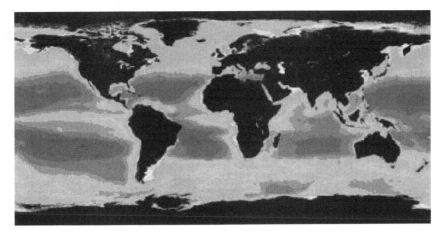

FIGURE 2.14
(See color insert following page 144.) The MODIS ocean color product. (Adapted from http://
www.nasa.gov/images/content)

satellite will be in a sun-synchronous orbit at an altitude of 824 km. The
spatial resolution will be 0.65 km with a swath of 3000 km.

2.4.3.2 Ocean Color Algorithms

The algorithms for converting the radiance into chlorophyll concentration
are fairly simple for first-order approximations. The algorithms have been
modified to many forms for different sensors and for different optical components in different waters. Experiences with SeaWiFS and MODIS show
that high-quality ocean color products for global open ocean (Case-1 waters)
can be achieved. Improvements in water color products are still needed for
inland and coastal regions (Case-2 waters), because turbid waters violate the
NIR black ocean assumption, and strongly absorbing aerosols violate the
non- or weakly absorbing aerosols.

In this section, we will introduce some basic algorithms for ocean color
retrieval. At the satellite's altitude, 90% of sensor-measured signals over the
ocean comes from the atmosphere and surface. Ocean color remote sensing
involves deriving the ocean water-leaving radiance spectra by accurately
removing the atmospheric and surface effects.

2.4.3.3 Physics of the Problem

Radiance received by a sensor at TOA:

$$L_t(\lambda) = L_{path}(\lambda) + T(\lambda)L_g(\lambda) + t(\lambda)L_{wc}(\lambda) + t(\lambda)L_w(\lambda) \qquad (2.16)$$

where $L_t(\lambda)$: radiance received by a sensor at TOA, $L_{path}(\lambda)$: radiance generated along the optical path, $L_g(\lambda)$: radiance reflected by surface (sun glitter),

$L_{wc}(\lambda)$: radiance reflected by whitecaps, $L_w(\lambda)$: water-leaving radiance (usually <10% of $L_t(\lambda)$), T and t are the direct and diffuse transmittances of the atmosphere.

After light enters the ocean, some of it eventually scatters back up and is the so-called "water-leaving radiance" $L_w(\lambda)$, which can be derived from satellite observations after removing atmospheric effects. The magnitude, spectral variation, and angular distribution of this radiance depend on the absorption and backscattering coefficients of the seawater, $a(\lambda)$ and $b_b(\lambda)$, respectively; the downwelling irradiance incident on the sea surface $E_d(\lambda,0^+)$; and the geometry of the light within the ocean. To make things easier, sea-water can be divided into three components, each one having distinct optical properties. These components are the seawater itself (water and salts), the particle fraction, and the dissolved fraction. Fortunately, $a(\lambda)$ is simply equal to the sum of the absorption coefficients for each component, and $b_b(\lambda)$ is equal to the sum of the backscattering coefficients. If we can accurately model each spectrally distinct component of the absorption and backscattering coefficients, then we can determine the magnitude of each one from measurements of $L_w(\lambda)$ and $E_d(0^+,\lambda)$, given some assumptions about the angular distribution of light in the water. The key here is to accurately model the spectral behavior of $a(\lambda)$ for each component. The spectral behavior of $b_b(\lambda)$ is not as dynamic as $a(\lambda)$ (Table 2.3).

The MODIS and VIIRS ocean color algorithms use water-leaving reflectance instead of water-leaving radiance:

$$\rho_t(\lambda) = \rho_{path}(\lambda) + T(\lambda)\rho_g(\lambda) + t(\lambda)\rho_{wc}(\lambda) + t(\lambda)\rho_w(\lambda) \qquad (2.17)$$

where $\rho_t(\lambda) = L_t(\lambda) \cdot \pi/F_o$, F_o is the solar irradiance, and π/F_o is constant for each λ. $T(\lambda)\rho_g(\lambda)$ is the sun glint contribution, which is usually avoided, masked, or corrected.

TABLE 2.3

Ocean Color and Other Useful Spectral Bands for VIRRS, MODIS, and SeaWiFS

VIIRS	MODIS		SeaWiFS	
Ocean Bands (nm)	Other Bands (nm)	Ocean Bands (nm)	Other Bands (nm)	Ocean Bands (nm)
412		412	645	412
445		443	859	443
488		488	469	490
—		531	555	510
555	SWIR Bands	551	SWIR Bands	555
672	1240	667	1240	670
746	1610	748	1640	765
865	2250	869	2130	865

The VIIRS does not retrieve ocean products when sun glint occurs. $\rho_{wc}(\lambda)$ is the whitecap reflectance, which is assumed to be only related to wind speed, W:

$$[\rho_{wc}(\lambda)]_N \sim 6.49 \times 10^{-7} \, W^{3.52}$$

The only unknown term is $\rho_{path}(\lambda)$:

$$\rho_{path}(\lambda) = \rho_r(\lambda) + \rho_a(\lambda) + \rho_{ra}(\lambda) \tag{2.18}$$

where $\rho_r(\lambda)$: reflectance resulting from Rayleigh scattering (can be calculated accurately when provided surface pressure and wind), $\rho_a(\lambda)$: reflectance resulting from aerosol scattering, $\rho_{ra}(\lambda)$: reflectance resulting from air–aerosol interaction.

The MODIS chlorophyll a algorithm (Carder et al. 2003) is based on a semi-analytical, bio-optical model of remote-sensing reflectance, $R(\lambda)$, where $R(\lambda)$ is defined as the water-leaving radiance, $L_w(\lambda)$, divided by the downwelling irradiance just above the sea surface, $E_d(\lambda,0^+)$.

The basic assumption of relating reflectance, $R(\lambda)$, seawater absorption coefficient, $a(\lambda)$, and seawater backscattering coefficient, $b_b(\lambda)$, is:

$$R(\lambda) = \text{const} \cdot \frac{b_b(\lambda)}{a(\lambda)} \tag{2.19}$$

where the "constant" is unchanging with respect to λ and θ_0. The value of the constant is not relevant to the algorithm. Both $b_b(\lambda)$ and $a(\lambda)$ will be divided into several separate terms. Each term will be empirically described.

The total backscattering is contributed by pure seawater and particulate suspended matter:

$$b_b(\lambda) = b_{bw}(\lambda) + b_{bp}(\lambda) \tag{2.20}$$

where the subscripts "w" and "p" refer to the water and particles, respectively. These terms can be derived using empirical methods. $b_{bw}(\lambda)$ is constant and well known (Smith and Baker 1981). $b_{bp}(\lambda)$ is modeled as:

$$b_{bp}(\lambda) = X \left[\frac{551}{\lambda} \right]^Y \tag{2.21}$$

The magnitude of particle backscattering is indicated by X, which is equal to $b_{bp}(551)$, whereas Y describes the spectral shape of particle backscattering.

$$X = X_0 + X_1 \, R(551)$$
$$Y = Y_0 + Y_1 \, R(488) \tag{2.22}$$

The total absorption coefficient is contributed by water, phytoplankton pigments, and dissolved organic matter (DOM):

$$a(\lambda) = a_w(\lambda) + a_{ph}(\lambda) + a_d(\lambda) + a_{dom}(\lambda) \tag{2.23}$$

where the subscripts "w," "ph," "d," and "dom" refer to water, phytoplankton pigments, detritus, and DOM, respectively. Here $a_w(\lambda)$ is taken from Pope and Fry (1997). Expressions for $a_{ph}(\lambda)$, $a_d(\lambda)$, and $a_{dom}(\lambda)$ need to be developed. The term $a_w(\lambda)$ is also empirically derived.

The shape of the $a_{ph}(\lambda)$ spectrum for a given water mass will change due to the pigment-package effect (i.e., the flattening of absorption peaks due to self-shading with increasing intracellular pigment concentration and larger cell size, Morel and Bricaud (1981) and changes in pigment composition. For MODIS, normalizing that measured $a_{ph}(\lambda)$ spectra to $a_{ph}(675)$ reduces the dynamic range and results in a smooth variation for $a_{ph}(\lambda)/a_{ph}(675)$ versus $a_{ph}(675)$ for the MODIS wavebands centered at $\lambda = 412$, 443, 488, and 551 nm. A hyperbolic tangent function was chosen to model this relationship in order to ensure that the value of $a_{ph}(\lambda)/a_{ph}(675)$ approaches an asymptote at very high or very low values of $a_{ph}(675)$. Logarithmic scaling for both axes results in the following equation for $a_{ph}(\lambda)$ as a function of $a_{ph}(675)$:

$$a_{ph}(\lambda) = a_0(\lambda)\exp\left[a_1(\lambda)\tanh\left[a_2(\lambda)\ln\left(\frac{a_{ph}(675)}{a_3(\lambda)}\right)\right]\right] * a_{ph}(675) \tag{2.24}$$

where the parameters $a_0(\lambda)$ to $a_3(\lambda)$ are empirically determined for each MODIS wavelength of interest. $a_0(\lambda)$ is the most important of these parameters, because it is directly proportional to $a_{ph}(\lambda)$. For simplicity, only $a_0(\lambda)$ and $a_1(\lambda)$ are varied to parameterize $a_{ph}(\lambda)$, with $a_2(\lambda)$ and $a_3(\lambda)$ being set to the constant values of −0.5 and 0.0112, respectively.

For VIIRS, the phytoplankton pigment absorption coefficient, $a_{ph}(\lambda)$, can be normalized with $a_{ph}(672\,\text{nm})$:

$$a_{ph}(\lambda) = a_{ph}(672\,\text{nm}) \cdot F(a_{ph}(l, 672\,\text{nm})) \tag{2.25}$$

Both the $a_d(\lambda)$ and $a_{dom}(\lambda)$ can be fit to a curve of the form $a_x(\lambda) = a_x(400)$ $\exp[-S_x(\lambda - 400)]$ where the subscript "x" refers to either "d" or "dom" (Bricaud et al. 1981; Carder et al. 1991; Roesler et al. 1989). Owing to this similarity in spectral shape, these terms cannot be spectrally separated with the MODIS channels, so the $a_d(\lambda)$ term is combined operationally with $a_{dom}(\lambda)$, and both detrital and DOM absorption are represented by $a_g(\lambda)$. The combined DOM and detritus absorption term is thus written:

$$a_g(\lambda) = a_g(400)\exp[-S(\lambda - 400)] \tag{2.26}$$

where S is empirically determined. Many researchers have reported that $S_d = 0.011$ nm^{-1}, on average (Roesler et al. 1989). An average value of 0.017 nm^{-1} was measured for S_{dom}.

2.4.3.4 Inverting the Model

All of the pieces of the reflectance model are now in place. Substitute all equations to the reflectance model Equation 2.18, and $R(\lambda)$ may be expressed solely as a function of the "constant" term, $R(443)$, $R(488)$, $R(551)$, $a_{ph}(675)$, and $a_{dom}(400)$, given values for the parameters for X, Y, $a_0(\lambda)$, $a_1(\lambda)$, and S. Then, for each pixel, the R model equation can be written for each of the five available MODIS wavebands yielding five equations written in three unknowns: the "constant" term, $a_{ph}(675)$, and $a_g(400)$. Using spectral ratios of R removes the "constant" term, because it is largely independent of wavelength. In principle, two spectral ratio equations can be used to solve the two remaining unknowns, $a_{ph}(675)$ and $a_{dom}(400)$. Based on the shape of the absorption curve for phytoplankton versus those for DOM and detritus, equations using spectral ratios of 412:443 (MODIS) or 412:445 (VIIRS) and 443:551 (MODIS) or 445:555 (VIIRS) for $R(\lambda)$ generally provide good separation of the two absorption contributions.

$$\frac{R(412)}{R(443)} = \frac{\left[b_b(412)a(443)\right]}{\left[b_b(443)/a(412)\right]}$$

$$\frac{R(443)}{R(551)} = \frac{\left[b_b(443)a(551)\right]}{\left[b_b(551)/a(443)\right]}$$

(2.27)

The right-hand side of each equation is a function of $a_{ph}(675)$, $a_{dom}(400)$, $R(443)$, $R(488)$, and $R(551)$. Since the R values are provided on input, we now have two equations with two unknowns. The equations can usually be solved algebraically to provide values for $a_{ph}(675)$ and $a_{dom}(400)$. Three channels (412, 443, and 551 nm) can be used to get two equations to solve for $a_{ph}(675$ nm) (MODIS) or $a_{ph}(672$ nm) (VIIRS) and a_0.

Pigment Algorithm for Semi-analytical Case. To evaluate variations of $a_{ph}(675)$ with [chl a] for subtropical to tropical waters, a data set is developed to explore the more limited variation in surface values of $a_{ph}(675)$ under highlight conditions. Linear regression of log ([chl a]) versus log $[a_\varphi(675)]$ yielded an equation of the form. Chlorophyll concentration (chl a) is then determined from empirical regression:

$$\log(\text{chl}\, a) = a + b * \log\left[a_{ph}(675\,\text{nm})\right]$$

(2.28)

where chl a is the chlorophyll concentration (mg/m^3), a and b are regression coefficients.

Pigment Algorithm for the Default Case. When the semianalytical algorithm does not return a value for $a_{ph}(675)$, an empirical, two-wavelength algorithm for [chl *a*] is used by default. Aiken et al. (1995) found that the $L_w(488)/L_w(551)$ ratio is best for empirical [chl *a*] determination. We use an equation of the form:

$$\log\left[\text{chl}\,a\right]_{emp} = c_0 + c_1 \log\left(r35\right) + c_2\left[\log\left(r35\right)\right]^2 + c_3\left[\log\left(r35\right)\right]^3 \quad (2.29)$$

$$r35 = \frac{R(488)}{R(551)} \quad (2.30)$$

where [chl *a*]$_{emp}$ is called the "empirically-derived" or "default" chlorophyll *a* concentration and c_0, c_1, c_2, and c_3 are empirically derived constants.

Weighted Pigment Algorithm. Another consideration is that there should be a smooth transition in [chl *a*] values when the algorithm switches from the semianalytical to the empirical method. This is achieved by using a weighted average of the [chl *a*] values returned by the two algorithms when near the transition border. When the semianalytical algorithm returns an $a_{ph}(675)$ value between 0.015 and 0.03 m^{-1}, [chl *a*] is calculated as:

$$\left[\text{chl}\,a\right] = w * \left[\text{chl}\,a\right]_{sa} + \left(1 - w\right)\left[\text{chl}\,a\right]_{emp} \quad (2.31)$$

where [chl *a*]$_{sa}$ is the semianalytically derived value, [chl *a*]$_{emp}$ is the empirically derived value, and the weighting factor is $w = [0.03\text{Ba}_\varphi(675)]/0.015$.

More algorithm developments for ocean color retrieval include:

- Arnone et al. (1998) and Siegel et al. (2000) account for the NIR ocean contributions for SeaWiFS and MODIS NIR bands.
- Hu et al. (1999) proposed an adjacent pixel method.
- Gordon et al. (1997) and Chomko et al. (2003) proposed the spectral optimization algorithm.
- Ruddick et al. (2000) proposed regional Case-2 algorithm using the spatial homogeneity of the aerosol in a given area.
- Lavender et al. (2004) developed a regional bio-optical model (suspended sediments) for SeaWiFS application.
- Wang and Shi (2005) derived NIR ocean contributions using the MODIS SWIR bands.
- Doerffer et al. (1999) developed Artificial Neural Network for coastal Case-2 waters (implemented for MERIS data processing).

- Wang et al. (2007) proposed atmospheric correction using the SWIR bands for the turbid coastal waters.

A comparison of ocean color derived from MODIS with atmospheric correction by using the standard NIR bands with those using the SWIR band and combined NIR and SWIR is shown in Figure 2.15.

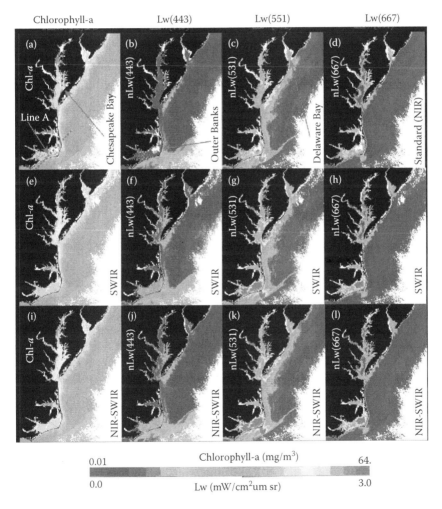

FIGURE 2.15
(**See color insert following page 144.**) Comparisons of MODIS ocean color products from NIR, SWIR, and NIR-SWIR Combined Methods. (Adapted From Wang, M. and W. Shi. 2007. *Optics Express*, 15:15722–15733.)

2.5 GeoSensor Networks and Sensor Web: The Move Toward Systems of Sensor Systems*

Recent advances in sensor technology and deployment strategies have greatly affected geoinformatics in terms of data collection and analysis. First, we have a substantial improvement in the ability to *collect* diverse types of georeferenced information. For example, traffic cameras provide continuous streams of georeferenced video at distributed locations in a neighborhood. Similarly, sensors distributed in an ecosystem may be monitoring local humidity or pollution measures. Further, we are witnessing a substantial change with regard to the manner in which this collected information is *analyzed*. Traditionally, such information was analyzed offline. However, with the advent of nanotechnology, it becomes feasible and economically viable to develop and deploy low-cost, low-power devices that are general-purpose computing platforms with multipurpose onboard sensing and wireless communications capabilities. In this manner, sensors not only collect information but also participate in the analysis of this information. These sensors may be acting collaboratively, functioning as nodes within broader network configurations that may range in scale and scope from a few cameras monitoring traffic to thousands of nodes monitoring an ecosystem. The resulting novel geocomputational paradigm integrates distributed sensing, analysis, and communications and is often referred to as a *sensor Web*. Since the area covered by the nodes decreases in scale and the nodes themselves are limited to zero- or low-altitude deployment (excluding, for example, satellites), the concept of the Sensor Web morphs into *wireless sensor networks* (WSN) or GSN, two terms that are often used almost interchangeably to refer to such integrated observation and computation infrastructures.

A Sensor Web can be defined as an autonomous, coordinated, reconfigurable, spatially distributed infrastructure comprised of sensing devices (also referred to as *nodes*) that *monitor* various conditions (e.g., temperature, illumination) at specific locations and transmit the collected information to either other nodes of this network or a remote central location. Inherent in this definition is the Sensor Web's ability to not just passively record data but also react to the phenomenon it monitors, for example, by selecting specific nodes to record information so as to better capture the underlying event. Typically, the information communicated in a Sensor Web includes sensor and/or footprint geolocation, readings, and accompanying metadata (Botts et al. 2007). Since the collected information may be diverse, ranging from seismic activity measurements to digital imagery and human reports, it is easy to understand that the term *sensor* within this context is used to refer to a wide variety of devices, such as thermometers and acoustic sensors or even satellite and human intelligence feeds.

---

* This section is contributed by Anthony Stefanidis.

Pioneering work on the Sensor Web was performed in the late 1990s by Kevin Delin at JPL (Delin et al. 1999), who pursued applications like climate and agricultural monitoring or the *in situ* exploration of gaseous biosignatures (Delin and Jackson 2000). During these early developmental stages, sensing and communication tasks were typically handled by different units with the network. As nanotechnology allowed the development of miniature sensors with sufficient computational and communicational capabilities, Sensor Web evolved. This evolution resulted in substantial increases in the number of sensors participating in the network and in raised levels of in-network collaboration, thus allowing us to embed higher levels of analysis within the network. This enables the network to function increasingly as an integrated sensing, analyzing, and reporting *reconfigurable infrastructure unit*, thus linking the concept of the Sensor Web with that of the WSNs (Akiyildiz et al. 2002). The defining characteristics of a WSN are

- Very high number of heterogeneous sensor nodes, often orders of magnitude higher than in traditional networks.
- These sensors are deployed over a limited geographic area, thus resulting in *dense coverage* that allows us to capture adequate detail of the observed phenomenon.

Advanced communication capabilities within the network are needed to support the effective and coordinated collaboration of individual sensors. Information is communicated within the network in-between nodes (typically from one node to one of its neighbors) or through select gateway nodes to end users (Figure 2.16).

- The majority of network sensors tend to be small-size observation units (often referred to as *motes* or *pods*) and typically have a limited energy supply, thus making *power consumption* a critical issue in sensor network operations. This also necessitates measures to ensure network robustness for failure of individual nodes.

  Sensor network applications are very broad, ranging, for example, from tactical surveillance and environmental monitoring to patient data monitoring.

  In geosensor networks (Nittel et al. 2008; Stefanidis and Nittel 2003), the geographic or spatial aspect is dominant in one or both of the following levels:
- *Data level*, as it may be the primary type of data collected by the sensors (e.g., sensors recording the movement or deformation of objects)
- *Analysis level*, as the spatial *distribution* of sensors may provide the integrative layer to support the analysis of the collected information, for example, when analyzing the dispersion of a pollutant as it is

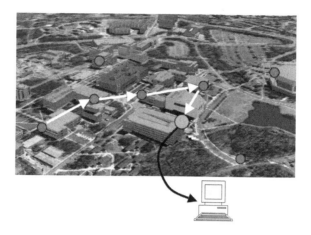

FIGURE 2.16
Sensors (marked by smaller, dark circles) distributed in a campus. Data traffic within the sensor network (white lines) and through a gateway node (larger, lighter circle) transmitted to an end user.

captured by sensors in a metropolitan area to determine the extent and source of a contamination (Cervone et al. 2009).

The geographic space covered by a sensor network, or analyzed through its measurements, may range in scale from the confined environment of a room or the infrastructure of a smart building to the highly complex dynamics of an ecosystem region.

2.5.1 Sensor Characteristics for WSN Applications

The sensors participating in a broadly defined Sensor Web design may be any of the wide variety of sensors that collect information for geoinformatics applications. Satellite feeds, RADAR, GPS measurements, surveillance camera feeds, thermometer readings, and even human reports are all examples of sensor input within the context of a sensor Web. However, the constraints imposed by WSN configurations dictate certain constraints in the design of the corresponding motes or pods. Motes or pods have to be capable of gathering data (sensing), communicating with neighboring nodes, and processing local sensor network information. Accordingly, each node comprises certain key elements (Figure 2.17):

- *One or more sensors:* Each node has one of more sensors to locally collect data to support the monitoring of a phenomenon via the senor network. These sensors may be active or passive, and their most critical characteristic from a sensor network point of view is the corresponding energy consumption and data transmission requirements. Although sensor network applications in the computer science community tend to focus almost exclusively on low-consumption passive

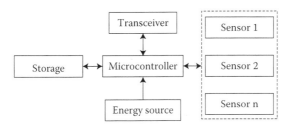

FIGURE 2.17
Components of a sensor node.

sensors (e.g., thermometers or sensors measuring the concentration of a particular gas), geoinformatics Sensor Web applications also consider sensors like cameras and radar.

- *A microcontroller or microprocessor:* This is the processing unit of the node that controls and executes its basic functionalities. Since energy consumption is often of particular interest, switching a sensor between *sleep* (inactive) and *awake* (active) states is a key part of a controller's activity.

- *A transceiver:* A two-way communication component is used to allow the interaction between neighboring sensors and the flow of information within the network. Typically, this is accomplished using a radio and antenna, though IR or laser transceivers are also available (but the latter are limited to line-of-sight applications). The range of such transceivers is commonly a few hundred yards.

- *Data storage:* Each node has some, rather limited, data storage capacity for future retrieval and data analysis.

- *An energy source:* All sensor operations require energy, which is typically provided by either batteries or dedicated capacitors. Although these energy sources are typically limited in their capacity, they may be renewed by tapping into ambient energy sources, such as solar- (e.g., accessible through solar panels) or vibration-derived energy. Communication within the network typically consumes much higher levels of energy than sensing itself.

2.5.2 Performance Requirements

Sensor Web performance requirements are typically dictated by the application itself; however, some popular representative requirements are easy to identify, and they include

- *Synchronized sampling:* Synchronization is often critical to better analyze the underlying phenomena corresponding to the distributed data collected by the network. In extreme applications, such as the above-mentioned volcanic monitoring, synchronization, and timestamping of the OASIS sensors with precise UTC, time with error in the order of milliseconds is required.

- *Real-time continuous data flow:* Although the network may be dormant when no activity is detected, it is desirable to have real-time continuous flow of data once an activity is detected.

- *Long-duration robust operation:* Although pods energy supply is typically limited, it is desirable that the network operation lasts as long as possible. This necessitates the development of efficient techniques to limit the active state of individual sensors to the minimum necessary.

- *Resilience to physical conditions:* Since sensor Webs are particularly useful for applications in remote and/or unfriendly environments, as is the case, for example, with the USGS monitoring of volcanic activity using the OASIS Sensor Web infrastructure (Xua et al. 2009), it is desirable that the sensors are able to survive extreme conditions (e.g., in terms of weather and chemical composition).

- *Ability to interact with external input and resources:* Since decision making that uses sensor Webs typically involves additional resources (e.g., access to additional data sources and computational capabilities), it is essential to support the interaction between the Sensor Web and these resources.

- *Fast and easy deployment or task assignment:* Optimally, a sensor network should be deployed in a fast manner requiring minimal human intervention, for example, air-dropped over a volcano (Song et al. 2009) or thrown as dust within the monitored area (Warneke et al. 2001). Fast deployment cannot be considered a realistic requirement when dealing with space-based sensors for rather obvious reasons, but the flexibility of such sensors to task assignments is the equivalent performance metric (e.g., the ability to orient sensors over a particular area of interest).

2.5.3 Sensor Web for Earth Sciences and GEOSS

A concerted effort to formalize the concept of the Sensor Web as it pertains to Earth Sciences was the focus of NASA's Earth Science Technology Office Sensor Web Technology in 2007. There, Sensor Web was defined as "a coherent set of heterogeneous, loosely coupled, distributed nodes, interconnected by a communications fabric that can collectively behave as a single dynamically adaptive and reconfigurable observing system" (ESTO 2007). The interesting aspect of this definition is the specific reference to a *communications fabric* as a key component of a sensor Web, in addition to its sensors. The term *communications fabric* refers to the aggregation of networking and communication technologies that allow the different nodes of a Sensor Web to communicate with each other and the user community to access the information contained and generated within a Sensor Web. The use of standards meets this goal and is of obvious importance in supporting interoperability.

The nodes of the Sensor Web may be *sensing, computational, storage,* or *composite* nodes, with the latter type being combinations of the first three (e.g., sensing *and* storage). The above-mentioned communications fabric allows the flow of information from sensing nodes to computational and storage nodes and back. These various components of a Sensor Web are shown in Figure 2.18. Sensing nodes allow the system to interact with the environment and handle the *data capturing* tasks of the system. The middle layer in this schema is the *information generation* layer, comprised of storage and computational operations. Computational operations make use of models that describe the function of the sensors and the complex environmental system that is monitored by the Sensor Web. The top layer of the system is *information delivery,* allowing the system to communicate with other systems of the user community at large. The communications fabric within the network enables in-network communication and controls its operations. As mentioned earlier, individual nodes may belong to one or more (in the case of composite nodes) of these components of the Sensor Web.

Although Figure 2.18 shows the various components of the system, it does not capture the hierarchical organization that is inherently assumed in this approach, which is a key issue for sensor Webs to support the dynamic reconfiguration of their resources and function. Individual Webs may interact to create a broader system, and an individual node may belong to more than one Sensor Web at any given time. This is bringing forward the notion of the Sensor Web as a system of systems (Van Zyl et al. 2009).

The Global Earth Observation System of Systems (GEOSS) is an example of a Sensor Web as a system of systems, where emphasis is placed primarily on

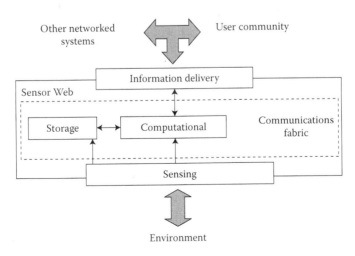

FIGURE 2.18
The different components of a Sensor Web.

interoperability (e.g., through the promotion of common standards) among its various components rather than classic WSN issues like energy conservation. The GEOSS is a concerted effort to link together existing and projected Earth observing capabilities and corresponding analysis tools, to better serve the needs of diverse science and decision-making communities. The potential applications of such a system of systems are numerous, ranging from drought prediction and disease monitoring to accurately forecasting weather and energy needs and disaster mitigation (Lautenbacher 2006).

Access to GEOSS, data sets are provided through *GEOPortal* (see e.g., www.geoportal.org), a Web-based interface for searching and accessing GEOSS data, information, imagery, services, and applications. Through the GEOPortal user, requests are directed to the *GEOSS Clearinghouse*, which accesses, collects, stores, and disseminates information, data, and metadata available through the various registered resources of GEOSS. Responses to user queries are thus identified and returned. Interoperability issues that relate to this operation are handled through the use of a *GEOSS Component and Service Registry*, wherein agencies and organizations that contribute data and services to GEOSS provide formal listings and descriptions of their contributions (data sets, models, and tools). The *GEOSS Standards Registry* supports system interoperability through the development and use of standards for the exchange of data and information among the various components of the system. Through its components and functionalities, GEOSS aims at optimizing data dissemination at global and multidisciplinary levels, minimizing unnecessary duplications, and fostering the exchange of reliable and up-to-date data sets.

References

Abdou, W.A., D.J. Diner, J.V. Martonchik et al. 2005. Comparison of coincident Multiangle Imaging Spectroradiometer and Moderate Resolution Imaging Spectroradiometer aerosol optical depths over land and ocean scenes containing Aerosol Robotic Network sites. *Journal of Geophysical Research*, 110(D10S07):11967–11976.

Abrams, M.J., R.P. Ashley, L.C. Rowan and A.F.H. Goetz. 1977a. Mapping of hydrothermal alteration in the Cuprite mining district, Nevada, using aircraft scanner images for the spectral region 0.46 to 2.36 µm. *Geology*, 5(12):713–718.

Abrams, M.J., R.P. Ashley, L.C. Rowan, A.F.H. Goetz and A.B. Kahle. 1977b. Use of imaging in the 0.46–2.36 µm spectral region for alteration mapping in the Cuprite mining district, Nevada. United States Geological Survey Open-File Report 77–585.

Acker, J., L. Chiu, P. Ardanuy et al. 2002. Remote sensing from satellites. In *Encyclopedia of Physical Science and Technology*, ed. R. Meyers, pp. 161–202. Academic Press, New York, NY.

Adams, J.B. and A.R. Gillespie. 2006. *Remote Sensing of Landscapes with Spectral Images: A Physical Modeling Approach*. Cambridge University Press, Cambridge, MA.

Adams, J.B., M.O. Smith, and A.R. Gillespie. 1993. Imaging spectroscopy: Interpretation based on spectral mixture analysis. In: *Remote Geochemical Analysis: Elemental and Mineralogical Composition. Topics in Remote Sensing 4*, eds. C.M. Pieters, and P.A.J. Englert, pp. 145–166. Cambridge University Press, Cambridge, MA.

Adler, R.F., G.J. Huffman, A. Chang et al. 2003. The Version-2 Global Precipitation Climatology Project (GPCP) monthly precipitation analysis (1979–present). *Journal of Hydrometeorology*, 4(6):1147–1167.

Akiyildiz, I., W. Su, Y. Sankarasubramaniam, and E. Cayirci. 2002. A survey on sensor networks. *IEEE Communications Magazine*, 40(8):102–114.

Anding, D. and R. Kauth. 1970. Estimation of sea surface temperature from space. *Remote Sensing of Environment*, 1(4):217–220.

Asner, G.P., D.E. Knapp, E.N. Broadbent, P.J.C. Oliveira, M. Keller, and J.N. Silva. 2005. Selective logging in the Brazilian Amazon. *Science*, 310:480–482.

Barrett, E. and D. Martin. 1985. *The Use of Satellite Data in Rainfall Monitoring*. Academic Press, London, UK.

Boose, E.R., D.R. Foster, and M. Fluet. 1994. Hurricane impacts to tropical and temperate forest landscapes. *Ecological Monographs*, 64:369–400.

Botts, M., G. Percivall, C. Reed, and J. Davidson. 2007. *OGC Sensor Web Enablement: Overview and High Level Architecture*. OGC White Paper. OGC 07–165.

Bowman, K.P. 2006. Transport of carbon monoxide from the tropics to the extratropics. *Journal of Geophysical Research—Atmosphere*, 111: D02107. doi:10.1029/2005JD006137.

Boyd, D. and F. Petitcolin. 2004. Remote sensing of the terrestrial environment using middle infrared radiation (3.0–5.0 μm). *International Journal of Remote Sensing*, 25(17):3343–3368.

Bricaud, A., A. Morel, and L. Prieur. 1981. Absorption by dissolved organic matter of the sea (yellow substance) in the UV and visible domains. *Limnology and Oceanography*, 26 (1):43–53.

Campbell, J. 2002. *Introduction to Remote Sensing*. Taylor & Francis, London.

Carder, K.L., F.R. Chen, Z. Lee, S.K. Hawes, and J.P. Cannizzaro. 2003. *Case 2 Chlorophyll a. MODIS Ocean Science Team Algorithm Theoretical Basis Document*, ATBD 19. http://modis.gsfc.nasa.gov/data/atbd/atbd_mod19.pdf (accessed April 26, 2009).

Carder, K.L., S.K. Hawes, K.A. Baker, R.C. Smith, R.G. Steward, and B.G. Mitchell. 1991. Reflectance model for quantifying chlorophyll *a* in the presence of productivity degradation products. *Journal of Geophysical Research*, 96(11): 599–611.

Cervone, G., A. Stefanidis, P. Franzese, and P. Agouris. 2009. Spatiotemporal modeling and monitoring of atmospheric hazardous emissions using sensor networks. *IEEE International Conference on Data Mining—Workshops (Spatial and Spatiotemporal Data Mining—SSTDM'09)*, Miami, FL, pp. 571–576.

Chang, C.I. 2003. *Hyperspectral Imaging. Techniques for Spectral Detection and Classification*. Kluwer Academic/Plenum Publishers, New York.

Chiu, L.S., 1988. Estimating areal rainfall from rain area. In *Tropical Precipitation Measurements*, eds. J. Theon and N. Fugono, 361–367. Deepak Co, Hampton, VA.

Chiu, L.S. and R. Chokngamwong. 2010. Microwave Emission Brightness Temperature Histograms (METH) rain rates for climate studies: SSM/I V6 results. *Journal of Applied Meteorology and Climatology*, 49(1):115–123.

Chiu, L.S., A.T.C. Chang, and J. Janowiak. 1993. Comparison of monthly rain rates derived from GPI and SSM/I using probability distribution functions. *Journal of Applied Meteorology and Climatology*, 32:323–334.

Chiu, L.S., R. Chokngamwong, and T.T. Wilheit. 2010. Modified Monthly Oceanic Rain Rate Algorithm to account for TRMM Boost. *IEEE Transactions on Geoscience and Remote Sensing*, 48(8):3081–3086.

Chiu, L.S., G.R. North, D.A. Short, and A. McConnell. 1990. Rain estimation from satellites: Effect of finite field of view. *Journal of Geophysical Research*, 28:2177–2185.

Chu, A. and L. Remer. 2006. MODIS observation of aerosol loading from 2000–2004. In *Earth Science Satellite Remote Sensing: Vol 1, Science and instruments*, eds. J. Qu, W. Gao, and M. Kafatos. Springer, New York, NY.

Clark, R.N., G.A. Swayze, A.J. Gallagher, T.V.V. King, and W.M. Calvin. 1993. *The U. S. Geological Survey, Digital Spectral Library: Version 1: 0.2 to 3.0 microns*. U.S. Geological Survey Open File Report 93–592, 1340pp.

Clark, R.N., G.A. Swayze, K.E. Live et al. 2003. Imaging spectroscopy: Earth and planetary remote sensing with the USGS Tetracorder and expert systems. *Journal of Geophysical Research*, 108(E12), 5131, doi:10.1029/2002JE001847.

Cracknell, A.P., S.K. Newcombe, A.F. Black, and N.E. Kirby. 2001. The ABDMAP (algal bloom detection, monitoring and prediction) concerted action. *International Journal of Remote Sensing*, 22:205–247.

DeFries, R.S., L. Bounoua, and G.J. Collatz. 2002. Human modification of the landscape and surface climate in the next fifty years. *Global Change Biology*, 8:438–458.

DeFries, R.S., M.C. Hansen, J.R.G. Townshend, A.C. Janetos, and T.R. Loveland. 2000. A new global 1-km dataset of percentage tree cover derived from remote sensing. *Global Change Biology*, 6(2):247–254.

Delin, K. and S.P. Jackson. 2000. Sensor web for *in situ* exploration of gaseous biosignatures. *IEEE Aerospace Conference*, Big Sky, Montana.

Delin, K., S.P. Jackson, and R.R. Some. 1999. Sensor Webs. *NASA Technical Briefs*, 23, 80.

DiGregorio, A.D. and L.J.M. Jansen. 2000. *Land Cover Classification System (LCCS): Classification Concepts and User Manual*. UN FAO, Rome.

Diner, D.J., B.H. Braswell, R. Davies et al. 2005. The value of multiangle measurements for retrieving structurally and radiatively consistent properties of clouds, aerosols, and surfaces. *Remote Sensing of Environment*, 97:495–518.

Diner, D.J., Di Girolamo, and A. Nolin, 2007 Multi-angle Imaging SpectroRadiometer (MISR) Special Issue—MISR Special Issue, *Remote Sensing of Environment*, 107(1–2):384.

Doerffer, R., K. Sorensen, and J. Aiken. 1999. MERIS potential for coastal zone applications. *International Journal of Remote Sensing* 20:1809–1818.

Drummond, J.R. and G.S. Mand. 1996. The measurements of pollution in the troposphere (MOPITT) instrument: Overall performance and calibration requirements. *Journal of Atmospheric and Oceanic Technology*, 13:314–320.

Elachi, C. and J. Van Zyl. 2006. *Introduction to the Physics and Techniques in Remote Sensing*. Wiley-Interscience, John Wiley & Sons.

ESTO. 2007. Report from the Earth Science Technology Office (ESTO) Advanced Information Systems Technology (AIST) Sensor Web Technology Meeting, NASA.

Friedl, M.A., D.K. McIver, J.C.F. Hodges et al. 2002. Global land cover mapping from MODIS: Algorithms and early results. *Remote Sensing of Environment* 83(1–2): 287–302.

Fry, J.A., M.J. Coan, C.G. Homer, D.K. Meyer, and J.D. Wickham. 2009. Completion of the National Land Cover Database (NLCD) 1992–2001 Land Cover Change Retrofit Product. U.S. Geological Survey Open-File Report 2008–1379, 18pp.

Fu, C. 2003. Potential impacts of human-induced land cover change on East Asia monsoon. *Global and Planetary Change*, 37:219–229.

Goetz, A.F.H., G. Vane, J.E. Solomon, and B.N. Rock. 1985. Imaging spectrometry for Earth remote sensing. *Science*, 228:147–1153.

Gurney, R.J., J.L. Foster, and C.L. Parkinson. 1993. *Atlas of Satellite Observations Related to Climate Change*. Cambridge University Press, Cambridge, MA.

Hackwell, J.A., D.W. Warren, R.P. Bongiovi, et al. 1996. LWIR/MWIR imaging hyperspectral sensor for airborne and ground-based remote sensing. Imaging Spectrometry II. In *Proceedings of the International Society for Optical Engineering*, 2819:102–107.

Hansen, M.C. and B. Reed. 2000. A comparison of the IGBP DISCover and University of Maryland 1 km global land cover products. *International Journal of Remote Sensing*, 21:1365 – 1373.

Hansen, M., R. DeFries, J.R.G. Townshend, and R. Sohlberg. 1998. UMD Global Land Cover Classification, 1 Kilometer, 1.0, Department of Geography, University of Maryland, College Park, Maryland, 1981–1994.

Hansen, M., R. DeFries, J.R.G. Townshend, and R. Sohlberg. 2000. Global land cover classification at 1km resolution using a decision tree classifier. *International Journal of Remote Sensing*, 21:1331–1365.

Hapke, B. 1993. *Theory of Reflectance and Emittance Spectroscopy*. Cambridge University Press, Cambridge, MA.

Hecht, E. 1987. *Optics*, 2nd Edition. Addison-Wesley Publishing Company, Reading, MA.

Homer, C., C. Huang, L. Yang, B. Wylie, and M. Coan. 2004. Development of a 2001 National Landcover Database for the United States. *Photogrammetric Engineering and Remote Sensing*, 70(7):829–840.

Houghton, R.A. 1999. The annual net flux of carbon to the atmosphere from changes in land use 1850–1990. *Tellus Series B—Chemical and Physical Meteorology*, 51: 298–313.

Houghton, R.A., D.L. Skole, C.A. Nobre, J.L. Hackler, K.T. Lawrence, and W.H. Chowmentowski. 2000. Annual fluxes of carbon from deforestation and regrowth in the Brazilian Amazon. *Nature*, 403:301–304.

Hsu, C., S.C. Tsai, M. King, and J. Herman. 2006. Deep blue retrieval of Asian aerosol properties during ACE-Asia, *IEEE Transactions on Geoscience and Remote Sensing*, 44(11):3180–3195.

Hu, Y., D. Winker, M. Vaughan et al. 2009. CALIPSO/CALIOP Cloud Phase Discrimination Algorithm. *Journal of Atmospheric and Oceanic Technology*, 26: 2293–2309, doi:10.1175/2009JTECHA1280.1.

Huete, A., K. Didan, T. Miura, E.P. Rodriguez, X. Gao, and L.G. Ferreira. 2002. Overview of the radiometric and biophysical performance of the MODIS vegetation indices. *Remote Sensing of Environment*, 83(1–2):195–213.

Huffman, G., R. Adler, P. Arkin et al. 1997. The Global Precipitation Climatology Project (GPCP) combined precipitation dataset. *Bulletin of the American Meteorological Society*, 78:5–20.

Hunt, W.H., D.M. Winker, M.A. Vaughan, K.A. Powell, P.L. Lucker, and C. Weimer. 2009. CALIPSO Lidar description and performance assessment. *Journal of Atmospheric and Oceanic Technology*, 26:1214–1228, doi:10.1175/2009JTECHA1223.1.

IEEE. 2002. *Transactions on Geoscience and Remote Sensing* (July 2002) Special issue on MISR, 40(7):1447–1655.

IOCCG. 1999. Historical Ocean-Colour Sensors, http://www.ioccg.org/sensors/historical.html, last accessed August 3, 2010.

IPCC. 2000. *Land Use, Land-Use Change, and Forestry. A Special Report of the IPCC.* Cambridge University Press, Cambridge, UK.

Jackson, C.R., and J.R. Abel. 2004. *Synthetic Aperture Radar Marine User's Manual.* US Dept of Commerce, NOAA/National Environmental Satellite, Data, and Information Services, Office of Research and Applications. http://www.sarusersmanual.com/

Janetos, A.C. and C.O. Justice. 2000. Land cover and global productivity: A measurement strategy for the NASA programme. *International Journal of Remote Sensing,* 21:1491–1512.

Ji, Y., P. Ardnauy, D.L. Sun, Q. Liu, and W. Yang. 2000. *NPOESS VIIRS Sea Surface Temperature ATBD Version 1–3.* Raytheon ITSS, Lanham, MD.

Joseph, G. 2005. *Fundamentals of Remote Sensing,* 2nd edition. Universities Press, Hyderabad, India, 477pp.

Justice, C.O., L. Giglio, S. Korontzi et al. 2002. The MODIS fire products. *Remote Sensing of Environment,* 83(1–2):244–262.

Justice, C.O., E. Vermote, J.R.G. Townshend et al. 1998. The Moderate Resolution Imaging Spectroradiometer (MODIS): Land remote sensing for global change research. *IEEE Transactions on Geoscience and Remote Sensing,* 36(4): 1228–1249.

Kidder, S.Q. and T.H. Vonder Haar. 1995. *Satellite Meteorology—An Introduction.* Academic Press, San Diego, CA.

Kramer, H.J. 2002. *Observation of the Earth and its Environment: Survey of Missions and Sensors,* 4th edition. Springer Verlag, New York, NY.

Krige, J. and A. Russo. 2000. *A History of the European Space Agency,* Vol. I. http://www.esa.int/esapub/sp/sp1235/sp1235v1web.pdf.

Kummerow, C., Y. Hong, W.S. Olson, S. Yang, R.F. Adler, J. McCollum, R. Ferraro, G. Petty, D-B. Shin, T.T. Wilheit. 2001. The evolution of the Goddard Profiling Algorithm (GPROF) for rainfall estimation from passive microwave sensors. *Journal of Applied Meteorology,* 40:1801–1820.

Landgrebe, D.A. 2003. *Signal Theory Methods in Multispectral Remote Sensing.* Wiley-Interscience, John Wiley and Sons, New Jersey.

Lautenbacher, C. 2006. The global Earth observation system of systems: Science serving society. *Space Policy,* 22(1):8–11.

Liu, D., Z. Wang, Z. Liu, D. Winker, and C. Trepte. 2008. A height-resolved global view of dust aerosols from the first year of CALIPSO lidar measurements. *Journal of Geophysical Research,* 113(D16214):15.

Loeb, N.G., W.B. Sun, W.F. Miller, K. Loukachine, and R. Davies. 2006. Fusion of CERES, MISR, and MODIS measurements for top-of-atmosphere radiative flux validation. *Journal of Geophysical Research—Atmospheres,* 111(D18), doi:D1820910.1029/2006jd007146.

Lu, G.Y., L. Chiu, and D. W. Wong. 2007. Vulnerability assessment of rainfall induced debris-flows in Taiwan. *Natural Hazards,* 43:223–244.

Mace, G.G., R. Marchand, Q.Zhang, and G. Stephens. 2007. Global hydrometeor occurrence as observed by Cloudsat; Initial observations from Summer 2006. *Geophysical Research Letters,* 34:doi: 10.1029/2006GL029017.

Mildrexler, D.J., M. Zhao, and S.W. Running. 2009. Testing a MODIS Global Disturbance Index across North America. *Remote Sensing of Environment*, 113:2103–2117.

Milman, A.S. 1999. *Mathematical Principles of Remote Sensing: Making Inferences from Noisy Data*. Ann Arbor Press, Chelsea, MI.

Morel, A. and A. Bricaud. 1981. Theoretical results concerning light absorption in a discrete medium, and application to specific absorption of phytoplankton. *Deep Sea Research*, 28:1375–1393.

Murphy, R. 2006. The NPOESS Preparatory Project. In *Earth Science Satellite Remote Sensing, I Science and Instrumentation*, eds. J. Qu., W. Gao and M. Kafatos, pp. 182–198. Springer, Berlin.

Nittel, S., A. Labrinidis, and A. Stefanidis. 2008. *Advances in GeoSensor Networks*, Vol. 4540. Lecture Notes in Computer Science, Springer, Berlin.

Pope, R.M. and E.S. Fry. 1997. Absorption spectrum (380–700 nm) of pure water. II. Integrating cavity measurements. *Appl. Opt.*, 36:8710–8723.

Potter, T.D. and B.R. Colman. 2003. *Handbook of Weather, Climate and Water—Atmospheric Chemistry, Hydrology, and Societal Impacts*. Wiley Interscience.

Potter, C., T. Pang-Ning, V. Mar et al. 2005. Recent history of large-scale ecosystem disturbances in North America derived from the AVHRR satellite record. *Ecosystems*, 8:808–824.

Prabhakara, C., G. Dalu, and V.G. Kunde. 1974. Estimation of sea surface temperature from remote sensing in the 11 μm to 13 μm window region. *Journal of Geophysical Research*, 79(12):1744–1749.

Qu, J.J., W. Gao, M. Kafatos, R. Murphy, and V.V. Salomonson. 2006a. *Earth Science Satellite Remote Sensing, Vol. 1, Science and Instruments*. Springer; Berlin, Heidelberg, New York/Tsinghua University Press; Beijing.

Qu, J.J., W. Gao, M. Kafatos, R. Murphy, and V. V. Salomonson. 2006b. *Earth Science Satellite Remote Sensing, Vol. 2, Data, Computational Processing, and Tools*. Springer; Berlin, Heidelberg, New York/Tsinghua University Press; Beijing.

Rao, P.K., S.J. Homes, R.J. Anderson, J.S. Winston, and P.E. Lehr. 1990. *Weather Satellites: System, Data and Environmental Applications*. American Meteorological Society, Boston.

Resmini, R.G., M.E. Kappus, W.S. Aldrich, J.C. Harsanyi, and M.E. Anderson. 1997. Mineral mapping with Hyperspectral Digital Imagery Collection Experiment (HYDICE) sensor data at Cuprite, Nevada, U.S.A. *International Journal of Remote Sensing*, 18(7):1553–1570.

Richards, J.A and X. Jia. 2006. *Remote Sensing Digital Image Analysis: An Introduction*. Springer, New York, NY.

Rodgers, C.D. 2000. Inverse method for atmospheric sounding: Theory and practice, *Series on Atmospheric, Oceanic and Planetary Physics*, Vol. 2. World Scientific, Singapore.

Roesler, C.S., M.J. Perry, and K.L. Carder, 1989. Modeling *in situ* phytoplankton absorption from total absorption spectra in productive inland marine waters, *Limnology and Oceanography*, 34:1510–1523.

Salisbury, J.W. 1993. Mid-infrared spectroscopy: laboratory data. In: *Remote Geochemical Analysis: Elemental and Mineralogical Composition*. Topics in Remote Sensing 4, eds. Pieters, C.M., and P.A.J. Englert, pp. 79–98. Cambridge University Press, Cambridge, MA.

Salomonson, V.V., W. Barnes, and E.J. Masuoka. 2006. Introduction to MODIS and an overview of associated activities. In *Earth Science Satellite Remote Sensing, Vol. I*, eds. J. Qu, W. Gao and M. Kafatos, pp. 12–31. Springer, Berlin.

Salomonson, V., W.L. Barnes, X. Xiong, S. Kempler, and E. Masuoka. 2002. An overview of the Earth observing system MODIS instrument and associated data systems performance. In *Proceedings of the International Geoscience and Remote Sensing Symposium (IGARSS 02)*, Sydney, Australia.

Schoeberl, M.R., A.R. Douglass, E. Hilsenrath et al. 2006. Overview of the EOS Aura Mission. *IEEE Transactions on Geoscience and Remote Sensing* 44(5):1066–1074.

Schott, J.R. 1997. *Remote Sensing: The Image Chain Approach*. Oxford University Press, New York.

Short, N.M. 2009. Remote Sensing Tutorial (online), http://rst.gsfc.nasa.gov/.

Smith, R.C. and K.S. Baker. 1981. Optical properties of the clearest natural waters (200–800 nm). *Applied Optics*, 61(20):177–184.

Solé, J.G., L.E. Bausá, and D. Jaque. 2005. *An Introduction to the Optical Spectroscopy of Inorganic Solids*. Wiley-Interscience, John Wiley and Sons, Hoboken, NJ.

Song, W., R. Huang, M. Xu, A. Ma, B, Shirazi, and R. LaHusen. 2009. Air-dropped sensor network for real-time high-fidelity volcano monitoring. In *Proceedings of the 7th ACM Conference on Mobile Systems, Applications, and Services*, pp. 305–318. Krakow, Poland.

Sorooshian, S., K.L. Hsu, X. Gao, H.V. Gupta, B. Imam, and D. Braithwaite. 2000. Evaluation of PERSIANN system satellite-based estimates of tropical rainfall. *Bulletin of the American Meteorological Society*, 81:2035–2046.

Stefanidis, A. and S. Nittel. 2004. *GeoSensor Networks*. CRC Press, Boca Raton, FL.

Stephens, G.L., D.G. Vane, R.J. Boain et al. 2002. The CloudSat mission and the A-TRAIN: A new dimension to space-based observations of clouds and precipitation. *Bulletin of the American Meteorological Society*, 83:1771–1790.

Stephens, G.L., D. G. Vane, S. Tanelli et al. 2008. CloudSat mission: Performance and early science after the first year of operation. *Journal of Geophysical Research*, 113(D00A18), doi:10.1029/2008JD009982.

Stocker, A.D., I.S. Reed, and X. Yu. 1990. Multi-dimensional signal processing for electro-optical target detection. In: *Signal and Data Processing of Small Targets 1990. Proceedings of the SPIE*, 1305:218–231.

Sun, D.L. and R.T. Pinker. 2007. Retrieval of surface temperature from the MSG_SEVIRI observations: Part I. Methodology. *International Journal of Remote Sensing* 28(23):5255–5272.

Swayze, G.A., R.N. Clark, A.F.H. Goetz, T.G. Chrien, and N.S. Gorelick. 2003. Effects of spectrometer band pass, sampling, and signal-to-noise ratio on spectral identification using the Tetracorder algorithm. *Journal of Geophysical Research*, 108(E95101): doi:10.1029/2002JE001975.

Townshend, J.R.G. and C.O. Justice. 1988. Selecting the spatial-resolution of satellite sensors required for global monitoring of land transformations. *International Journal of Remote Sensing*, 9:187–236.

Tucker, C.J. 1979. Red and photographic infrared linear combinations for monitoring vegetation. *Remote Sensing of Environments*, 8:127–150.

Tucker, C.J., J.E. Pinzon, M.E. Brown, D. Slayback, E. Pak, R. Mahoney, E. Vermote, and N. Saleous. 2005. An extended AVHRR 8-km NDVI data set compatible with MODIS and SPOT vegetation NDVI data. *International Journal of Remote Sensing*, 26(20):4485–4498.

Tucker, C.J., J.E. Pinzon, and M.E. Brown. 2004. *Global Inventory Modeling and Mapping Studies*. Global Land Cover Facility: University of Maryland.

Turner, B.L.II., W.B. Meyer, and D.L. Skole. 1994. Global land-use/land-cover change: Towards an integrated program of study. *Ambio*, 23(1):91–95.

Van Zyl, T.L., I. Simonis, and G. McFerren, 2009. The Sensor web: Systems of sensor systems. *International Journal of Digital Earth*, 2(1):16–30.

Vogelmann, J.E., S.M. Howard, L. Yang, C.R. Larson, B.K. Wylie, and J.N. Van Driel. 2001. Completion of the 1990's National Land Cover Data Set for the conterminous United States. *Photogrammetric Engineering and Remote Sensing*, 67:650–662.

Wang, M. and W. Shi. 2005. Estimation of ocean contribution at the MODIS near-infrared wavelengths along the east coast of the U.S.: Two case studies. *Geophysical Research Letters*, 32(13):1–5.

Wang, M. and W. Shi. 2007. The NIR-SWIR combined atmospheric correction approach for MODIS ocean color data processing. *Optics Express*, 15:15722–15733.

Wang, M., S. Bailey, C. McClain, C. Pietras, and T. Riley. 1999. Remote sensing of aerosol optical thickness from SeaWiFS in comparison with *in situ* data. In *Proceedings of ALPS Symposium*, Meribel, France, January 18.

Wang, M., K.D. Knobelspiesse, and C.R. McClain. 2005. Study of the Sea-Viewing Wide Field-of-View Sensor (SeaWiFS) aerosol optical property data over ocean in combination with the ocean color products: Global aerosol system. *Journal of Geophysical Research*, 110(10):1–14.

Wang, W., J.J. Qu, X. Hao, Y. Liu, and J. Stanturf. 2010. Post-hurricane forest damage assessment using satellite remote sensing. *Agricultural and Forest Meteorology*, 150(2010):122–132.

Warneke, B., M. Last, B. Liebowitz, and K. Pister. 2001. Smart dust: Communicating with a cubic-millimeter computer. *Computer*, 34(1):44–51.

Wielicki, B.A., B.R. Barkstrom, B.A. Baum, T.P. Charlock, R.N. Green, D.P. Kratz, and R.B. Lee III. 1998. Clouds and the Earth's Radiant Energy System (CERES): Algorithm overview. *IEEE Transactions on Geoscience and Remote Sensing*, 36(4):1127–1141.

Wilheit, T.T., A.T.C. Chang, and L.S. Chiu. 1991. Retrieval of monthly rainfall indices from microwave radiometric measurements using probability distribution functions. *Journal of Atmospheric and Oceanic Technology*, 8:118–136.

Wilson, R. 1997. Total solar irradiance trend during solar cycles 21 and 22. *Science*, 277(5334):1963–1965.

Wolfe, W.L. 1997. *Introduction to Imaging Spectrometers*. An SPIE Press Book, v. TT25, ISBN: 9780819422606, http://spie.org/x648.html?product_id=263530, 164pp.

Xua, M., W.Z. Song, R. Huang, Y. Peng, B. Shirazi, R. Lahusen, and A. Kiely. 2009. Design of smart sensing components for volcano monitoring. *Pervasive and Mobile Computing*, 5(5):639–653.

Young, S.A. and M.A. Vaughan. 2009. The retrieval of profiles of particulate extinction from Cloud Aerosol Lidar Infrared Pathfinder Satellite Observations (CALIPSO) data: Algorithm description. *Journal of Atmospheric and Oceanic Technology*, 26(6):1105–1119.

Zhang, X., M.A. Friedl, and C.B. Schaaf. 2006. Global vegetation phenology from Moderate Resolution Imaging Spectroradiometer (MODIS): Evaluation of global patterns and comparison with *in situ* measurements. *Journal of Geophysical Research*, 111:G04017, doi: 10.1029/2006JG00217.

Zhang, W., J. Xu, C. Dong, and J. Yang. 2006. China's current and future meteorologi-
    cal satellite systems. In *Earth Science Satellite Remote Sensing, I Science and
    Instrumentation*, eds. J. Qu, W. Gao, M. Kafatos, R. Murphy, and V. Salomonson,
    pp. 392–413. Springer, Berlin.
Zhang, X., M.A. Friedl, C.B. Schaaf, A.H. Strahler, J.C.F. Hodges, and Gao, F. 2003.
    Monitoring vegetation phenology using MODIS. *Remote Sensing of Environment*,
    84:471–575.
Zhang, X., M.A. Friedl, C.B. Schaaf, and A.H. Strahler. 2004. Climate controls on
    vegetation phenological patterns in northern mid- and high latitudes inferred
    from MODIS data. *Global Change Biology*, 10:1133–1145.

3

*Geoinformation Computing Platforms**

Qunying Huang, Chaowei Yang, Wenwen Li,
Huayi Wu, Jibo Xie, and Ying Cao

CONTENTS

* Supported by NASA projects (NNX07AD99G and SMD-09-1448) and FGDC project (G09AC00103).

79

Earth observation, ground sensing, and other data collection techniques have exponentially accelerated our ability to collect a large amount of data. Recent research has produced computing intensive algorithms and models for analyzing data and simulating scientific phenomena. Geoinformation science requires ongoing research and development to leverage different computing platforms to address these data and computing intensive problems. This chapter discusses how to utilize distributed computing, high performance computing (HPC), grid computing, and cloud computing to address the requirement.

3.1 Introduction*

Geoinformation science is an applied science that deals with vast amounts of observations, experiments, data analysis and information extraction, and phenomena simulations. One of its focal objectives is to address how computing technology can be efficiently and effectively utilized to solve Geoinformation science problems. Two basic questions need to be answered: (1) Can a Geoinformation science problem be solved with available computing facilities, considering the computing requirements, such as CPU cycles and memory size? (2) If yes, how much time is needed for computers to solve a Geoinformation science problem? For example, we cannot spend one week to forecast a hurricane that may occur during the second day of forecasting. Geoinformation computing platform research (1) leverages the latest advancements in computing technology to help expand the frontiers of Geoinformation science and (2) poses newer and higher requirements for computing technology to drive advancements of computing technology.

Geoinformation science problems include analyses and simulations of phenomena within geography, weather, atmosphere, climate, ocean, and global systems studies. A simulation of such phenomena involves a large volume of data and parameters. Computing technology applied to simulations helps improve our understanding of Geoinformation science phenomena

---

* This section is contributed by Huayi Wu and Chaowei Phil Yang.

and enables us to predict timely severe weather, pollution, and climate events from seconds, hours, days, months, seasons, decades, and up to century time scales and from global, regional, to local space scales. For example (NRC 2008), in the short term (1–5 years), 103–104 processors could help us address scientific questions of immediate interest in atmospheric science. In the long term (5–10 years), the exploration advancements will require 104–105 or more processors per run on a routine basis, thus requiring breakthroughs in computing technology.

The advancements of data acquisition technology, such as high-resolution sensors and active sensors, help collect PB of data on a daily basis. For example, GeoEye's GeoEye-2 will be an advanced, third-generation satellite capable of discerning objects on the Earth's surface as small as 0.25 m (9.75 in) in size, that is, the image size of a single square kilometer is 16 million pixels (ASIA-ASM 2008; Martin and Chris 2009). Besides these unprecedentedly advanced passive optical sensors, active sensors, such as SAR and LiDAR (light detection and ranging), have also been developed and widely used. The Shuttle Radar Topography Mission (SRTM) obtained digital elevation models (DEM) on a near-global scale from 56°S to 60°N and generated the most complete high-resolution digital topographic database of Earth to date (Jakob 2001) in up to 1 Tb in 10 m resolution. The data collecting capability and data volume continue to increase exponentially.

With the size of EO data constantly growing, more storage space is required for data archiving and dissemination. For example, the NASA's GES DISC is one of the NASA's DAACs tasked with archiving and distributing remote sensing data from the NASA's EOS. The center archives data from various Earth observing satellite missions along with data from various field campaigns. The GES DISC processes more than 1 TB of data per day and archives 1.75 PB of data (Leptoukh 2005). These data are geographically distributed to different centers and institutions. A distributed computing platform is required to process this data for Geoinformation science problems.

The improved data availability creates the potential for us to solve Geoinformation science problems that could not be tackled in the past through higher spatial and temporal resolution in a multidimensional context. For example, earlier 5–10 km spatial resolution and 2-h temporal resolution of dust storm prediction and simulation in 36 h were considered a high-level demand. Scientists are requested to achieve 1 km spatial resolution and minute temporal resolution in 120 h. This increased requirement poses a grand demand for computing technology (Xie et al. 2009).

The recent advancements of network infrastructure facilitate the integration of distributed computers. Although data and computing resources are geographically dispersed, distributed computing technologies provide the possibility to integrate and utilize all the resources in a seamless fashion. In recent years, computing technology has also been developed at a much faster pace, thus providing greater capacity for processing data. In addition, the development of computing technology has gone beyond merely increasing

CPU frequency and the size of RAM. Distributed computing, HPC, grid computing, cloud computing, and other emerging computing paradigms have become dominant approaches with novel mechanisms and, therefore, increased opportunities for expanding computing capabilities. Hardware and software platforms have been developed to leverage these computing paradigms and are being applied to specifically solve Geoinformation science problems (Xie et al. 2009).

Geoinformation science problems and modern computing resources are correlated in that Geoinformation science deals with geographically dynamic phenomena, whereas computing resources are also geographically dispersed. Conversely, utilizing the intrinsic spatial principles of Geoinformation to leverage the distributed resources greatly helps deploy efficient computing resources to solve scientific problems. Complex scientific problems that cannot be solved by a single computer can be tackled through the collaboration of distributed computers. Geoinformation computing leverages spatial principles, such as space and time connections and constraints, in computing arrangements, selection, and utilization to enable the computability of Geoinformation science problems (Yang et al. 2010). Computing platforms for Geoinformation science include the integration of hardware, models, algorithms, software, intellectual capacity, and computational infrastructure so as to enable the desired computations.

Instead of increasing only CPU frequency and network bandwidth to improve computer performance, HPC cluster improves computing capability by increasing the number of CPUs and the memory size in one computer. On one hand, the more CPUs are integrated, the better computing capacity we will have; on the other hand, the communication overhead among the CPUs also increases. Usually, there is an optimum number of CPUs that can be efficiently leveraged for a specific scientific application (Erich et al. 2005).

Grid computing is an emerging technology that coordinates resource sharing and usage in a dynamic, scalable context for common tasks that require a large number of CPU cycles to process large amounts of data (Foster et al. 2001). The size of grid computing may vary from being small—for example, confined to a network of computer workstations within a corporation—to being a large public collaboration across many companies and networks. What distinguishes grid computing from conventional cluster computing is that grid computing tends to be more loosely coupled, heterogeneous, and geographically dispersed. Although a computing grid may be dedicated to a specialized application, it is often constructed with the aid of general-purpose grid software or middleware.

Cloud computing is a new distributed computing paradigm that is driven by economies of scale, in which a pool of abstracted, virtualized, dynamically scalable, managed computing power, storage, platforms, and services are delivered on demand to external customers over the Internet (Foster et al. 2008). Cloud computing allows consumers and businesses to use applications without installation and to access their personal files at any computer that has Internet access. This technology allows much more efficient computing with distributed storage, memory, processing, and bandwidth. It typically involves the provision of dynamically scalable and often virtualized resources as a service over the Internet.

This chapter addresses how computing platforms could be used to advance Geoinformation science. Four recent computing paradigms, such as distributed computing, HPC, grid computing, and cloud computing, are introduced. These technologies share the purpose of organizing distributed computing resources, including hardware, software, data, and other functions, to solve large-scale problems, including emerging Geoinformation science problems.

3.2 Distributed Geoinformation Computing*

3.2.1 Introduction

Distributed computing refers to delivering computing resources through a large and global network. It encompasses a wide range of the latest computer systems, ranging from VLSI (Very Large Scale Integration) chips, tightly coupled shared memory multiprocessors, and local-area cluster of workstations, to the Internet (Attiya and Welch 1998). A distributed computing system achieves a much better performance than a single computer by harnessing more CPU cycles and storage space in tens to thousands of networked systems for a computing intensive problem. Distributed computing is essential to Geoinformation science, because the proliferation of large volume of remote sensing imageries, cartographic or photogrammetric data, and other spatial data makes Geoinformation computing very time consuming. Geoinformation data sets, such as demographic data sets and socioeconomic data sets, are highly distributed and make centralized computing impossible.

Research has been conducted to investigate Distributed Geospatial Computing (DGC) (Yang 2008) and Distributed Geospatial Information Processing (DGIP) (Yang et al. 2008) for Geoinformation science and applications. The history of DGC (Figure 3.1) can be traced back to the idea of an "intergalactic computer network," introduced in the early 1960s by Licklider, who enabled the development of the Advanced Research Projects Agency Network (ARPANET) in 1969. His vision was for everyone on the globe to be interconnected and have access to programs and data at any site, from anywhere. Other experts attribute the concept to computer scientist John McCarthy, who proposed the idea of computation being delivered as a public

---

* This section is contributed by Wenwen Li and Chaowei Phil Yang.

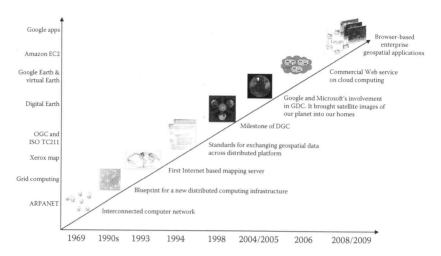

FIGURE 3.1
Evolution of DGC.

utility, which was similar to the concept of service bureaus. In Geoinformation science, Xerox's mapping server is recognized as the first system for processing Geoinformation across the Internet (Plewe 1997). The Xerox PARC Map Viewer was created in June 1993 by Steve Putz at Xerox Corporation's Palo Alto Research Center, as an experiment in providing interactive information retrieval via the World Wide Web (WWW). The term *grid computing* originated in the 1990s as a metaphor for making computer power as easy to access as an electric power grid in Ian Foster's and Carl Kesselman's seminal work *The Grid: Blueprint for a New Computing Infrastructure*. In 1994, FGDC (FGDC 1994) was established to share Geoinformation across distributed platforms, and OGC (1994) and ISO/TC211 (1994) were established to define a set of standardized interfaces. In 1995, MapQuest and other DGC applications were released and eventually gained great success by leveraging popular Geoinformation applications, such as routing, to serve the public. In 1996, ESRI, Intergraph, and other GIS companies began to participate in the DGC effort by fully implementing spatial components in the Internet environment (Peng and Tsou 2003).

A major milestone of DGC is the Digital Earth vision vocalized by Vice President Al Gore in 1998. The aim of the Digital Earth vision is to integrate all Geoinformation resources to support a virtual environment that could enhance human existence from the research and development fields to everyday life. Since then, distributed computing has spawned many innovative technologies, including grid computing, utility computing, and, more recently, cloud computing. In 2004, Google Earth was launched and became another milestone in the implementation of such a vision. In 2005, Microsoft started Virtual Earth. Only some limited functions of Geoinformation applications are addressed in these two implementations. These applications focus on massive data and friendly user interaction and thus solved many problems associated with massive simultaneous users by deploying thousands to millions of computers (Tao 2006). In 2006, Amazon launched its Elastic Compute Cloud (EC2) as a commercial web service that allowed small companies and individuals to rent computers on which to run their own applications. In 2008 and 2009, Google and others started to offer browser-based enterprise applications through services like Google mapping applications.

All the developments just mentioned have been enabled by the maturing of virtualization, high-speed Internet, and computing interoperability.

3.2.2 DGC Architecture

Figure 3.2 illustrates the three-layer DGC architecture: (1) The bottom layer is the computing pool, which includes a dedicated computing management service and a number of distributed computing clients with lightweight

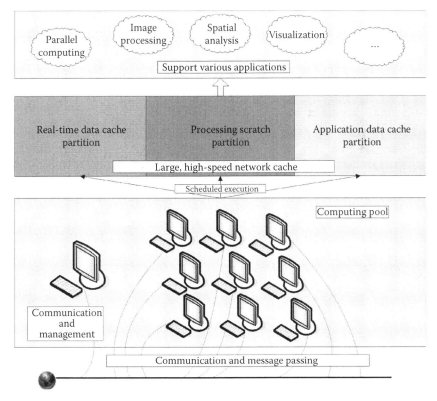

FIGURE 3.2
A three-layer DGC architecture.

software agents. The agents are responsible for notifying the management service when the client system is idle. The client then receives more computing tasks and sends back the results. The communications among these agents are supported by messages in different languages, such as eXtensible Markup Language (XML). The message can be transmitted through pipe, a computing level communication mechanism, and the client to web service communication protocol can be HyperText Transportation Protocol (HTTP) (Yang et al. 2006). Since Geoinformation computing is also data intensive, computing clients copy large portions of a data set from the distributed locations to a local disk before performing a task. (2) The middle layer is the data handler model. Once a job is submitted, it is divided into several sub-jobs, including real-time data partition, processing scratch partition, and application data partition. All these subtasks are scheduled to be executed by the computing nodes at the bottom layer. To increase the data transmission rate and to provide high-speed access, an online cache system is applied to the data handler model (Tierney et al. 2000). The cache system provides standard interfaces to a large, application-oriented, and distributed online storage system. (3) The top layer is the application layer and it includes various DGC applications, such as spatial analysis, image processing, and geographic visualization accelerated by the DGC platform. In this way, the DGC platform works as a supercomputer, and the computing topology is transparent to the application submitters.

This general architecture can be adapted and customized to popular computing platforms, such as HPC, grid computing, and cloud computing. When executing computing tasks, if they rely on a tightly coupled cluster or supercomputers, the DGC can be considered as HPC; when the computing devices in the computing pool are a combination of low-end personal computers, clusters, and supercomputers, the DGC works as a grid computing platform; and when the entire infrastructure is made transparent to end users and the computing platform is provided as a service, the DGC can be treated as a cloud computing platform.

3.2.3 Methodologies, Algorithms, and Strategies in DGC

To organize geographically dispersed computing resources to work harmoniously and efficiently, systematic mechanisms serve as the key to integrate them. Currently, there are three major mechanisms: message passing, shared memories, and fault tolerance.

3.2.3.1 Message Passing

The message-passing mechanism is important for distributed systems, because the mechanism greatly facilitates component communication and controls system concurrency. The exchange of messages may be conducted synchronously or asynchronously.

Synchronous message passing involves two entities: a client for sending messages and a server for receiving messages. The message-passing process occurs only when both sender and receiver processes are ready for the communication. After the receiver receives the message and before it finishes the processing and returns the results, the sender pauses until the response arrives. A typical synchronous message-passing problem is the atomic commitment problem (Skeen 1982).

The *atomic commitment problem* refers to the issue of ensuring that all distributed clients of a transaction make the same decision (to either commit or abort the transaction) after an action is taken. If the decision is made as "Commit," all the clients should update their status or data permanently. If the decision is to "Abort," the transaction takes no effect on any of the participating clients. This problem becomes very difficult to solve if some clients involved in the execution of the transaction commitment fail (Lemlouma and Badache 2007). The atomic commitment problem is very common in the DGC context, for example, when distributed clients are editing the same Geoinformation (adding/deleting features or revising feature attributes) simultaneously. Each of them has a local copy of the data sets and once any of them finishes editing, he or she wants to commit the changes to the server copy by synchronizing so that the other clients can view the changes. The atomic commitment problem comes up when one of the clients fails to update the changes. In this situation, to make the local copy on each client consistent, the specific client should send out a message to notify all the other clients to rollback to previous state. Thus, an algorithm is needed to ensure that the message is successfully received by other clients and the rollback actions are correctly executed.

Currently, the popular algorithms to solve the atomic commitment problem include the two-phase commit protocol (2PC) (Bernstein et al. 1987) and the three-phase commit protocol (3PC) (Dale 1983). The 2PC uses a coordinator to decide whether the clients should commit or abort the transaction. The decision is made through a voting process, and the coordinator is notified once a consensus is reached. The disadvantage of 2PC is that it is a blocking protocol. The node is blocked and its local resource for the specific transaction is locked. Therefore, the other thread that needs the resource has to wait until the resource is released. If the coordinator fails, the resource will be tied up forever. In contrast to 2PC, 3PC is an unblocking protocol. The 3PC sets a time-out variable, which indicates the longest time required before making a decision on the locked transaction. This variable ensures that the locked resources of a given transaction are released, no matter what, when the time is out. However, the 3PC assumes that the time-out (failure) event can be accurately detected. Therefore, it is not adaptable to asynchronous communication or a complex network partition.

Asynchronous message passing does not require the entities among which the message is passing to be closely coordinated. The entity can engage in other tasks while waiting for a response. For example, after sending a message to the service entity, a client entity is not blocked and is able to conduct other computational intensive tasks it manages. An asynchronous communication has no restriction on the sequence of messages; thus, the implementation of such an algorithm is difficult, because it needs to handle all the sequencing of the messages. A simple algorithm is First-In-First-Out (FIFO), meaning that the messages are processed in the sequence of time that they have been added to the queue. One entity always maintains two queues: one queue stores all the sending messages, and the other one stores all the received messages. Different from FIFO, casual ordering (Raynal and Helary 1990; Raynal et al. 1990) requires that a single message should not be overtaken by a sequence of messages (Murty and Garg 1993).

In practice, it is often possible to mix synchronous and asynchronous message passing mechanisms in a single distributed system. For example, the Message Passing Interface (MPI) supports various types of synchronization, such as point-to-point send or receive operations, and computation in combining partial results and detecting network-related properties (Hampel 1994). The selection of synchronization mode is up to the particular requirement of specific applications.

3.2.3.2 Shared Memories

A distributed shared memory system provides a shared memory among distributed processors for sharing and scheduling distributed data and computing resources. According to Tanenbaum (1994), a distributed shared memory system can be classified into three broad categories: page-based distributed shared memory, shared-variable-based distributed shared memory, and object-based distributed shared memory.

The aim of the page-based distributed shared memory approach is to use fixed-size blocks, called pages, to share address space on top of multicomputers (Li 1986; Li and Hudak 1989). The translation from virtual to physical memory is through the Memory Management Unit (MMU). Once a requested page is not found in the local memory, the address translation hardware triggers a page fault exception. Then, the exception handler transfers the page from the owner in the distributed environment to the node requesting it. There are usually two approaches to find the owner of the missing page: broadcasting and page manager. Broadcasting interrupts each processor and forces them to process the request packets. It is both time consuming and resource consuming (a big portion of bandwidth will be used up by the broadcasting). The other approach is through a central manager who maintains the ownership information for each page. This method saves resources for communication; however, the page manager may have a heavy load due to handling all of the incoming requests. A drawback of this approach is that page-based distributed shared memory exchanges information at a page granularity, resulting in poor performance especially when the size of the requested resource is less than a page.

The aim of the shared-variable distributed shared memory approach is to avoid the page-based system's drawback by only sharing variables or data structures that are used by distributed processes. This is accomplished on software level rather than on hardware level. This approach (Bershad and Zekauskas 1991; Carter et al. 1991) provides optional declaration and synchronization mechanisms for multiple processors to share information at different levels.

The aim of the object-based distributed shared memory approach is to share the encapsulated information as an object (Carriero and Gelernter 1986; Bal et al. 1992). Different from a shared-variable system, the object-based approach not only shares data but also shares the methods that invoke and process the data. Since the methods can define different access permission by different processors to the data, it can avoid violated intrusion and increase the performance of the distributed system.

3.2.3.3 Fault-Tolerance

Fault tolerance is used to provide, by redundancy, service complying with the reliability in spite of faults having occurred (Laprie 1985). Research in fault-tolerant distributed computing aims at making distributed systems more reliable and continuously satisfactory (Gartner 1999). User statistics show that web end users lose patience if the response time of a service is longer than 8 s (Kim 2002). That means that the web servers must meet high-availability requirements and be able to respond to massive concurrent requests in a timely fashion. Nowadays, providing a robust and reliable system is a critical research topic, especially when the Geoinformation computing in a distributed environment becomes the mainstream computing for Geoinformation science in the twenty-first century.

According to Jalote (1994), the term *fault* refers to a defect at the lowest level of abstraction, for example, a memory cell that always has return value 0. Instead, an error is a system state that could be caused by a fault. An error may lead to a failure, meaning that the system fails to stay in its correct specification. The fault usually has many types, such as (1) transient, intermittent, or permanent hardware faults; (2) software and hardware design errors, such as incorrect design and design flaws; and (3) operating errors or externally induced upsets or physical damage (Nelson 1990; Somani and Vaidya 1997). These faults need to be addressed at the hardware and system software level. In general, hardware fault tolerance refers to the practice of designing computing equipment capable of recovering from random faults. Techniques used for fault tolerance include (1) fault masking, a structural redundancy technique by which a number of identical modules are used to execute the same function. The outputs of the modules are compared to remove the errors in hardware computing. Fault masking does not require recovering faults but is able to contain faults. (2) Dynamic recovery utilizes special mechanisms to detect faults in modules and move the computation from the faulty module to a spare module to continue the task. Dynamic recovery is more space efficient than a voted system and it is better to be used in a resource-contained system or in a high-performance scalable system in which the active computing power should be maximized. However, dynamic recovery is not as time efficient as fault masking, because fault recovery delays the computational process. Hardware fault tolerance techniques have been significantly improved over the past decades, and the reliability concerns in terms of hardware have been diminished. Thus, researchers pay more attention to software fault tolerance.

Software fault tolerance is used to describe the ability of software to detect and recover from a fault that is happening or has already happened when the software is running in order to provide service in accordance with the specification (Inacio 1998). Although the fault will not cause permanent faults (Reis et al. 2005), it may alter the processor's state, signal transfers or stored values on register, and affects a program's normal execution. Approaches for software fault tolerance are mostly based on traditional hardware fault tolerance, including N-version programming (NVP), recovery blocks (RcB), N self-checking programming (NSCP), and others. The RcB is a simple method introduced by Randell (Horning et al. 1974), and it operates with an adjudicator, which determines the correctness of various blocks tried. In such a system, the overall system view is broken into several recoverable blocks. Each block is responsible for certain subtasks and is recursively dividable. The system will run based on atom system blocks, and these blocks contain primary, secondary, and exceptional codes. The adjudicator will choose the software block that has a higher priority to run and if the result is validated as wrong, it will run the block with the next highest priority. If none of these results is correct, the exceptional codes will be executed and the errors will be reported back to the executors. The idea of NVP is similar to N-way redundant hardware fault tolerance, but it is an implementation attempt on software. In a system with NVP, each module consists of N different ways of implementation with the same functionality. The voting procedure then decides the correct output and returns that as the final result of the module. The NVP approach requires more development efforts, as each module should be implemented as diversely as possible. Another software fault tolerant approach is NSCP, which has the aim of adding extra checking points and rollback recovery methods to eliminate as many possible faults to achieve high reliability in a safety critical system. Systems with NSCP have shown to be surprisingly effective in industry (Inacio 1998), but there are very limited studies in the literature in comparison to the other two approaches.

Although a solid technical foundation has been built for fault tolerance in DGC over the past decades, there are still many issues that need to be addressed. In the future, researchers should emphasize the quantitative treatment of design techniques and protocols, including a scientific assessment of fault source models, to further accelerate advancements in fault-tolerant DGC technologies (Kim 2002).

In summary, message passing, shared memory, and fault tolerance have been widely used in distributed computing systems. Geoinformation science application requires (a) tremendous computing and storage resources, (b) good communication networks, and (c) good job scheduling. The message-passing strategy fits into the characteristics. The shared memory strategy can also be used in a multiprocessor computer system. It is also easier for a job dispatcher to distribute a computational task to multiple processors and reassemble the results into an integrated solution. Moreover, to assure the reliability of distributed systems, fault tolerance strategy must be adopted to detect and handle any possible system errors in case a system failure happens. Due to the significance of the strategies just described, any design and implementation of a distributed system should carefully select the appropriate strategy for better performance.

3.2.4 Applications

The technical advancements in DGC have made it possible to transfer global climate modeling from traditional supercomputers to distributed platforms. Climateprediction.net, the world's largest climate forecasting experiment in the twenty-first century, depends on such a DGC platform called BOINC (Berkeley Open Infrastructure for Network Computing) to produce predictions of the Earth's climate up to 2080 (Allen 1999). Currently, there are almost 100,000 PCs from 138 countries participating in the project. To accomplish the just-in-time access, asynchronous data request and delivery between the participating nodes and donated servers are adopted. Benefiting from the "shared" virtual memory strategy, this platform can successfully handle the asymmetric data requirement of multiple participants (Stainforth et al. 2002). Moreover, it enables the distributed visualization modules to access data from the model when it is running (Figure 3.3). The climateprediction.net project has brought the concept of DGC to the public and, at the same time, improves public understanding of the nature of uncertainty in climate prediction.

Besides climate simulation, DGC is also used in many other domains that include the processing of distributed Geoinformation resources. In geography, DGC can help integrate widely, geographically dispersed Geoinformation resources to provide a comprehensive overview of the Earth surface (EOP 1994). In oceanography, DGC can be used to help integrate the *in situ* and satellite observation systems with the modeling system to monitor tsunamis, sea level changes, and coastal disasters (Mayer et al. 2004). In public health, DGC can help integrate the health information and environmental observations to find a correlation between environmental changes and human health. In transportation, DGC can be used to integrate road network data sets, dispatching a routing request to different servers to select the

FIGURE 3.3
(See color insert following page 144.) A 3D global view of model cloud fields. (Adapted from Stainforth, D. et al. 2002. *Computing in Science and Engineering*, 4(3):82–89.)

shortest or fastest path. This can be used in (1) driving directions, such as Mapquest or Yahoo map, (2) rapid response, such as routing planning for emergency response, and (3) operation planning, such as coordinating the super shuttle or scheduling a FedEx package pick up.

In addition, NASA identified application areas of interest at the national level (Birk et al. 2006), whereas GEO identified nine application areas of interest at the global level. All these areas require the use of DGC to integrate distributed EO and Earth system models to improve decision support tools hosted by government agencies or other organizations.

3.2.5 Future Development

Many Geoinformation applications, if not all, will be extended to DGC platforms throughout the global networks. New technologies, such as grid computing, HPC, and cloud computing, will be leveraged to achieve better performance. The DGC should be developed to be more capable in handling (1) massive data management, such as PB data obtained from EO system on a daily basis, (2) intensive computing, such as Earth science model simulations, (3) quality of services, such as on-the-fly reliable application building, (4) interoperability, such as plug-and-play integration of DGC components in a real-time fashion, and (5) intelligent spatial service searching engines (Li et al. 2008).

3.3 Geoinformation High-Performance Computing*

3.3.1 Introduction

For decades, CPU development has been following Moore's Law and continues to grow with smaller transistors, faster frequency, and rapidly increasing processing power. Along with improvements of the CPU manufacturing processes, memory technologies have also evolved with faster speed, growing from dynamic random access memory (DRAM), to synchronous dynamic random access memory (SDRAM), and finally to Double Data Rate (DDR) SDRAM. The improvement of CPU memory along with the development of the Internet contributes to the rapid growth of HPC.

The study of HPC originated in the scalability of the cluster system. Improving the CPU frequency is initially the primary strategy to improve computer performance. However, it is very difficult and sometimes impossible to improve only the performance of a single CPU to satisfy computing requirements. Consequently, researchers begin to improve computing power by increasing the number of CPUs and memory capacity in the same computer, such as the vector machines and symmetric multiprocessor (SMP). However, when the number of CPUs exceeds a certain threshold, these multiprocessor systems (e.g., SMP) are not good in scalability. For example, computers with the current Intel architecture can be extended to a maximum of eight CPUs. The main bottleneck is bus bandwidth, which is used to access memory and cannot be increased as fast as the number of CPUs. Parallel computing cluster methods emerge to satisfy those high computational demands of scientific tasks. The famous Beowulf cluster system is one of the implementations of parallel computing clusters made by assembling Intel processors running Linux (Hargrove et al. 2001).

Generally, a HPC cluster system includes a management node and multiple computing nodes connected through Ethernet (or other network). The management node is usually the gateway to access the computing nodes, and it is responsible for monitoring and controlling computing nodes. In a large-scale cluster system, the master node management functionalities may also be shared by multiple nodes. Cluster systems are equipped with common hardware devices and rarely include specially customized equipment. Those low-cost and widely disseminated software, such as the Linux operating system, parallel virtual machine (PVM), and MPI, are often used in the HPC cluster systems.

Many spatial analyses are very time consuming, especially when large data sets are involved. For instance, spatial interpolation for large geographic areas could become a problem in online applications such as terrain visualization, where a fast response is required and computational demands exceed

---

* This section is contributed by Qunying Huang, Jibo Xie, and Chaowei Phil Yang.

the capacity of a traditional single processing unit. Utilization of HPC for Geoinformation enables the solution for spatial problems that could not be tackled in the past. For example, HPC is used to satisfy the demand of weather forecasting modeling for computing and storage resources (Xie et al. 2009). High performance computing is also used to satisfy the ever-increasing demands in transportation engineering for supporting transport vehicle crash, safety manufacturing simulation (Haug et al. 1994), and real-time routing (Cao 2007).

Through parallel computing, such as that provided by HPC, we are able to process larger data sets, at a higher resolution, and to process more scenarios in less time (Clarke 2003). Parallel computing partitions serial computation task into subtasks and dispatch each subtask onto different processors. The subtasks can be independent when no communication between tasks is needed. However, more than often, the tasks require communication, wherein decomposed domains or functions need synchronization. In domain decomposition, data is partitioned and distributed across computing nodes. Each node is given a subset of the data. In functional decomposition, each node executes different portions of the code simultaneously. Both methods suit spatial data, yet domain decomposition is used more for Geoinformation applications (Wang and Armstrong 2003; Xie et al. 2009). However, domain decomposition is known to have problems working with random numbers and at map and image edges (Mower 1996).

Generally, there are two approaches to achieve communication between parallel programs that are not independent and require synchronization: (1) use MPI and (2) use the shared memory model (the operating system threads, OpenMP). Although there are some other methods available, these two methods are the most used. For applications using a MPI model, multiple processes are invoked. Although the programs using shared memory models employ only one process, after invocation, multiple threads are created to participate in the computing. During the computing, both of the threads employed by the OpenMP programs and the processes invoked by the MPI programs will communicate to exchange the data and synchronize. The MPI model-based programs need to copy data between the CPU, whereas the thread is able to share data between the CPUs. Data transfer delay and efficiency of message passing are the most critical factors affecting the performance in the MPI model.

3.3.2 Architecture

A cluster system is generally composed of multiple commodity computers and computing nodes, which can run independently and are connected by a high-speed network. A typical HPC cluster includes the following major system components (Figure 3.4).

Network: An HPC cluster usually uses three types of network simultaneously, which include (1) Inter-process communication (IPC) network, (2)

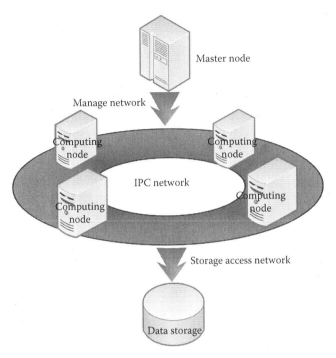

FIGURE 3.4
The basic architecture of an HPC cluster.

management network, and (3) storage access network. The IPC network is a fast private network used for parallel task execution, high throughput, and very low latency (Barrera 1991). Management network is used to gather cluster information, monitor, and perform maintenance tasks for the cluster. Storage access network is the private network for accessing public storage. Presently, the connection among the nodes in the cluster commonly uses fast Ethernet, Gigabit Ethernet, Myrinet, and others. Fast Ethernet can provide 100 Mbps bandwidth; Gigabit Ethernet can provide 1 Gbps bandwidth, both of which can connect a switch or router directly to the cluster node; and Myrinet is formed by a series of switching devices. The internal switches use a pipelining mechanism, and the current bandwidth has achieved more than 2 Gbps. The protocol for the network communications can be TCP/IP or the equivalent of TCP/IP, such as active message, fast message, and VIA (Virtual Interface Architecture). According to specific needs and characteristics of selected nodes, these networks can use a variety of media and protocols. Usually, the IPC network (Barrera 1991) would use Gigabit Ethernet or Myrinet network for less communication delay, whereas management network and storage network would use a fast Ethernet network.

Client node provides a user interface, receives tasks from the users, and sends the tasks to the master node, which will assign the tasks to the computing nodes and return the results to the client node. Sometimes, a client node and a master node could be the same node. To control access to the cluster, users are only permitted to access the client node and are blocked from accessing other nodes.

Master (or control) node is responsible for distributing tasks to computing nodes through scheduling or queuing software, such as Condor, Portable Batch System (PBS), or LSF (Boukerche 2006). The master node provides management functionality for the cluster and enables an administrator to conduct monitoring, operations, and handle all logs and alarm information in the cluster.

Storage Node is a node for providing shared storage. In order to execute tasks in parallel, each computing node must be able to access the same data resources. Storage nodes usually utilize a network share (NFS) or other methods to ensure the synchronization of data access.

Computing nodes execute computing tasks and include most of the computers in the cluster. Computing nodes running the parallel applications use MPI to complete inter-process communication. Usually, the computing nodes are tightly connected through an IPC network.

Open source software is often used to manage the cluster systems. In the TOP500 list, the cluster systems have accounted for 42%, whereas of the top 10 of the TOP500 list, there are seven clusters.

3.3.3 Case Study: Dust Storm Simulation

Simulations of dust storms and potential forecasting are of significant interest to public health, environment sciences, and GEOSS. To support improved public health decision making with higher resolution of dust storm forecasting, HPC needs to be leveraged to increase the simulation or prediction resolution to the zip code level. This poses significant computational challenge for dust storm simulations. This includes enhancing dust storm forecasting to (a) reduce the computing time, (b) support high-resolution simulations, and (c) lengthen the period of forecast.

In this example, WRF-NMM and NMM-dust (Janjic et al. 2001) based dust storm simulation models are utilized to demonstrate HPC for dust storm forecasting. Parallelized using the MPI programming model, they are able to obtain higher-resolution dust simulation results up to about 1 km. The study utilized the GMU CISC cluster (with 28 computing nodes and 224 CPU cores) and its environment to provide HPC support.

Data decomposition: Numerical weather and climate prediction involves solving a system of coupled nonlinear partial differential equations with appropriate boundary conditions (Purohit et al. 1999). Atmosphere is modeled by dividing the studying area into three-dimensional regions or cells. The calculations of each cell are repeated many times to model the phenomena of evolution. The computational cost of an atmospheric model is a function of the number of cells in the domain and of the time step (Baillie et al. 1997). Since atmospheric models perform essentially the same set of computations in each cell, an SPMD (single program, multiple data stream) data-domain decomposition approach to parallelism is preferred, and nearest-neighbor communication in the physical domains is required (Nanjundiah 1998). The process of data decomposition should define the regions assigned to each processor and also define a virtual array of processors used to execute these regions, thereby creating neighboring relations between regions.

Performance analysis: The NMM-dust model (with 3 km resolution) was tested by using different number of CPU cores on the CISC HPC cluster for performance comparison. Excluding preprocessing and postprocessing time (Postprocessing is used to quilt the separate tiles into a composite result. It is time consuming, but it can only run in a serial computing mode. Therefore, it is fixed in performance.), as illustrated in Figure 3.5, the near best performance (cost of 125 s with a speedup factor of 30.6) was obtained when 40 CPU cores were used, and the model is in an efficiency of 30.6/40 = 76.6% in utilizing the CPU cores.

High spatial resolution: This experiment is to analyze the relationship between the spatial resolutions of the dust storm prediction model and the number of CPU numbers involved in predictions. Short-range weather prediction is computation intensive and a time critical task that has to be completed in less than 2 h (Lenz et al. 2002). Thus, only HPC centers are able to execute such forecasts.

It is observed that one core is sufficient for successfully completing the 10 km resolution simulation in 2 h (Figure 3.6). At least 16 cores are needed for the cluster to finish the task of 4 km simulation, and around 2 h are needed while 8 CPUs are required for the 5 km simulation. The computation is greatly increased with increases of resolution, and the computing time of 4 km

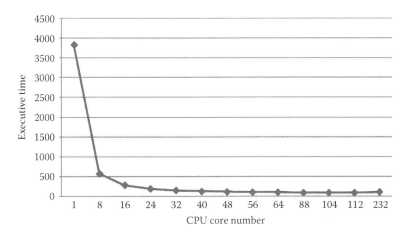

FIGURE 3.5
Performance of HPC-enabled NMM-dust model high-resolution simulations.

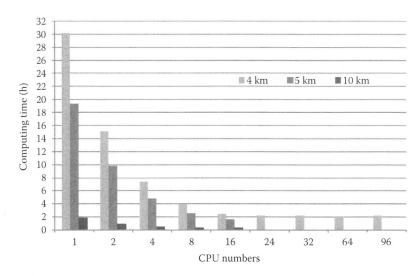

FIGURE 3.6
CPU number and spatial resolution analysis.

resolution and 5 km resolution simulation increases by a factor of 16.5 and 10.5, respectively, compared with 10 km resolution simulation when both use only one core. The experiment also shows that a resolution which increases to 3 km in each dimension of the model will not run in the cluster for the domain of 2000 km × 2000 km × 37 levels. It requires either (1) a redesign of the existing algorithms, codes, and data structures or (2) an increase of the speed of the CPU and the network connection. Thus, this application poses a grand challenge to both physical and computer science.

Long-term prediction: The capability of HPC support to long-term dust-storm prediction has also been exploited. It is observed that 4 CPUs can successfully complete a 1-day simulation in 2 h (Figure 3.7). However, it is impossible to complete the 5 days or 10 days in 2 h when using 28 computing nodes (224 cores), and the peak performance is obtained using around 20 CPU cores, which can predict dust storm in 6 h for 10 days and in 3 h for 5 days. The limitation is partially caused by cache and memory contention (27–28), thus again this poses a grand challenge to both physical and computer sciences.

The case study demonstrates that the capabilities of HPC are widely needed in supporting Earth science applications. The experiments illustrate that increasing of spatial resolution and temporal scope would greatly increase the computation, and only HPC can meet such computationally intensive demands.

3.3.4 Future Development

The development of HPC is faced with enormous challenges, such as the scalability of a cluster system. Cluster nodes require relatively large power,

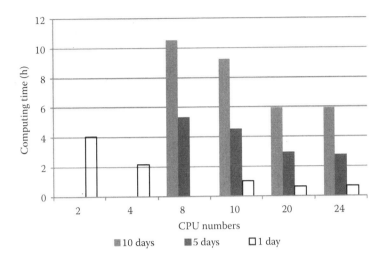

FIGURE 3.7
CPU number and temporal resolution analysis.

space, and heat dissipation to complete the computation. When the number of nodes reaches a few hundred to several thousand, the cumulative effect of these factors will become most apparent. If a goal is to further expand the cluster size, originally secondary factors will become a major problem.

Another problem with cluster system expansion is reliability and manageability issues. The reliability of a parallel system is inversely proportional to the number of nodes. If the number of nodes is increased by ten times, reliability will be decreased by ten times. An original system can last for a month without fault and now it can fail in 3 days after expanding ten times. Therefore, larger-scale parallel systems or more efficient use of computing resources becomes one of the main directions of research HPC systems.

3.4 Geoinformation Grid Computing*

3.4.1 Introduction

The concept of a grid was first introduced in the 1960s and it was given a concrete form by Grid Pioneers in the 1990s (Foster and Kesselman 1999). A grid is a hardware and software infrastructure that provides dependable, consistent, pervasive, and inexpensive access to high-end computational capabilities. A grid resource could include computers, clusters, computer pools, devices, equipments, sensors, storage devices, data, software, and

---

* This section is contributed by Wenwen Li, Qunying Huang, Chaowei Phil Yang, and Ying Cao.

other entities, and the related software, data, and equipments. In general, grid computing is considered a special type of parallel computing that relies on complete computers with onboard CPU, storage, power supply, network interface, and so on connected to a network by a conventional network interface, such as Ethernet. This aspect makes it greatly different from the traditional notion of a supercomputer, which has multiple processors connected by a local high-speed bus.

Grid computing combines grid resources so that they will have a similar computing power as a multiprocessor supercomputer but at a lower cost. However, grid resources are generally geographically distributed and have low-speed connections, which will introduce communication latency. This characteristic makes grid computing well applicable to applications that can be partitioned into subtasks without internal connections among subtasks.

Research in recent years proved that grid computing is efficient in addressing computing demands (Armstrong et al. 2005). With the continuing price dropping in computer hardware and networks, it becomes practical for any laboratory even with limited funding to deploy a grid computing platform. In order to utilize a grid computing environment to support various applications, a middleware is required to enable heterogeneous resources to be shared within the communities. An efficient middleware with the capability to discover, allocate, negotiate, monitor, and manage the use of network-accessible computing resources is fundamental to the success of grid computing.

Currently, a variety of middleware systems are available. For example, Globus Grid Toolkit (The Globus Project, http://www.globus.org), which has become a standard for grid computing, is a low-level middleware and requires a substantial amount of technical expertise to set up and use (Hawick et al. 2003). Condor is a relative high-level distributed batch computing system that provides facilities such as job management, scheduling policy, priority scheme, resource monitoring, and management (Thain et al. 2005).

3.4.2 Grid Architecture

A grid has a layered architecture or, as we call it, fabric, in which higher layers are more user-centric in comparison to lower layers, which are more hardware-centric, focusing on computers and networks. Figure 3.8 demonstrates the architecture of a grid. The lowest layer is the network layer, which is used to connect grid resources. Two groups of protocols are utilized in a network layer: (1) one is for interoperable data exchange, such as XML, Simple Object Access Protocol (SOAP), and Web Service Description Language (WSDL); (2) the other group of protocols facilitates data transmission over Internet or Ethernet wherein grid resources are located, such as HTTP, Transmission Control Protocol (TCP), File Transfer Protocol (FTP), and Simple Mail Transfer Protocol (SMTP).

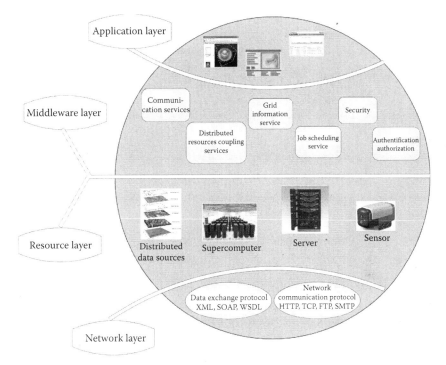

FIGURE 3.8
Grid architecture.

Above the network lies the resource layer, where both the data resources and computing resources are connected to the network. For general grids, data resources can be of any type, such as statistic data for businesses and numerical data for mathematical analysis. For Geoinformation grids, the data resources are mainly vector or raster data sets, such as remote sensing imageries and surveying and digitalized vector data. These data are always in large volumes, thus there is a great need in the grid-enabled platform to achieve high performance in computing. The computing resources in a grid could be supercomputers, which have high frequency multiprocessors, multicores, and large memories. These resources could also be a cluster of servers, sensor networks, and even personal computers. This is one of the innovative ideas of a grid: to share the idle CPU cycles of many types of computing devices without disturbing the user's work by seamlessly plugging them into grid platforms (Berman et al. 2003). In this way, the utilization of computing resources could be maximized.

A middleware enables, coordinates, and manages the computing and data resources to achieve the functions of a grid. As shown in Figure 3.8, the middleware layer provides a series of services including (1) communication services, (2) distributed resource coupling services, (3) grid information services, (4) job scheduling services, (5) grid security control, and (6) authentification and authorization.

The topmost layer of a grid is the application layer, which includes Geoinformation applications, such as real-time disaster monitoring, dynamic transportation routing, and geology evolutional simulation, supported by lower-layer grid resources as well as portals and client customized tools. This is the layer where users interact with the grid resources.

3.4.2.1 Resource Management

Grid computing involves the sharing and coordination of dynamic, multi-institutional, geographically distributed, and heterogeneous resources in a distributed environment. Grid resource management is the process of identifying requirements, matching resources to applications, allocating those resources, and scheduling and monitoring grid resources over time in order to run grid applications as efficiently as possible (Nabrzyski et al. 2004). It does not concern what functions of the grid resources and services are provided to the clients; rather, it is concerned with the manner in which these functions are performed (Foster and Kesselman 1999). Generally, the goal of resources management is how to effectively and efficiently communicate with the grid resources, such as applications, jobs, files, directories, schedulers, managers, special class of scientific instruments (such as radiotelescopes), authentication objects (representations of users in a grid), databases, visualization devices, tools, and others to satisfy different computational needs by a great variety of users.

For a grid environment, in which the managed resources are geographically distributed, much valuable work concerning the resources management issues has been done. In a decentralized grid environment, peer-to-peer technique is usually used to organize and manage the grid resources, and it enables the grid node to freely join and leave (Chen et al. 2005). Recently, agent-based methods were widely used for resources management in grid computing, because these methods are capable of solving some key problems involved with resources management in a grid computing environment (Kesselman 2004). Li et al. (2006) present a novel agent-based dynamic grid resource management model considering both the grid resource management and job scheduling as coalition integrity.

The process of resource management can be generally represented in Figure 3.9. A user submits a job to the job control center, which will invoke the grid resource manager to parse the request and determine the required resources for the job. The resource manager will then discover the current state and availability of resources. If all of the required resources are available, it will pass the job to a scheduler, which will put the job into a ready queue. Protocols are used to determine the priority of the jobs according to the parameters like importance, emergency, profit, and relationship. Finally, the jobs can be executed within the resources through the network.

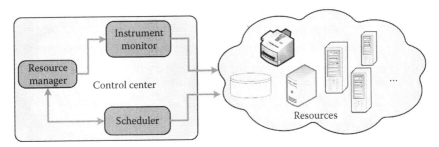

FIGURE 3.9
General process of resource management.

3.4.2.2 Data Management

Data management of a grid is one of the central challenges of grid computing (Plantikow et al. 2009). The infrastructure that focuses on management of distributed application data is commonly labeled as a *data grid* (Chervenak et al. 2000). Data grids provide the mechanisms needed to manage distributed data, the tools that simplify automation of data management processes (Moore et al. 2004). Generally, performance, security, and reliability are critical issues required to be considered for accessing and managing large amounts of data in a grid environment (Allcock et al. 2001). More specifically, Foster and Kesselman (1999) mentioned that high-performance and distributed data-intensive applications require two fundamental services: (1) secure, reliable, and efficient data transfer and (2) the ability to register, locate, and manage multiple copies of data sets.

The Earth science applications and spatial processing are generally data intensive with large area and time dimensions involved, such as global climate change, environmental modeling, and hydrological modeling. Therefore, Geoinformation grid computing first requires efficient data management to enable the sharing and transfer of terabytes or even PB of EO data or spatial data over wide-area and distributed computing environments. The data management environment must provide security services, such as the authentication of users and control over who is allowed to access the data. In addition, once multiple copies of files are distributed at multiple locations, researchers need to be able to locate copies and determine whether to access an existing copy or to create a new one to meet the performance needs of their applications.

There are a variety of solutions proposed for efficient management of grid data. Plantikow et al. (2009) presented a conceptual data management system architecture that separates application, community, and resource concerns, using three layers of addressing, thus providing a highly adaptable architecture for different community grids. The Storage Resource Broker data grid is an example of a system that has been successfully applied to a wide variety of scientific disciplines for managing massive data sets

(Rajasekar et al. 2003). Another example is GridFTP (Allcock et al. 2001), which implements extensions to FTP that provide Grid Security Infrastructure (GSI) security and parallel, striped, partial, and third-party data transfers, whereas the Globus replica management architecture supports the management of complete and partial copies of data sets. Performance studies of both strategies provide promising results.

3.4.2.3 Job Management

Job monitoring and management are the central part of grid computing. Its main goal is to maximize the utilization of resources such that computers with high computing capabilities should do more work, and a submitted job would be assigned resources for running the job. In a grid, computing resources are transparent to users. When a job submitter writes a job description file, it is not necessary for the job submitter to know on which node the job will be executed. Instead, what the job submitter only needs to do is to point out the running requirements (e.g., what is the CPU frequency or minimal memory?) for its job. Once a job is submitted, the job manager would select the best-fit resources for it from a list of candidate resources. *Best-fit* is a strategy that represents to minimize the cost of job submitters, meaning the waiting time is the shortest or the labor that the job submitter provides is minimal. Meanwhile, a job manager should also satisfy the requirement of users on grid service quality.

Specifically, a job manager has the following subtasks:

Manage the lifecycle of the job and be responsible for the whole process from the user submitting the job to returning the desired results to users.

Find the best-fit computing resources and assign available resources to users by automatically matching job requirements.

Manage input and output of the jobs. The input and output of a job is usually done on remote grid nodes, but this might not be specified in the job description file. For example, job input might be from keyboard input, and output might be to another file or to screen. So, the job manager should be able to read and output the data to the right place on the right node.

Job migration is needed to move the job running on one resource to another by keeping the current running status. It is hard to tell the running status of a job, for example, when a job in a running queue has some exceptions and the node is blocked, the other jobs behind it will be delayed to be executed. In this situation, these blocked jobs should be migrated to other available resources for load balancing.

The main function a job manager is job scheduling. It is composed of two parts: match scheduling and sequence scheduling. Match scheduling is to find the proper computing resource for a certain job in a waiting queue, whereas sequence scheduling is to determine the execution sequence for jobs in the waiting queue. Match scheduling handles the relationship between a job and multiple resources, whereas sequence scheduling handles the relationship among multiple jobs. Suppose there are n jobs $J = \{j_1, j_2, \ldots j_n\}$ scheduled on m resources $S = \{s_1, s_2, \ldots s_m\}$, the execution time for job j_i on resource s_k is c_{jk} and wait time is a_k (a_k is the earliest complete time of the jobs that are already scheduled on s_k before assigning j_i). The scheduling strategy is to find a match between a resource and a job such that $\sum_{\substack{i\in\{1,2,\ldots n-1\} \\ j\in\{1,2,\ldots m-1\}}} (c_{ik} + a_k)$ is minimized.

3.4.2.4 Grid Security

Security is one of the most challenging issues in grid computing. It needs to make sure that any user is the one he or she claims to be (authentication) and has the proper right to access certain resources to perform some tasks (authorization). Meanwhile, a grid system should be able to ensure the integrity of grid resources that no unauthorized modification is allowed. It should also balance between privacy and availability control, namely some information should be solely controlled by a role, whereas others should maintain enough openness for access from authorized users. Since most grid services are encapsulated into web services, the delivery of secure, integrated, and interoperable solutions is critical for open grid architecture and grid security. According to Nagaratnam et al. (2003), a secure grid model should satisfy requirements in the following aspects:

Binding security: Messages for transferring between grid nodes are always in the format of XML. Due to this, the security in service binding is very important. In a grid environment, XML are always encrypted by Secure Sockets Layer (SSL) and Transport Layer Security (TLS) techniques. This is performed on a whole XML document and now a new trend is to encrypt only on essential fragments of an XML document. The secured information will be sent through SOAP, which can be confidentially protected by using XML.

Secure association: In order for messages to be securely exchanged, a mechanism is required to establish a mutual authentication context between service requestors and service providers. There are several protocols and mechanisms to support the secure context. For example, WS-Security is used to realize the secure information exchange between SOAP nodes by adding identity authentication, digital signature, and encryption. The WS-license is introduced in the SOAP extended model, which describes how to encode trust to be used in WS-Security. Another example is Internet Inter-ORB Protocol (IIOP), where CSIv2 specification is used for context establishment.

Trust: Trust relationships are dynamic due to the dynamic nature of grid. It is essential for services accessed between the members to traverse network firewalls and satisfy authorization policies associated with a service achieved by translating credentials from one domain to another and mapping identities across security domains (Nagaratnam 2003). Within a trustable infrastructure, the message could be securely transferred in the same grid layer and across multiple grid layers.

Besides the above-mentioned three aspects, secure logging, privacy, and authorization enforcement are also important in securing a grid environment.

3.4.3 Grid Computing Applications

Robust middleware services are of great importance to support applications and facilitate collaborations in sharing of data and computing resources. The NSF Middleware Initiative (NMI) was launched in 2001, aiming to "develop, deploy and sustain a set of reusable and expandable middleware functions that benefit many science and engineering applications in a networked environment" (Barton et al. 2002). A suite of leading software is provided to the grid computing community, such as Globus (The Globus Project. http://www.globus.org), Legion, DOCT (The Distributed Object Computation Testbed, http://www.sdsc.edu/DOCT/), and Punch (http://punch.ecn.purdue.edu/CeHub/).

Since the development of grid computing techniques, a much wider variety of applications have been explored. With the characteristics of accessing, querying, and manipulating massive spatial data sets (e.g., satellite imagery) from distributed spatial databases and distribution of the services, models, and data, Earth science applications have become a great potential candidate for grid computing. Especially, for those applications requiring large-scale, long-term data sets, for example, global climate change modeling or real-time or near real-time information, emergency services decision support, and quick response system, grid computing power is needed.

The Earth System Grid II (ESG, http://www.earthsystemgrid.org/) is a new research project to address the formidable challenges associated with enabling the analysis of and knowledge development from global Earth System models (ESG, http://esg-pcmdi.llnl.gov/). Incorporating with grid technologies and emerging community technology, ESG with distributed supercomputers and large-scale data and analysis servers is able to provide a seamless and powerful environment for the climate change research. In the following real-time routing example, we illustrate how grid computing supports Geoinformation application.

Traffic management centers must operate in real time, as the traffic data such as speed, volume, and incident are changing at every minute, especially during peak periods. The time taken to collect data, process data, and broadcast the resulting information may constitute a possible information bottleneck. To avoid this bottleneck, the Intelligent Transport Systems (ITS) models and algorithms must run in a time much faster than real time (Chabini 1998), and grid computing support is required (Cao 2007). The MITSIMLab (Yang 1997), especially its simulator MITSIM, is utilized for travel time estimation and prediction.

To meet the computing needs of traffic simulation by MITSIMLab, we deployed a CISC grid at GMU (George Mason University), which is used to support Geoinformation research and development. The grid computing pool has a configuration as illustrated in Figure 3.10.

In this computing pool, Condor is used as a middleware to dispatch and execute jobs. Condor implements job scheduling by ClassAds/Match mechanism, a strategy to match the resource of condor pool with the specified job descriptions. Each execution node advertises all of the information regarding the machine, such as capabilities, usage, preferences, machine name, CPU type, memory size, operating system, workload, and others, as a ClassAd to the central manger. Based on the standards of Condor, the job description file, which includes the job running environment CPU core number, specific CPU cores, and memory required, can be created by users. After submitting jobs, Condor will match the computing resources to execute tasks according to these requirements and ClassAds. The grid-enabling near real-time traffic simulation processes scheduled by Condor are conducted in six steps as follows:

Data decomposition: The entire traffic data can be decomposed into different sections. Data redundancy is needed, because there is spatial correlation

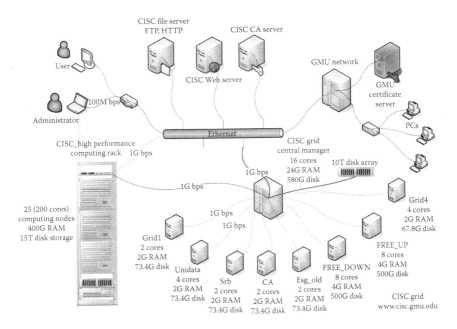

FIGURE 3.10
Grid computing pool utilized for the example.

FIGURE 3.11
Traffic data decomposition method.

between neighboring sections; each section with a MITSIMLab program is submitted as a job to the grid computing environment with a job description file illustrating job resource requirement and running command information; a central manager of the Condor matches the jobs to computing nodes by analyzing the job description file information and ClassAds and uses matched computing nodes to execute jobs; computing nodes separately and concurrently return results to the central manager and merge separate results received into a complete traffic information table; as a real case study of near real-time traffic simulation experiment, part of the traffic network in Washington DC, a region with over 50,000 roads, is used as test data. The study area is divided into 25 regular sections with an overlap of 1000 m for each section as shown in Figure 3.11. The overlap between sections is used to reduce the boundary errors induced by data decomposition. Each section is sent to a computing node matched by the job scheduler in the condor pool and computed separately.

To test the performance, 2, 4, 8, and 22 CPU cores in the condor pool are used. Figure 3.12 illustrates how the grid platform effectively improves the performance of traffic simulations. As the CPU cores were added, the executing time for traffic simulation decreased correspondingly. When 22 CPU cores are used, the execution time is more reasonable for the near real-time pattern.

3.4.4 Discussions and Future

The research on Geoinformation grid computing evolved from concept infrastructure to today's matured grid applications. It successfully enables and accelerates performance, improves productivity and collaboration, and optimizes the resiliency of the Cyberinfrastructure. However, with the global

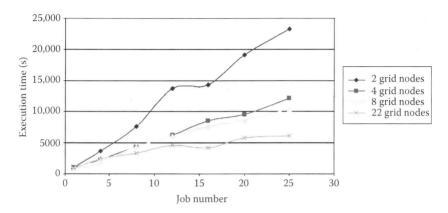

FIGURE 3.12
Performance of grid-enabled traffic simulation.

availability of high speed network (speed range from 100 Mbps to 1 Gbps), PB data resources from various types of devices (such as sensors, Personal Digital Assistant [PDA], and health monitors) as well as petaflop computational resources will bring unprecedented heterogeneity and performance challenges to grid computing. It requires the next generation grid to be more sophisticated and capable of supporting unprecedented diversity, scale, globalization, and adaption (Berman et al. 2003). In addition, grid computing software will be developed to be more intelligent to achieve the goals of automatic and adaptive computing. We believe that through the technical development and the integration with various scientific applications, for example, emergency services decision support and quick response system, Geoinformation grid computing platform will advance the research and education for Earth science community.

3.5 Geoinformation Cloud Computing*

3.5.1 Introduction

For decades, Earth scientists continually investigate techniques that can be used to facilitate the accessing and sharing of massive Geoinformation in a transparent manner through interoperability. Cao et al. (2009) proposed an interoperable framework to disseminate Earth science data to different application domains. Within the framework, different Earth science data products and raster snapshots over time can be efficiently managed and handled through the use of relevant metadata information. In addition, a variety of

---

* This section is contributed by Qunying Huang and Chaowei Phil Yang.

international organizations, such as FGDC, ISO/TC211, and OGC, are also working to advance interoperability.

It is also of great research interest to utilize grid computing and other HPC technologies to support spatial decision support within global to local initiatives (Yang and Raskin 2009). In the Information Technology (IT) field, with the emergence of technological advancements, such as multicore processors and networked computing environments, computing paradigms have been shifted from cluster computing, grid computing (Foster and Kesselman 1998), and P2P computing (Oram 2001) toward market-oriented computing (Buyya et al. 2008). After similar paradigm development in the IT field, the computing paradigm for supporting Geoinformation applications has been developed from Geoinformation distributed computing (Yang et al. 2005), Geoinformation HPC (Clematis et al. 2003), and Geoinformation grid computing (Wang et al. 2002) to, most recently, cloud computing (Armbrust et al. 2009).

Cloud computing is a computing model, where all servers, networks, applications, and other relevant parts of the data centers are connected through the Internet and provided to organizations and individuals. From one perspective, it is new, because it changes how we invent, develop, deploy, scale, update, maintain, and pay for software and the infrastructure on which software run. From another perspective, nothing is new, because cloud computing uses approaches and concepts that have already been established.

The basic concept of cloud computing is the use of web or software applications, data storage, and computing powers through the Internet. It is a way of computing in which typically scalable resources are provided as a transparent service for users over the Internet. This computing model integrates Web 2.0, on-demand deployment, Internet delivery of services and open source software, and visualization, which is emphasized by identifying it as a key component (Buyya et al. 2008), and other recent well-known trends of technologies to satisfy the computing needs of the individuals and users. Cloud services are transparent and easily usable, hiding the deployment details from the users (Vaquero et al. 2009). Users need only to purchase their own specific type and quantity of computing services. Customers who connect to the "cloud" can access infrastructure services, platform (operating system) services, or software services and the "cloud" will then provide the same functionality as the internal data centers or computers. For example, with technology virtualization, Infrastructure as a Service (IaaS), one type of cloud computing, is able to split, assign, and dynamically resize computing resources to create ad hoc systems as demanded by individual users or organizations.

Utilizing Geoinformation cloud computing paradigms and emphasizing interoperability, computing and transparent services can facilitate the research of Earth sciences in many aspects:

- Discover access and use Earth science data and research results for Earth sciences, such as climate change.

- Facilitate Research to Operations (R2O).
- Efficient and convenient to tap into vast varieties of distributed resources for decision support.
- Create new opportunities for national, international, state, and local partners to leverage research easily.
- Leverage government investments and capabilities.
- Contribute to an Enterprise Architecture for bridging Earth science and mainstream IT.

3.5.2 Cloud Computing Types

The consulting company Accenture gives a practical, concise definition of a cloud as the third-party provider that provisions and configures IT functions (hardware, software, or services) through the network dynamically. Armbrust et al. (2009) refers to cloud computing as both the applications delivered as services over the Internet and the hardware and systems software in the datacenters that provide those services. In summary, all those definitions emphasize the concept of services, with hardware, software, applications, and APIs all considered as services.

Generally, it is very easy to confuse cloud computing with grid computing. The distinctions between them are not clear. Maybe it is because clouds and grids share similar visions: to reduce computing costs and increase flexibility and reliability by using third-party operated hardware (Vaquero et al. 2009). The Google Trends diagram (Figure 3.13) reveals that people especially in IT have now become more interested in cloud computing than grid computing.

Currently, a variety of companies provide cloud services. Google and Yahoo provide web-based e-mail service; Carbonite and MozyHome provide

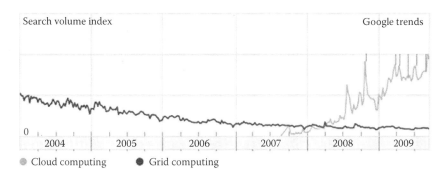

FIGURE 3.13
The Google search trend of cloud computing and grid computing. (Courtesy of http://www. google.com/trends.)

backup service; Salesforce.com (http://www.salesforce.com) provides customer resource management applications; and America Online (AOL), Google, Skype, Vonage, and other companies offer instant messaging and VoIP services. All these are cloud computing services, which are hidden behind an abstraction layer so that end users do not need to know the complexity and details of the cloud computing architecture.

Cloud computing encapsulates many aspects of computing complexities from hardware to software. Generally, there are three basic types of cloud computing:

Infrastructure as a Service (IaaS) delivers the computer infrastructure, for example, grid or cluster virtual server, network, storage, and system software, as standardized services over the network. The end user has full control over the virtualized computer instance and can customize the instance accordingly. The virtualization technology is used to provide multitenancy and isolation to the users, as different virtual instances may be allocated to a single physical machine. Unlike purchasing the physical servers, IaaS is charged on a utility basis, depending on the consumption of the resources, such as memory, storage space, and CPUs.

The IaaS introduces a variety of benefits for applications and enterprises: (1) Elasticity. Users can connect to and purchase the computing power on demand according to their needs, for example, applications can dynamically acquire more resources to host their services to handle peak workloads and release when the load decreases. (2) Inexpensive. Enterprises do not have to invest hardware or software utilities, but they can be provisioned as many resources as they need and when they need. Moreover, by moving their IT infrastructure into the cloud, enterprises reduce their administration and maintenance costs. This model is also convenient for service providers that aim to maximize the revenue from their physical infrastructure. (3) Reliability. The cloud providers are often able to utilize more advanced and improved network infrastructures and datacenters.

The most notable examples are Amazon's Elastic Compute Cloud (EC2, http://aws.amazon.com/ec2/), Joyent (Joyent, http://www.joyent.com/), and Amazon's Simple Storage Services (S3, http://aws.amazon.com/s3/). The IBM and other traditional IT vendors are also providing such services, and the Verizon Business, which provides telecommunications and more services, also initiates such activities.

Platform as a Service (PaaS) delivers a computing platform as a service. It encompasses a layer of software and provides it as a service that can be used to build higher-level services. Users can run existing applications or develop new applications on such a platform and do not need to consider maintaining the operating system, server hardware, load balancing, or computing capacity. It hides all the complexity of managing the underlying hardware and provides all the facilities required to support the complete lifecycle of building and deploying web applications and services entirely from the Internet. The most notable PaaS examples include the Microsoft "Azure" (http://www.microsoft.com/azure/), Google's Google App engine, and Salesforce's Force.com.

Software as a Service (SaaS) is the most famous and widely used type of cloud computing, providing all kinds of capabilities such as sophisticated traditional applications, installed locally but through the web browser, that are made available to end users. From the users' perspective, SaaS saves costs on the servers used to host the software and software licenses. From the providers' point of view, they need only to maintain one program, which also reduces costs. The most notable examples are Salesforce.com, Google's Gmail and Doc, America Online, Yahoo and Google's instant messaging, and VoIP of Vonage and Skype.

3.5.3 Cloud Computing Architecture

The success of cloud computing is largely based on the effective implementation of its architecture. Through proper implementation of cloud computing architecture, services provided to users could be accessed in a very fast fashion. Developers who design the architecture for cloud computing have to remember that it might be composed of different types of cloud computing that are integrated for the consistent computing of the online application. Figure 3.14 shows the architecture of the cloud computing with four layers from infrastructure layer to application layer.

The first layer is the SaaS layer. It provides all kinds of application services through the Internet. The application in cloud computing will call on the assistance of the middleware layer to deploy computing resources to respond to the users' requests.

The second layer is middleware, on which the applications are deployed in the Cloud and computing powers are relied. It should provide the functionality of management, including user authorization, authentication, log management, and computing resources management. Middleware allows networked computers to communicate with each other. It receives the requests from users and forwards them to the appropriate corresponding procedures. Middleware is also used to monitor and calculate the usage of cloud system resources, in order to make a quick response to complete the node synchronization configuration, load balancing configuration, and resource monitoring and to ensure that resources are well allocated to the appropriate users. Therefore, it can deploy and configure the resources intelligently and dynamically.

The bottom layer is the IaaS layer, including Computing Resource, Data Center, and Data Storage Server, which are managed by the middleware layer. Computing clusters are virtual or physical servers and are responsible for the high volume and concurrent requests. They can implement a large amount of computing operations and web application services.

If a cloud computing company has a lot of clients, there is likely to be a high demand for a lot of storage space. Some companies require hundreds of

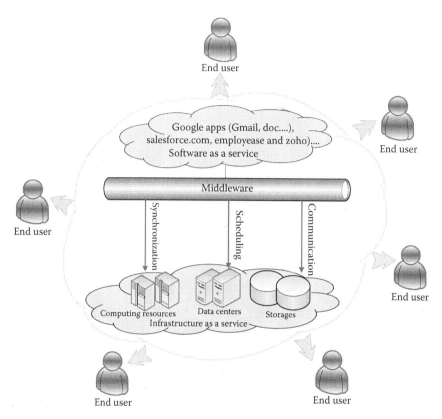

FIGURE 3.14
General architecture of cloud computing.

digital storage devices. Cloud computing systems need at least twice the number of storage devices they require to keep all their clients' information stored. That is because these devices, like all computers, occasionally break down. A cloud computing system must back up all its clients' information and store it on other devices. The copies enable the central server to access backup machines to retrieve data that otherwise would be unreachable. Making copies of data as a backup is called redundancy. Very recently, McFedries (2008) described the data center (conceived as a huge collection of clusters) as the basic unit of the cloud, offering huge amounts of computing power and storage by using spare resources. Keeping up with the demands of the application are the data centers and server farms, which are implemented to ensure that the application would have as many backup plans as it could have.

The outside cloud client layer provides the interaction interface for users to request cloud services, including IaaS, PaaS, and SaaS services, which is also entrance for users accessing the clouds. It enables users to utilize and operate cloud-based services like using traditional desktop systems. The user can register log and custom services through the web browser. Cloud users after obtaining the appropriate permissions (paid or other restrictions) are allowed to choose or customize the list of services and can also cancel the existing services in the cloud interface.

Figure 3.15 shows an extensible and open portal architecture enabling Earth science utilizing cloud computing. Within the Earth Science Components (ESC), Earth science data and metadata are stored in different data servers that are geographically distributed. These data and information are kept cross a number of categories and format types and are provided as accessible services based on standards and protocols proposed by FGDC, OGC, and ISO/TC211. The ESC is the counterpart of the bottom layer, IaaS layer of the general cloud computing architecture, which functions as data centers, providing the capability of preprocessing spatial data and making the data ready to be accessed by Geoinformation Gateways, for example, projection transformation and data format transfer. The Geoinformation Gateway provides an online discovery and access to Geoinformation resources (data, applications, and Web sites) through the interoperable way by the Web Map Service (WMS), Web Catalogue Service (WCS), Web Feature Service (WFS), and catalogue services. Based on the Geoinformation Gateway, all of the available services, data, and procedures can be brought together and utilized by any Earth science application. Through the Geoinformation Gateway, the complexity of data processing

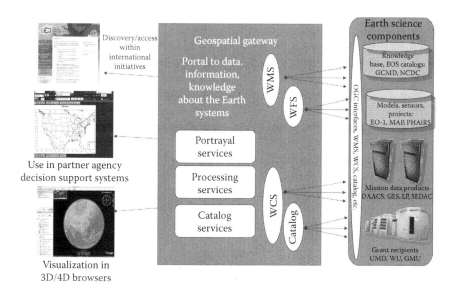

FIGURE 3.15
Extensible, open portal architecture.

and assessing are hidden from the Earth scientists without knowing where and how data come from.

3.5.4 Applications

3.5.4.1 *Current Cloud Computing Initiatives*

The IaaS solutions provide users with physical or virtual resources that satisfy the requirements of the user applications in terms of CPU, memory, operating system, and storage. Amazon.com launched EC2 3 years ago and is now the king of offering a pay-as-you-go elastic hosting service. Other examples of IaaS are Amazon EC2 (http://aws.amazon.com/ec2/), Google App engine, Amazon S3 (http://aws.amazon.com/s3/), Mosso (http://www.mosso.com/), Nirvanix, and Go Grid. Nirvanix has a subscription-based service, and GoGrid provides infrastructure as a utility but similar to the enterprise datacenter.

PaaS inherits characteristics of both clusters and grids while developing specific capabilities such as strong support for virtualization; dynamic provisioning of web services; and support for creating third-party, value added services by building on cloud compute, storage, and application services. The infrastructure on which clouds are hosted is hidden from the users. It is important for the platform to be scalable, reliable, flexible, easily customizable, and secure. The PaaSs need to provide the capability to remotely control, monitor, and dynamically change single and groups of nodes, which is important not only to simplify the management but also to identify and remove bottlenecks.

Many of the existing middleware platforms are proprietary and tied to their infrastructure or applications' service such as Google's App Engine (http://code.google.com/appengine/) and Force.com, which offer an on-demand development platform that allows developers to build multitenant applications using Salesforece.com's infrastructure as a service (www.salesforce.com). Most of the nonproprietary PaaSs are developed based on Java/Linux environments, with a few on .NET framework. PaaS solutions, such as Google App Engine and Microsoft Azure, provide users with a development platform for creating distributed applications that can automatically scale on demand.

Some popular PaaSs are 3Tera (http://www.3tera.com/), Apprenda (http://apprenda.com/), RightScale (http://www.rightscale.com/), Univa UD (http://www.univaud.com/), DataSynapse, Manjrasoft (http://www.manjrasoft.com/), Elastra, and Enomaly. The 3Tera's AppLogic provides a cloud computing platform that enables infrastructure solutions which adapt to changing needs at the speed of business. Apprenda's SaaSGrid is a distributed SaaS Application Server that eliminates the difficulties of building and delivering SaaS for .NET enterprise applications. RightScale offers a cloud computing management platform for managing cloud infrastructure from multiple public clouds. Univa UD provides software products for cloud computing management (e.g., Univa UD' UniCluster bundles the Sun Grid Engine scheduler and monitoring and HPC capabilities and Univa UD 'Grid MP an advanced job scheduler and application provisioning platform). Similarly, Manjrasoft's Aneka virtualizes and accelerates applications and dynamically provisions resources across private, public, or hybrid networks for .NET environments.

The SaaS enables clients to be able to access their applications and data from anywhere at any time. Clients could access the cloud computing system using any cloud devices linked to the Internet. Data would not be confined to a hard drive on one user's computer or even a corporation's internal network. Cloud computing systems give these organizations company-wide access to computer applications. The companies do not have to buy a set of software or software licenses for every employee. Instead, the company could pay a metered fee to a cloud computing company.

Some SaaS applications include custom resource management (CRM), enterprise resource planning (ERP), accounting, scheduling, automated billing, content management, human resource management, and so on. Examples of SaaS are Salesforce.com (www.salesforce.com) and Clarizen. com (www.Clarizen.com), which respectively provide on line CRM and project management services. Google and Zoho provide desktop applications that are hosted in the cloud and are replacing the traditional desktop-based Microsoft Office software.

3.5.4.2 Supporting Geoinformation Applications through UniPortal

A large amount of WMS resources are available on the Internet. Challenges exist when trying to utilize WMS for supporting geospatial sciences: (1) WMSs are widely dispersed and hard to be found; (2) An intuitive and efficient client is needed to integrate the services to enable service composition and chaining; and (3) A multidimension visualization tool is crucial to different Geoinformation applications. UniPortal(http://eie.cos.gmu.edu/c/portal/layout?p_l_id = PUB.1.417) is a unified Spatial Web Portal (SWP) addressing the three challenges by providing powerful functions to discover, integrate, and finally utilize the WMS resources (Li et al. 2010). UniPortal can help users to (1) search multiple catalogues synchronously; (2) seamlessly integrate tens of thousands of WMS, navigate and overlay the services with an intuitive interface for customizing new applications; (3) find more accurate records by using a layer-based search engine with spatial, temporal, and performance criteria; (4) visualize these services in a 2D/3D/4D (time) manner; and (5) share their own maps to others based on the Web Map Context (WMC) and KML standards.

With the support of the UniPortal, we can easily compose an Earth science application. For example, in order to analyze the relationship between the global land use or land cover changes and forest fires, we can overlay forest

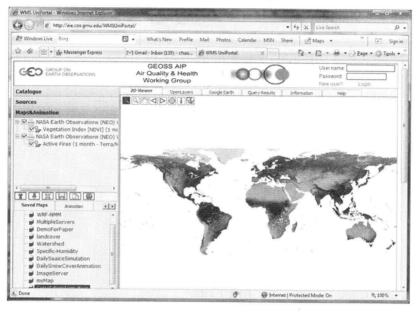

FIGURE 3.16
(See color insert following page 144.) Land use and forest fire correlation.

fire WMS on land use or land cover changes WMS, which are fetched from multiple servers (Figure 3.16). By exploring the overlay maps in different years, we can understand the places that are most vulnerable to the forest fires, global migration of forest fires, and the reasons triggering more and more forest fires every year.

3.5.5 Future Development

Cloud computing is at its beginning and changes the network from a text, community-oriented platform into one that can provide unlimited applications and bring together various kinds of services at anytime and anywhere.

In summary, Earth science cloud computing can provide smart or broad discovery, enhanced access, on-the-fly integrations, and transparent platforms for Earth scientists so that they can focus on research. Interoperability through standard protocols and service interfaces allow (1) Earth science data to be applied to many priorities, (2) data served up once and used many times, and (3) contributor to flexible or extensible global framework. The IT and Earth science collaborations can develop the cloud computing platform to enable Earth science discoveries. The current cloud computing platform is also facing many challenges:

Data security: The security of user data is considered a computing platform security issue, which is an important issue for the security of cloud computing. Computing platform security issues are generally more complicated than the network security. The current solutions use a number of different methods of computer science, for example, using trusted computing (Trusted Computing) and virtualization (Virtualization) technology to achieve computing platform security issues. The cloud storage security also includes data security, fault tolerance, continuous data protection, and other aspects of issues.

Personal privacy: In the cloud computing platform, each person is in an open environment to provide or receive services in the cloud computing platform, which makes personal privacy of users insecure. If a client can log in from any location to access data and applications, it is possible that the client's privacy could be compromised. Cloud computing companies will need to find ways to protect client privacy. One way is to use authentication techniques, such as user names and passwords. Another is to employ an authorization format—each user can access only the data and applications relevant to his or her job. Therefore, the privacy protection is also a challenge of cloud computing.

Service Interoperability: The support for the interoperability of services of current cloud computing is not enough. There are still many obstacles for users to obtain cross-platform services and different services in one platform. This is also a challenge of cloud computing research in the future.

How will cloud computing affect other industries? There is a growing concern in the IT industry about how cloud computing could impact the business of computer maintenance and repair. If companies switch to using streamlined computer systems, they will have fewer IT needs. Some industry

TABLE 3.1

Geoinformation Computing Platforms Characteristics

Platform	Applicability	Time	Scalability	Usability
GDC	Generic Geoinformation sharing and Spatial processing, e.g., spatial interpolation	1994 (Yang et al. 2007)	Depends	Normal
GHPC	Data and memory computing intensive applications, such as weather forecasting	2000 (Strohmaier et al. 2005)	Not good	Limited access
GGC	Data-intensive applications, which rarely require communication and data transfer between computing nodes	2002 (Wang et al. 2002)	Good	Normal
GCC	Applications serve massive users and require opaque or transparent access, such as Uniportal (Li et al. 2010)	2008 (Armbrust et al. 2009)	Very good	Very good

experts believe that the need for IT jobs will migrate to the back end of the cloud computing system.

3.6 Conclusion and Discussion

This chapter introduces the leveraging of the four computing platforms for Geoinformation science problems. As illustrated through the four examples, Geoinformation science problems have to be solved by leveraging the new computing platforms. Special considerations should be taken to specially locate, select, and organize computing resources and Geoinformation resources to achieve needed performance due to the spatial and time constraints of Geoinformation and the scientific phenomena. According to different applications, the platforms have different usage (Table 3.1). The increasing demands of Geoinformation science for computing to answer fundamental scientific and complex application questions will not only benefit from the advancement of computing technology but also provide opportunities to advance computing science.

References

Allcock, W., J. Bester, J. Bresnahan et al. 2001. Data management and transfer in high-performance computational grid environments. *Parallel Computing*, 28(5): 749–771.

Allen, M. 1999. Do it yourself climate prediction. *Nature*, 401(6754):642.

Armbrust, M., A. Fox, R. Griffith et al., 2009. *Above the Clouds: A Berkeley View of Cloud Computing*. University of California, Berkeley, Berkeley, CA.

Armstrong, M.P., M. Cowles, and S. Wang, 2005. Using a computational grid for geographic information analysis. *Professional Geographer*, 57(3):365–375.

Attiya, H. and J. Welch. 1998. *Distributed Computing: Fundamentals, Simulations, and Advanced Topics*. McGraw-Hill, New York.

Baillie, C., J. Michalakes, and R. Skilin. 1997. Regional weather modeling on parallel computers. *Parallel Computing*, 23(14):2135–2142.

Bal, H.E., M.F. Kaashoek, and A.S. Tanenbaum. 1992. Experience with distributed programming in Orca. *IEEE Transactions on Software Engineering*, 18:190–205.

Barrera, J.S. 1991. A fast Mach network IPC implementation. In *Proceedings of the Second USENIX Mach Symposium*, pp. 1–11, Monterey, CA, November 1991.

Barton, T., N. El-Khoury, M. Gettes et al. 2002. NSF Middleware Initiative Draft, work-in-progress, expires November 2002. http://middleware.internet2.edu/video/drsaftdocs/draft-nmiedit-vidmid_vc-commObject_White_Paper-1.0.html (accessed April 26, 2009).

Berman, F., G. Fox, and T. Hey. 2003. The grid: Past, present, future. In *Grid Computing: Making the Global Infrastructure a Reality*. ed. F. Berman, G. Fox, and T. Hey, Wiley, West Sussex.

Bernstein, P.A., V. Hadzilacos, and N. Goodman. 1987. *Concurrency Control and Recovery in Database Systems*. Addison Wesley, Reading, MA, ISBN 0-20110-715-5.

Bershad, B.N. and M.J. Zekauskas. 1991. Midway: Shared memory parallel programming with entry consistency for distributed memory multiprocessors. *Research Report CMU-CS-91-170*, Carnegie-Mellon University, Pittsburgh.

Birk, R., M. Frederick, L.C. Dewayne, and M.W. Lapenta. 2006. NASA's Applied Sciences Program: Transforming research results into operational success. *Earth Imaging J.*, 3(3): 18–23.

Buyya, R., C.S.Yeo, and S.Venugopal. 2008. Market-oriented cloud computing: Vision, hype, and reality for delivering it services as computing utilities. *CoRR*, vol. abs/0808.3558.

Cao, Y. 2007. *Transportation Routing with Real-Time Events Supported by Grid Computing*. Ph.D. dissertation, George Mason University.

Cao, Y., C. Yang, and D. Wong. 2009. An interoperable spatiotemporal weather radar data disseminating system. *International Journal of Remote Sensing*, 30:1313–1326.

Carriero, N. and D. Gelernter. 1986. The S/Net's Linda Kernel. *ACM Transaction on Computer Systems*, 4:110–129.

Carter, J.B., J.K. Bennett, and W. Zwaenepoel. 1991. Implementation and performance of Munin. In *Proceedings of 13th ACM Symposium on Operating Systems Principles*, pp. 152–164.

Chabini, I. 1998. Discrete dynamic shortest path problems in transportation applications: Complexity and algorithms with optimal run time. *Transportation Research Record: Journal of the Transportation Research Board*, 1645:170–175.

Chen, S., X. Du, F. Ma, and J. Shen. 2005. A grid resource management approach based on P2P technology. *Proceedings of the Eighth International Conference on High-Performance Computing in Asia-Pacific Region (HPCASIA'05)*, 2005, pp. 362–369.

Chervenak, A., I. Foster, C. Kesselman, C. Salisbury, and S. Tuecke. 2000. The data grid: Towards an architecture for the distributed management and analysis of large scientific datasets. *J. Network Comput. Appl.*, 23(3):187–200.

Clarke, K.C. 2003. Geocomputation's future at the extremes: High performance computing and nanoclients. *Parallel Computing*, 29(10):1281–1295.

Clematis, A., M. Mineter, and R. Marciano. 2003. High performance computing with geographical data. *Parallel Computing*, 29(10):1275–1279.

Dale, S. 1983. A formal model of crash recovery in a distributed system. *IEEE Transactions on Software Engineering*, 9(3):219–228.

ESRI, 2002. *ArcXML Programmer's Reference Guide, ArcIMS 4*. (2002).

Executive Office of the President (EOP), 1994. Coordinating geographic data acquisition and access: The National Spatial Data Infrastructure. http://www.fgdc.gov/publications/documents/geninfo/execord.html (accessed August 20, 2009).

FGDC, 1994. This website has information for clearinghouse, NSDI, GOS portal, and funding opportunities. http://www.fgdc.gov/ (accessed August 20, 2009).

Foster, I. and C. Kesselman. 1998. *The Grid: Blueprint for a New Computing Infrastructure*. Morgan Kaufmann Publishers Inc, San Francisco, CA.

Foster, I. and C. Kesselman. 1999. *The Grid: Blueprint for a New Computing Infrastructure*. Morgan Kaufmann, Los Altos, CA.

Foster, I., C. Kesselman, and S. Tuecke. 2001. The anatomy of the grid: Enabling scalable virtual organizations. *The International Journal of High Performance Computing Applications*, 15(3):200–222.

Foster, I., Y. Zhao, I. Raicu, and S. Lu. 2008. Cloud computing and grid computing 360-degree compared. In *Grid Computing Environments Workshop, 2008. GCE '08*, pp. 1–10, 12–16 Nov. 2008, ISBN: 978-1-4244-2860-1.

Gartner, F.C. 1999. Fundamentals of fault-tolerant distributed computing in asynchronous environments. *ACM Computing Surveys*, 31(1):1–26.

Google app engine, http://code.google.com/appengine/docs/whatisgoogleappengine.html (accessed April 16, 2009).

Google Trend, http://www.google.com/trends (accessed April 16, 2009).

Hampel, R. 1994. The MPI synchronous to asynchronous communication. In *Specification and Verification of Concurrent Systems*, ed. C. Rattray, pp. 247–252. Springer, Berlin, LNCS 797.

Hargrove, W.W., F.M. Hoffman, and T. Sterling. 2001. The Stone Soupercomputer: ORNL's first Beowulf-style parallel computer: The do-it-yourself supercomputer. *Scientific American*, 265(2):72–79.

Hawick, K.A., P.D. Coddington, and H.A. James. 2003. Distributed frameworks and parallel algorithms for processing large-scale geographic data. *Parallel Computing*, 29:1297–1333.

Haug, E., J. Dubois, J. Clinckemaillie, S. Vlachoutsis, and G. Lonsdale. 1994. Transport vehicle crash, safety and manufacturing simulation in the perspective of high performance computing and networking. *Future Generation Computer Systems*, 10(2&3):173–180.

Horning, J.J., H.C. Lauer, P.M. Melliar-Smith, and B. Randell. 1974. A program structure for error detection and recovery. *Lecture Notes in Computer Science*, 16:177–193.

Inacio, C. 1998. *Software Fault Tolerance*. Carnegie Mello University, http://www.ece.cmu.edu/~koopman/des_s99/sw_fault_tolerance/ (accessed November 11, 2009).

International Standards Organization (1994) Technical Committee 211: Geoinformation, Geomatics. http://www.isotc211.org/ (accessed November, 2009).

Jakob, J.Z. 2001. The Shuttle Radar Topography Mission (SRTM): A breakthrough in remote sensing of topography. *Acta Astronautica*, 48(5–12): 559–565.

Jalote, P. 1994. *Fault Tolerance in Distributed Systems*. Prentice-Hall, Inc., Upper Saddle River, NJ.

Kesselman, C. 2004. Applications of intelligent agent technology to the grid. In *Proceedings of the IEEE/WIC/ACM International Conference on Web Intelligence (WI'04)*, Sept. 20–24, 2004, pp. 4–4.

Kim, K.H. 2002. Fault-tolerant distributed computing: Evolution and issues. *IEEE Distributed Systems Online*, 3(7):1–4.

Laprie, J.C. 1985. Dependable computing and fault tolerance: Concepts and terminology. In *Proceedings of Twenty-Fifth International Symposium on Fault-Tolerant Computing*, pp. 2–11.

Leptoukh, G. 2005. NASA remote sensing data in Earth sciences: Processing, archiving, distribution, applications at the GES DISC. *31st International Symposium on Remote Sensing of Environment*, June 20–34, Saint Petersburg.

Li, F., D. Qi, L. Zhang, X. Zhang, and Z. Zhang. 2006. Research on novel dynamic resource management and job scheduling in grid computing, *First International Multi-Symposiums on Computer and Computational Sciences, 2006. IMSCCS '06*, Vol. 1, pp. 709–713.

Li, K. 1986. Shared virtual memory on loosely coupled multiprocessors. PhD thesis, Yale University.

Li, K. and P. Hudak. 1989. Memory coherence in shared virtual memory systems. *ACM Transactions on Computer Systems*, 7(4):321–357.

Li, Z., C. Yang, W. Li, and L. Miao. 2010. An optimized framework for seamlessly integrating OGC web services. *IJGIS*, 2010 (forthcoming).

Li, W., C. Yang, and R. Raskin. 2008, A semantic enhanced search for spatial web portals. In *Technical Report of Semantic Scientific Knowledge Integration of AAAI Spring Symposium*, pp. 47–50, AAAI Press, Menlo Park, CA.

Mayer, L., K. Barbor, P. Boudreau et al. 2004. *A Geospatial Framework for the Coastal Zone: National Needs for Coastal Mapping and Charting*. National Academies, Washington, DC.

McFedries, P. 2008. The cloud is the computer. *IEEE Spectrum Online*, August 2008. Electronic Magazine, http://www.spectrum.ieee.org/aug08/6490 (accessed December 20, 2009).

Moore, R.W., A. Jagatheesan, A. Rajasekar, M. Wan, and W. Schroeder. 2004. Data grid management systems. *Proceedings of the 21st IEEE/NASA Conference on Mass Storage Systems and Technologies (MSST)*, April 13–16, 2004, College Park, MD.

Mower, J.E. 1996. Developing parallel procedures for line simplification. *International Journal of Geographical Information Systems*, 10(6):699–712.

Murty, V.V. and V.K. Garg. 1993. Synchronous message passing. In *Proceedings of International Symposium on Autonomous Decentralized Systems*, Phoenix, AZ, pp. 208–214.

Nanjundiah, R.S. 1998. Strategies for parallel implementation of a global spectral atmospheric general circulation model. *5th International Conference on High Performance Computing, 1998. HIPC '98*, pp. 452–458, December 17–20, 1998.

Nabrzyski, J., J.M. Schopf, and J. Weglarz. 2004. *Grid Resource Management: State of the Art and Future Trends*. Kluwer Academic Publishers, Norwell, MA.

National Research Council (NRC). 2008. *The Potential Impact of High-End Capability Computing on Four Illustrative Fields of Science and Engineering*, The National Academies Press, Washington, DC.

Nelson, V.P. 1990. Fault-tolerant computing: Fundamental concepts. *Computer*, 23(7):19–25.

Open Geospatial Consortium (OGC). 1994. This website has information on WMS, WFS, WCS, CS-W, and other OGC specifications; http://www.opengeospatial.org (accessed April 26, 2009).

Peng, Z.R. and M.H. Tsou. 2003. *Internet GIS: Distributed Geographic Information Services for the Internet and Wireless Networks*. Wiley, Hoboken.

Plantikow, S., K. Peter, M. Högqvist, C. Grimme, and A. Papaspyrou. 2009. Generalizing the data management of three community grids. *Future Generation Computer Systems*, 25:281–289.

Plewe, B. 1997. *GIS Online: Information Retrieval, Mapping, and the Internet*. OnWord, Santa Fe.

Purohit, S., A. Kaginalkar, I. Jindani, J.V. Ratnam, and S.K. Dash. 1999. Development of parallel climate/forecast models on 100 GFlops PARAM computing systems. *Proceedings of the Eight ECMWF Workshop on the Use of Parallel Processors in Meteorology*. eds. W. Zwieflhofer and N. Kreitz, pp. 125–132. World Scientific, Singapore.

Raynal, M. and J.M. Helary. 1990. *Synchronization and Control of Distributed Systems and Programs*. Wiley & Sons, New York.

Raynal, M., A. Schiper, and S. Toueg. 1990. The causal ordering abstraction and a simple way to implement it. *Information Processing Letters*, 39(6):343–350.

Rajasekar, A., M. Wan, R.W. Moore et al. 2003. Storage resource broker—Managing distributed data in a grid. *Computer Society of India Journal*, special issue on SAN, 2003.

Reis, G.A., J. Chang, N. Vachharajani, R. Rangan, D.I. August, and S.S. Mukherjee. 2005. Software—controlled fault tolerance. *ACM Transactions on Architecture and Code Optimization*, pp. 1–28, December 2005.

Skeen, D. 1982. Nonblocking commit protocols. In *Proceedings of the ACM SIGMOD Conference on Management of Data*, pp. 133–147, Orlando, FL, June 1982.

Somani, A.K. and N.H. Vaidya. 1997. Understanding fault tolerance and reliability. *IEEE Computer*, 30(4):45–50.

Stainforth, D., J. Kettleborough, M. Allen, A. Collins, A. Heaps, and J. Murphy, 2002. Distributed computing for public-interest climate modeling research. *Computing in Science and Engineering*, 4(3):82–89.

Strohmaier, E., J.J. Dongarra, H.W. Meuer, and H.D. Simon. 2005. Recent trends in the marketplace of high performance computing. *Parallel Computing*, 31(3–4): 261–273

Tanenbaum, A.S. 1994. Distributed shared memory. In *Distributed Operating Systems*, ed. A.S. Tanenbaum, pp. 289–372. Prentice-Hall, Englewood Cliffs, NJ.

Tao, V. 2006. Microsoft virtual Earth: New horizons in on-line mapping. Paper presented at *GeoInformatics'2006*, Wuhan, China.

Thain, D., T. Tannenbaum, and M. Livny. 2005. Distributed computing in practice: The Condor experience. *Concurrency—Practice and Experience*, 17(2–4):323–356.

Tierney, B., W. Johnston, J. Lee, and M. Thompson. 2000. A data intensive distributed computing architecture for "grid" applications. *Future Generation Computer Systems*, 16(5):473–481.

Vaquero, L.M., L. Rodero-Merino, J. Caceres, and M. Lindner. 2009. A break in the clouds: Towards a cloud definition. *ACM SIGCOMM Computer Communication Review*, 39(1):50–55.

Wang, S., M.P. Armstrong, and D.A. Bennett. 2002. Conceptual basics of Middleware design to support grid computing of geographic information. In *Proceedings of Second International Conference on Geographic Information Science*, Boulder, CO, pp. 197–200, September 25–28, 2002.

Wang, S. and M.P. Armstrong. 2003. A quadtree approach to domain decomposition for spatial interpolation in grid computing environments. *Parallel Computing*, 29(10):1481–1504.

Xie, J., C. Yang, B. Zhou, and Q. Huang. 2009. High performance computing for the simulation of dust storms. *Computers, Environment, and Urban Systems*, http://dx.doi.org/10.1016/j.compenvurbsys.2009.08.002.

Yang, C., D. Wong, and B. Li. 2005. Introduction to computing & computational issues of distributed GIS. *Geographic Information Sciences*, 11(1):1–3.

Yang, C., D. Wong, M. Kafatos, and R. Yang. 2006. Implementing computing techniques to accelerate network GIS. In *Proceedings of SPIE the International Society for Optical Engineering*. Vol. 6418, 64181C, *Geoinformatics 2006: GNSS and Integrated Geospatial Applications*, eds. D. Li and L. Xia. Bellingham, WA.

Yang, C., H. Wu, Q. Huang, Z. Li, and J. Li. 2010. Spatial computing for supporting physical sciences, *Proceedings of the National Academy of Science* (in press).

Yang, C., W. Li., J. Xie, and B. Zhou. 2008. Distributed geospatial information processing: Sharing earth science information to support Digital Earth. *International Journal of Digital Earth*, 1(3):259–278.

Yang, C. and R. Raskin. 2009. Introduction to distributed geographic information processing. *International Journal of Geographic Information Science*, 23(5):1–8.

Yang, Q. 1997. A simulation laboratory for evaluation of dynamic traffic management system. Unpublished PhD dissertation, Massachusetts Institute of Technology.

4

*Data Access and Data Systems**

Ruixin Yang, Hampapuram K. Ramapriyan, and Carol B. Meyer

CONTENTS

4.1 Introduction

The twenty-first century is marked by an unprecedented, constantly increasing volume of data in most of the scientific disciplines. In particular, there has been a phenomenal growth in the volume and variety of spatial and spatiotemporal data available for research and applications in geoscience and geoinformatics. A large number of spaceborne, airborne and *in situ* instruments have been accumulating Petabytes (PB 10^{15}) of Earth Observation (EO), and mathematical models have been contributing comparable amounts as well. The typical data sets handled in analyses have grown in size from Gigabytes (GB 10^9) to Terabytes (TB 10^{12}). Data holdings are in the PB 10^{15} range with expected growth to the Exabyte (EB 10^{18}) level (Williams 2009).

To use the rapidly growing volumes of data efficiently and effectively, one cannot rely on the traditional method wherein a user downloads data and uses local tools to study the data residing on a local storage system. Instead, a cyber-infrastructure for supporting users to find, access, and analyze data is needed so that users can focus on their applications without expending significant effort in handling data with issues such as data locations, data formats, and data models. Such a cyber-infrastructure would consist of several data, information, and computational systems communicating with each other.

---

* The authors thank Karl Benedict for helpful input to this chapter.

Depending on the scope of a given data and information system, there are many issues that need to be considered for its successful operation. Common issues include a long-term stewardship of data and data systems; type and degree of interoperability; metadata handling such as standard and semantic; and seamless data search, access, and analysis (Yang and Kafatos 2001).

The need for data, recognition of importance of preserving and providing access to data, and the existence of data centers are not really new. For example, NOAA's NCDC has existed since 1951. The U.S. National Aeronautics and Space Administration (NASA) has preserved data and derived products from its missions since its inception in 1958. However, the nature of the data and information systems has changed in an evolutionary, sometimes in a revolutionary, manner since the 1970s, given the significant changes in information technology as well as policies over time. Currently, there are many efforts in building cyber-infrastructure as well as data and information systems to provide distributed geoscience data to research and education communities. Generally speaking, there are two complementary high-level approaches for building such systems. One is a top-down approach for large data centers emphasizing data processing, archiving, and distributing a core set of standard data products, which may also provide tailored services for specific user communities. Given the emphasis in this case to handling large quantities of data, the resulting systems can be referred to as "data centric systems." The other is a bottom-up approach, by which user communities integrate the tools including data, application software, and even domain knowledge to form a new infrastructure for individuals to search, access, and analyze data in a new paradigm. Due to the collaborative governance structure of such systems, they can be referred to as "federated systems." One example of a data centric system is the Earth Observing System Data and Information System (EOSDIS), designed for the processing and archiving of data from NASA's EO missions and their distribution as well as provision of specialized services to users. The Federation of Earth Science Information Partners (ESIPs), initially sponsored by NASA, with a current membership from NASA, other federal agencies, and commercial and nonprofit organizations, is an example of a federated system.

In the next section of this chapter, we will introduce EOSDIS, a data-centric system. Then, we will provide an overview of the ESIP Federation as an example of a bottom-up approach. After that, we will discuss a use case as an example of an end-to-end system developed under the ESIP Federation. We will then conclude this chapter by discussing the future strategies for spatial data infrastructure.

4.2 Data Centric System: EOSDIS

In this section, we present a brief discussion of an example of a data-centric system defined earlier. NASA's EOSDIS has been in operation since 1994, and it is a large distributed system that processes, archives, and distributes over 3700 distinct types of data products needed for research and applications in a variety of Earth science disciplines. Its scope includes all of the data from NASA's EOS Program and a significant portion of the pre-EOS data as well as data from non-U.S. partners according to international agreements between NASA and the corresponding foreign agencies.

The EOS Program was NASA's contribution to the U.S. Global Change Research Program (USGCRP). The USGCRP was initiated by the U.S. Congress in 1990 as a part of the Global Change Research Act. NASA's Earth Science Enterprise (currently called the Earth Science Division in the Science Mission Directorate) established the EOS Program with the objective of launching a series of satellite missions to help answer fundamental questions such as "How is the Earth changing?" and "What are the consequences for life on Earth?" The development of EOSDIS was initiated in 1990 to support the EOS Program with the capabilities to process, store, and disseminate to a broad user community, the data from the instruments aboard EOS satellites as well as those from other missions as indicated earlier.

The major components of EOSDIS are 12 Data Centers, 14 Science Investigator-led Processing Systems (SIPSs), and the EOS Clearing House (ECHO). These components are geographically distributed across the United States. The Data Centers (some of which are also called DAACs) archive, distribute, and provide users with access to data and services. Each Data Center focuses on one (or a few) Earth science disciplines and serves mainly users in those disciplines. The SIPSs, generally collocated with the Principal Investigators or Team Leaders for the various EOS instruments, process the data from the respective instruments into standard digital data products. The products are at various "levels": calibrated radiances (Level 1); various geophysical parameters in the satellite-acquired coordinate systems and resolutions (Level 2); the Level 2 products spatially and temporally mapped into standard grids (Level 3); and model outputs or results of analyses of lower level data (Level 4) (NASA 2006). Most of these standard products are generated at the SIPSs, whereas a few are generated at some of the Data Centers. Each of the standard products generated at the SIPSs is sent to a specific, assigned Data Center for archiving and distribution.

Figure 4.1 shows the context in which EOSDIS exists and operates. From the left to the right, this figure shows the capture and early processing of data from satellites (handled by the Mission Operations, managed by the Earth Science Mission Operations [ESMO] Project at NASA Goddard Space Flight Center [GSFC]) and the higher levels of processing, archiving, and dissemination to a broad and diverse user community (handled by the Science Operations, managed by the Earth Science Data and Information System [ESDIS] Project at NASA GSFC).

The distribution of the EOSDIS Data Centers and the SIPSs across the United States is shown in Figure 4.2. Indicated with each Data Center's name is the set of Earth science disciplines addressed by its data holdings. Indicated

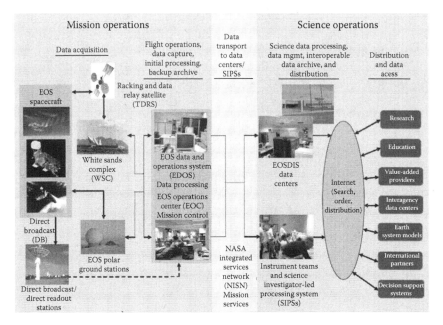

FIGURE 4.1
(See color insert following page 144.) Context of EOSDIS.

with each of the SIPSs locations is the set of the names of instruments whose data are processed at the location. It is to be noted that even though there may be several instrument names shown at a given geographic location, the SIPSs for each instrument are distinct from the others at the same location. More details about these can be found at NASA's EOSDIS Web site (http://www.esdis.eosdis.nasa.gov/).

As of this writing (October 2009), the volume of the holdings of EOSDIS is about 4 PB and the growth rate is about 1 TB per day. The growth in volume depends on the processing and ingestion as well as on the deletion rates. Processing instrument data into scientific products involves the generation of multiple versions over time, with each new version resulting from reprocessing the entire data set from the beginning of the mission. As newer versions are produced, the older versions are deleted after a certain period of overlap when both versions are maintained in the archives. The processing rate in EOSDIS has varied between about 1 and 5 TB/day. The distribution of data from EOSDIS to the external user community (excluding the science instrument teams responsible for production and quality assurance of data products) has varied between 3 and 6 TB/day. In the year ending on September 30, 2009, EOSDIS had delivered over 300 million products (data files) to external users; over 900,000 users had obtained data and/or services from the EOSDIS Data Centers; and there were over one million visits (counting only those that exceeded 1 min) to the EOSDIS Data Center Web sites.

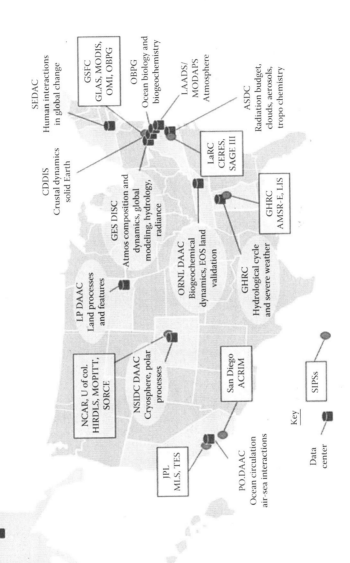

FIGURE 4.2
Geographic distribution of EOSDIS Data Centers and SIPSs.

The data in EOSDIS are currently stored in mixed archives containing online disks and robotic tape silos (it is expected that in early 2010, all the data will be on online disks with tape backups). The data stored online are easily downloadable via ftp. The data stored in robotic silos are staged for ftp pickup on being ordered by users. In either case, users need to search for the data of interest and download or order them. NASA has an open data policy, whereby the Earth science data held by EOSDIS are available to all users on a nondiscriminatory basis. The data are provided at no cost except in a few cases where international agreements dictate that nominal charges be made (not to exceed a marginal cost of fulfilling a user's request).

Given the diversity of users and the number of Earth science disciplines served by EOSDIS, a variety of methods is provided for searching and accessing the data. The Global Change Master Directory (GCMD) (Olsen and Major 1996; http://gcmd.nasa.gov/) provides pointers to locations of data in various Earth science disciplines. The access here is at the "directory level," that is, the pointers indicate where the data collections of interest are held, but to search for and obtain specific files or instances of data needed to cover a given region and/or a time interval, a user would have to go to the respective data centers' sites. However, GCMD provides a very broad coverage of sources of data in the world. Its database holds more than 25,000 descriptions of Earth science data sets and services covering all aspects of Earth and environmental sciences. Data and service providers from all over the world enter their data set descriptions using the standard Directory Interchange Format (DIF) and service descriptions using the standard Service Entry Resource Format (SERF). Standard terminology in the form of a hierarchical set of Earth science keywords that is widely accepted in the community facilitates maintenance and the usage of GCMD.

There are several methods to search for and access data at the "granule" or "file" level. These methods help users locate and order or access the specific data files relevant to their work. The ECHO consists of a metadata repository for all of the data sets handled by the EOSDIS Data Centers (except the Ocean Biology Processing Group). The ECHO is also the middleware with which many clients can interface (using well-defined Application Programmer Interfaces [APIs]) to provide different views of the data set inventory and different, discipline-specific, or other specialized clients can facilitate user access to data. The ECHO is based on a Service Oriented Architecture (SOA) and consists of a set of infrastructure services that enable the fundamental SOA functions: publish, discover, and access Earth science resources. It also provides additional services such as user management, data access control, and order management. The ECHO system has a data registry and a services registry. The data registry enables organizations to publish EOS and other Earth-science related data holdings to a common metadata model. These holdings are described through metadata in terms of data sets (types of data) and granules (specific instances of those types). The ECHO also supports browse images, which provide a visual representation of the data. The published metadata can be mapped to and from existing standards (e.g., FGDC, ISO 19115). With ECHO, users can find the metadata stored in the data registry and then access the data either directly online or through a brokered order to the data archive organization using client software (Mitchell et al. 2009). Currently, there are three clients in operation and four under development (Client Partners at http://www.echo.nasa.gov/). The operational clients are the Warehouse Inventory Search Tool (WIST), Search 'N Order Web Interface (SNOWI), and European Space Agency's (ESA) client. The WIST provides an accessing or ordering mechanism for users without needing to know at which Data Centers the data sets of interest are held. Users can submit cross-Data Center queries using spatial and temporal criteria, examine search results for relevance using built-in tools, and submit orders via ECHO to the appropriate data providers. The SNOWI tool is a Web-based client used to search a selected set of National Snow and Ice Data Center (NSIDC) data holdings. It is intended as a lightweight tool for quick access to data for its user community. The ESA's client EOLI allows access to ECHO holdings as well as ESA, DLR (Deutsches Centrum für Luft- und Raumfahrt-German Aerospace Center), and other catalogs. This client is a part of ESA's eoPortal. In addition to ECHO clients, there are several clients developed by the individual Data Centers that are part of EOSDIS. Such clients are more tailored to the specific discipline communities and provide search and access capabilities for holdings at the respective Data Centers. Although the implementation details vary, the basic capabilities include specification of spatial and temporal bounds, and dataset/parameter/product/mission/campaign names for the data of interest. In addition, the Data Centers provide online services such as subsetting, reprojection, and visualization for their data holdings and also make tools available for downloading and installation at users' sites. Several Data Centers offer data through OPeNDAP servers wherein appropriate, several data sets are provided through OGC standard-based Web services. The online services can be invoked as post-processing operations as a user searches for and selects data.

The EOSDIS operates in a larger context of NASA's Earth Science Data Systems (ESDS). NASA's ESDS consist of core and community capabilities. The core capabilities provide the basic infrastructure for robust and reliable data capture, processing, archiving, and distributing a set of data products to a large and diverse user community. The EOSDIS is an example of a core capability. Community capabilities provide specialized and innovative services to data users and/or research products offering new scientific insight. Such systems are generally supported by NASA through peer reviewed competitions. Examples of community capabilities are projects under the three Programs: Research, Education, and Applications Solutions Network (REASoN); Advancing Collaborative Connections for Earth System Science (ACCESS); and Making Earth System Data Records for Use in Research Environments (MEaSUREs). The core and community capabilities complement each other. NASA ensures community inputs to the ongoing evolution of its ESDS through the ESDS Working Groups (ESDSWG) with representation from developers of both the core and community capabilities. There are currently four working groups addressing the following four areas, respectively: Metrics, Software Reuse, Standards' Processes, and Technology Infusion. Most of the ESDSWG members also participate in the Federation of ESIP (see Section 4.3) to obtain perspectives of the broader community and convey them to NASA.

4.3 Federation of Earth Science Information Partners

The Federation of ESIPs (ESIP Federation) is a community-driven consortium of Earth science data and technology providers whose partners span a wide range of interests, including government data centers, research universities, technology developers, research laboratories, commercial enterprises, nonprofit organizations, and educators. In 1998, the ESIP Federation was created in response to a 1995 National Research Council (NRC) recommendation which determined that NASA's EOSDIS was too centralized and would not achieve all it was capable of achieving without additional involvement from the external science community and private sector (NRC 1995). The NRC provided the following guidance:

> To meet these expectations, we must now embrace a revolutionary expansion of the conceptual model that governs the management and operation of the system by affording the scientific community full partnership with shared responsibility.

In response to the NRC report, NASA created a competitive program that would enable product generation and publication and provide user services for NASA's EOSDIS program. The awardees, or ESIPs, were funded to create higher-level data products, tools, and commercial applications for EOSDIS data. As a condition of winning an award, each of the 24 recipients was required to federate, leading to the creation of the ESIP Federation. The ESIP Federation's original partners were distributed institutionally and geographically, a network made possible by the ubiquity of the Internet and associated increases in bandwidth capabilities. Further, all were connected to Earth science, but the organization itself was discipline-independent.

In contrast to the centralized approach for managing EOSDIS, the ESIP Federation embraces openness and community decision making, which results in innovation and new collaborations across the public, academic, and private sectors. The ESIPs are classified by Type:

Type 1 being primarily distributors of remotely sensed and ground-based data sets as well as standardized products derived from those data.

Type 2 being principally engaged in the scientific development, provision, and support of data and information products, technology, or services aimed primarily at the Earth science and research communities.

Type 3 being principally engaged in the development and provision of Earth science applications.

These Types correspond to functions along a value chain, along which the resulting products become more highly refined and potentially usable by a broader range of Earth science data and information consumers. ESIP Federation partners produce products for researchers, decision makers, and the public.

The ESIP Federation has evolved in many ways during its history. Created by NASA with NASA-funded projects as its original partners, the ESIP Federation began to grow from the beginning. Shortly after its formation, NASA DAACs joined the ESIP Federation. As its potential began to be realized, additional NASA-funded projects and nonfunded entities joined the ESIP Federation. An important indicator of the ESIP Federation's success was when the first NOAA entity, the National Climatic Data Center (NCDC), joined as a partner. The NCDC's partnership led two other NOAA data centers to join. In recognition of the ESIP Federation's potential, NOAA became the ESIP Federation's second financial sponsor in 2005. During the time when NOAA became more involved in the ESIP Federation, the international Group on Earth Observations (GEO) was solidifying its intent to build a Global Earth Observation System of Systems (GEOSS), a catalyst for the collaboration on the international stage, much like the ESIP Federation was a catalyst for the collaboration in the United States. The goal of GEO is to create a coordinated and distributed EO system. The model for GEO allows member countries and participating organizations to make their unique contribution to GEOSS. The model for GEO is akin to what the NRC envisioned for NASA and the ESIP Federation.

The work of GEO is premised upon the goal that Earth observations will benefit society. The GEO has offered several societal benefit areas in which applied Earth observations offer context—disasters, health, energy, climate, water, weather, ecosystems, agriculture, and biodiversity. As GEO considers how to meet the needs of these distinct user communities, the ESIP Federation is supporting those efforts in the areas of Health and Air Quality. The ESIP Federation formed an Air Quality Cluster (now Working Group), which has pulled together multiple federal agencies, state air quality managers, private sector companies, and academics to unite the many air quality data systems already in existence. The Air Quality Working Group has developed implementation pilot projects for GEO, resulting in uniting a disparate community behind its efforts. The most compelling storyline about the Air Quality Working Group's effort is that it has been done as a voluntary effort, for the good of the order. Participants in the pilots leveraged existing projects to support a greater community effort with the expectation that a community-wide product would be better than any single product from an individual provider. The Air Quality Working Group's success is a clear indicator of the ESIP Federation's community concept at work.

The ESIP Federation is unique; it does not produce data systems on its own. Rather, the ESIP Federation's unique contributions to Earth science data systems stem from its interconnectedness across institutions, disciplines, functions, and sectors and from the open collaboration that drives its networking. The ESIP Federation provides a forum for the diverse Earth science data and information community to meet so that ESIP partners can leverage expertise from the community to advance the field collectively. In recent years, the ESIP Federation has become a neutral venue in which multiple agencies can communicate with the Earth science data and information community, while collectively considering community-wide issues. The neutrality of the ESIP Federation is valuable in its linking federal agencies to a community of experts who guide both policy and technical decisions.

At their core, many of the activities of ESIP Federation member organizations include the acquisition, generation, management, and delivery of data and information; facilitation of data discovery by diverse individual and organizational end users; and delivery of analytic products and capabilities. The diverse ESIP member types typically focus on different activities, but there is substantial overlap between member organizations in terms of the actual activities that are undertaken.

The data holdings of Federation member organizations include those already described in the previous section related to EOSDIS, as NASA's EOSDIS Data Centers have been members of the ESIP Federation since its inception, but this is only a partial representation of the collective data available from Federation members. The more recent addition of Federation members from NOAA and increasing involvement from US EPA further expand the number of U.S. Federal data providers and products engaged with the Federation. Exact numbers are not available, but the total data holdings of Federation members include those described above as part of EOSDIS, those associated with NOAA Federation members (including representatives from NOAA's National Geophysical Data Center and NCDC), and many university and private sector Federation member organizations.

Discovery and delivery of data and products derived from data is a core activity of many ESIP Federation members. In many instances, Federation member data and products are made available through open standards such as the Web and data exchange standards developed by the W3C Consortium (http://www.w3.org/), Internet standards developed by the Internet Engineering Task Force (IETF http://www.ietf.org/), geospatial interoperability standards developed by the Open Geospatial Consortium (OGC http://www.opengeospatial.org/), spatial metadata standards from the International Organization for Standardization (ISO http://www.iso.org/iso/home.htm), and the U.S. Federal Geographic Data Committee (FGDC http://www.fgdc.gov/). Specific standards that are in use by Federation members include OGC's WMS, WFS, WCS, and Catalogue Service (CSW); ISO's Geographic Information—Metadata (ISO 19115:2003); and FGDC's Content Standard for Digital Geospatial Metadata (FGDC-STD-001-1998).

In addition to the above-mentioned standards that are broadly supported by Federation members, several *de facto* standards or protocols are also in common use by Federation members, such as Open-source Project for a Network Data Access Protocol (OPeNDAP) and Representational State Transfer (REST) (Fielding 2000; OPeNDAP 2009).

The common use of standards, de facto standards, and protocols with the Federation yields a high degree of interoperability between Federation member systems and with applications used by external data and information consumers. One successful example rooted in Federation activities is the GrADS Data Server (GDS) and an end-to-end system based on GDS for supporting Earth scientists.

4.4 An Example of Integrated Data Search, Access, and Analysis

Scientists prefer to work with data from their familiar environment for both data analysis and data support to others. Therefore, an infrastructure that can help scientists make their data available to other scientists without much extra effort and also to seamlessly search, access, and analyze distributed data will be very helpful to scientists. Such a system should be evolved from existing data handling components through a federated bottom-up approach. One such system is the GDS, which is created by integrating two well-known components, OPeNDAP and GrADS (Grid Analysis and Display System), with NASA support for the Seasonal to Interannual Earth Science Information Partnership (SIESIP) program (Kafatos et al. 1998).

The OPeNDAP is a robust, client-server data transport protocol based on the HyperText Transfer Protocol (HTTP) (OPeNDAP 2009). With an OPeNDAP-enabled application program such as Matlab, a scientist can open a data set with a Uniform Resource Locator (URL) instead of a local file name. In other words, OPeNDAP-enabled application programs can be considered special Web browsers. The special browsers receive data though HTTP and handle the data based on capabilities of the programs such as data analysis and data visualization. The simple design and ease of use of OPeNDAP has led to its widespread adoption to distribute online digital data by many data providers in geosciences and other disciplines.

The GrADS is another widely used program that provides an integrated user interface for accessing, analyzing, and displaying geoscience data (Doty et al. 1997). Geoscience data are described by GrADS control files with information about spatial coverage and resolutions in latitude, longitude, and vertical levels as well as start and end times and temporal resolutions such as hourly, 6-hourly, daily, and so on. The OPeNDAP and GrADS were tightly integrated to enable the application of the analysis power of GrADS over the Internet under the SIESIP project (Wielgosz et al. 2001; Yang et al. 2003b). This GDS extends the OPeNDAP to enable on-the-fly, server-side data analysis and manipulation. Scientists who use GDS can retrieve data and metadata, and they can transfer the data to their application programs through the Internet. A GDS user can also insert data processing commands in the requesting URLs to manipulate and/or filter distributed data in remote servers and receive the processed data only. Another benefit of GDS is that scientists who are using GrADS to manage, access, and analyze their data can put their data on the Internet without much extra effort.

The OPeNDAP and GDS construct the cyber-infrastructure for scientists to conduct research without spending much time in resolving data interoperability issues. A client can access distributed geoscience data supported by this infrastructure without knowing the logistic details about the data, such as physical disk locations and data formats.

The GDS is a powerful data infrastructure but it lacks metadata support for easy data search. To supplement the metadata, a Distributed Metadata Server (DIMES) supporting metadata navigation, nearest neighbor search based on links between metadata nodes, and other traditional searches for geoscience data were developed under SIESIP and were tailored to support metadata from GDS servers (Yang et al. 2001, 2002). Further, to give scientists a more powerful system moving toward an end-to-end system of data search, access, and analysis, a Metadata Integrated Data Analysis Server (MIDAS) was developed by combining DIMES for metadata support with GDS for data support, initially named an Enhanced Server (ES) (Yang et al. 2003b) and a Scientific Data and Information Super Server (SDISS) (Yang et al. 2003a). The power of the integrated server is reflected by its support of interactive access to both metadata and data. In addition, a GDS URL generator is included in the system to help users build the relatively complex GDS URLs.

4.4.1 A Use Case

Scientists are interested in searching for answers to science problems. They often need large numbers of multiple dimensional data sets to generate and to verify scientific hypotheses. However, scientists do not really care much about the details of data format, location, storage media, and so on, as long as they can access and use the data effectively and with ease. Without a "seamlessly operating" distributed data system, scientists must find, download, and spend more time in handling the various data formats and other details before finally being able to use the data. A data system should free scientists from such nonscientific tasks and provide data interoperability. The MIDAS makes one significant step in that direction. Here, we describe

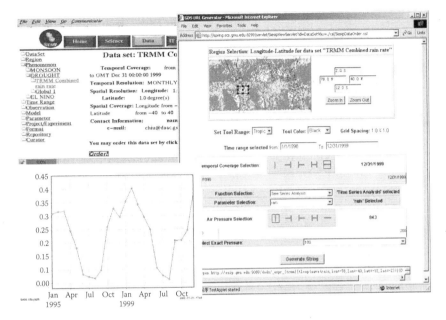

FIGURE 4.3
MIDAS use case. (Upper-left panel): DIMES metadata navigation interface for metadata search and browse; (right): GDS URL generator interface; (bottom-left): A time series of rainfall rate over Amazonian area.

a "use case" of MIDAS to demonstrate the usage and the usefulness of the system. Major components used for this case are shown by the three panels in Figure 4.3.

To access MIDAS, a user should first visit a Web page with a specific URL to a MIDAS/DIMES page. Since the DIMES metadata model connects different metadata concepts by links and those concepts can be considered as different metadata dimensions, the user could browse the metadata through different metadata dimensions until finding the data sets of interest. Metadata dimensions may match data dimensions directly, such as spatial or temporal resolution and coverage, or may contain other information about the data. For example, data sources such as a satellite instrument or a numerical model from which the data is retrieved are useful metadata information. Since DIMES is very flexible, scientific knowledge, such as what natural phenomenon is related to a specific data set, can also be integrated into DIMES. A DIMES user can find data through any metadata dimension with a metadata navigation system based on the DIMES metadata model and the nearest neighbor search. Suppose in our use case, the user browses the "Phenomenon" folder and then the "DROUGHT" phenomenon due to his or her interest. From this browsing point, the user can find "TRMM Combined rain rate" data set, as shown in the upper-left panel of Figure 4.3, which gives users basic information about the data set, such as spatial or temporal coverage and resolutions. Since the user is in the data page already, MIDAS allows the users to order the found data by simply clicking the "Order" button, which leads the user into the GDS URL generator Web page, as shown by the right panel of Figure 4.3.

The GDS allows users to fully leverage the power of GrADS to manipulate data on the server side before ordering the final products. The GDS URLs are relatively more complex than the plain OPeNDAP URLs. For GrADS users, although it is straightforward to write a GDS URL, the process is very tedious and error-prone. However, for non-GrADS OPENDAP client users, the GDS URLs could be too complex to be easily used. For such non-GrADS users, the GDS URL generator with supports of plain data access and predefined data processing functionalities is very helpful. It is expected that the generator will help GDS users to get familiar with the GDS URLs and then modify the sample URLs to create more specific URLs. One advantage of MIDAS is that the URL generator page is preloaded with the metadata data from the search result to better guide users to create their data requests. The most prominent preloading information is the browse image from the corresponding data set. The user not only can view the browse image for an overview of the data but also can use the image as a background to make further spatial selections. Similarly, a user can make selections for a time or a time period in the binding temporal coverage range. The interface also allows users to select a physical parameter in this data set and select a function.

In this use case, suppose the user is interested in high rain rate area. By viewing the rain rate image of January 1998, the user uses the rubber band to select the Amazonian area (between 12°S–2°S latitude and 78°W–40°W longitude), as displayed in the area selection panel. The user further selects the temporal coverage from January 1998 to December 1999 and the time series of rain rate. After the user makes all the selections, the user clicks on the "Generate String" button in the interface. The GDS URL would be created based on the user's selections. In this example, the GrADS command for opening data through OPeNDAP protocol was included although it is not part of the URL. With most computers right now, the user can cut and paste the generated URL to his or her OPeNDAP client and handle the data sets opened in this way as any other data sets opened locally. In this example, the resulting time series with 24 values is sent back to the client and is displayed as the result shown in the left-bottom panel of Figure 4.3.

The above use case shows how easy it is to use MIDAS to find and access data. Metadata for search, metadata for use, and data access mechanisms have been integrated to provide users a seamless way from the beginning to the end. It is expected that a scientist with basic knowledge of geoscience does not need anything else, such as downloading the data, reading data description documents, and reformatting the data before really using the remote data products for research.

An example of using the EOSDIS is given in Section 9.1 of this book, Global Agriculture Information System and Its Applications.

4.4 Future Strategies of Data Infrastructure

The Federation of Earth Information Partners has grown from the original 24 NASA-sponsored partners to a broad-based community with over 100 members. The Federation will continue supporting Earth-related data, information, and knowledge collection, management, access, and analysis to serve data providers, data users including scientists, educators, decision makers, and the public communities. The strategies of the Federation include coordinating with NASA data systems groups and international Earth data groups to enhance the use of Earth science data and information. The Federation will not only continue promoting the community-grown data information systems but also advocate service-oriented architecture for furnishing data collections from observations and data uses for scientific researches and other applications. The Federation will also create mechanisms for Earth data quality assessment and for identifying Earth science community data needs (ESIP Federation 2004, 2009).

As a community-grown integrated data system under the Federation, MIDAS can be potentially enhanced to have a real end-to-end system to support Earth science researchers. In an ideal case, scientists can search, access, and analyze data in the same familiar working environment. That is one of the ultimate goals for ESDS supporting Earth science research, as described by Lynnes (2008). The MIDAS is possibly one of the best candidates for achieving this goal through the Federation approach.

NASA's EOSDIS has undergone significant changes during 2005–2009 as a part of its ongoing evolution. These changes were carried out by (1) establishing a vision for 2015 through collaboration between an external "EOSDIS Elements Evolution Study Team" and internal "EOSDIS Elements Evolution Technical Team" (NASA 2005); (2) developing an implementation plan for the first step of evolution; and (3) completing the activities in the plan during 2006–2009. As a result, the system has been simplified, reducing the amount of software to be maintained and using commodity hardware. This has resulted in significant reductions in costs for ongoing maintenance and operations. Most of the data are available online, thus facilitating access and enabling provision of on-the-fly processing and analysis services, based on a service-oriented architecture (Ramapriyan et al. 2009).

Most of the missions being currently supported by EOSDIS will be completed or will be in extended operations by 2015. NASA will ensure the safe stewardship of the EOS data and the scientific research based on EOSDIS data holdings enabled uniquely by NASA. Therefore, the EOS data will be available to research and applications user communities beyond the EOS era, and users can access and analyze the data more efficiently with new technology such as on-the-fly data processing based on service-oriented architecture from different data sources. NASA is planning several spacecraft missions to be flown during the next decade as recommended by the National Research

Council in its Decadal Survey Report (NRC 2007). These are generally referred to as "Decadal Survey Missions." As a part of this planning, NASA is assessing how best to take advantage of its existing ESDS infrastructure and evolve it to meet the needs of the scientific and applications communities of the future.

The NOAA has created the Comprehensive Large Array-data Stewardship System (CLASS), a Web-based archive and distribution system for supporting one-stop shopping and data access for NOAA environmental data and products. The CLASS is NOAA's premier online facility for the archiving and distribution of the U.S. operational environmental satellite data, satellite-derived data products, data from NOAA's *in situ* environmental sensors, and other environmental data. The satellites supported by CLASS include the Geostationary Operational Environmental Satellite (GOES) series, the Polar Operational Environmental Satellites (POES) such as the DMSP series, and the future NPOESS and, its bridge mission, the NPP. It is expected that CLASS will be a large, permanent, long-term data archiving and distribution system which may also hold data from other agencies and institutions (NOAA 2005, 2009).

References

Doty, B.E., J.L. Kinter III, M. Fiorino, D. Hooper, R. Budich, K. Winger, U. Schulzweide, L. Calori, T. Hol, and K. Meier. 1997. The grid analysis and display system (GrADS): An update for 1997. In *Proceedings of 13th Conference on Interactive Information and Processing Systems for Meteorology, Oceanography, and Hydrology*, pp. 356–358. American Meteorological Society, Boston.

ESIP Federation. 2004. Strategic Plan 2004–2007. http://wiki.esipfed.org/images/c/cb/StrategicPlan_FINAL.doc (accessed November 23, 2009).

ESIP Federation. 2009. 2009–2013 Strategic Plan. http://wiki.esipfed.org/images/9/95/StrategicPlan_2009–2013_FINAL.pdf (accessed November 23, 2009).

Fielding, R. 2000. Architectural styles and the design of network-based software architectures. PhD thesis, University of California, Irvine. http://www.ics.uci.edu/~fielding/pubs/dissertation/top.htm (accessed December 28, 2009).

Kafatos, M., X. Wang, Z. Li, R. Yang, and D. Ziskin. 1998. Information technology implementation for a distributed data system serving earth scientists: Seasonal to interannual ESIP. In *Proceedings of the 10th International Conference on Scientific and Statistical Database Management*, eds. M. Rafanelli and M. Jarke, p. 210, Capri, Italy.

Lynnes, C. 2008. Earth Science Data Usability Vision. A document recorded after a discussion during the *7th NASA Earth Science Data Systems Working Group Meeting*, Philadelphia, PA.

Mitchell, A., D. Lowe, and H. Ramapriyan. 2009. Evolution of web services in EOSDIS—Search and Order Metadata Registry (ECHO). *IEEE International Geoscience and Remote Sensing Symposium (IGARSS)*, Cape Town, South Africa.

NASA. 2005. Evolution of EOSDIS Elements. http://esdswg.eosdis.nasa.gov/evolution/ppt/Final_EEE_NASA_Feb_2005_mm.ppt (accessed November 24, 2009).

NASA. 2006. *Earth Science Reference Handbook: A Guide to NASA's Earth Science Program and Earth Observing Satellite Missions*, eds. C. Parkinson, A. Ward, and M. King, p. 31, Washington, DC.

NOAA. 2005. CLASS Background. http://www.osd.noaa.gov/class/docs/CLASS_Background.ppt (accessed November 25, 2009).

NOAA. 2009. Comprehensive Large Array-data Stewardship System. http://www.osd.noaa.gov/class/ (accessed November 25, 2009).

NRC (National Research Council). 1995. *A Review of the U.S. Global Change Research Program and NASA's Mission to Planet Earth/Earth Observing System*. National Academy Press, Washington, DC.

NRC. 2007. *Earth Science and Applications from Space: National Imperatives for the Next Decade and Beyond*. National Academy of Sciences, Washington, DC.

Olsen, L.M. and G.R. Major. 1996. Global Change Master Directory Enhances Search for Earth Science Data. *Transactions of the EOS*, American Geophysical Union. http://www.agu.org/eos_elec/95127e.html (accessed October 16, 2009).

OPeNDAP. 2009. What is OPeNDAP?. http://www.opendap.org/ (accessed October 16, 2009).

Ramapriyan, H.K., J. Behnke, E.J. Sofinowski, D.R. Lowe, and M.A. Esfandiari. 2009. Evolution of the earth observing system (EOS) data and information system (EOSDIS). *Standard-Based Data and Information Systems for Earth Observation*, Chapter 7, eds. L. Di and H. K. Ramapriyan. Springer-Verlag, Berlin.

Wielgosz, J., B.E. Doty, J. Gallagher, and D. Holloway. 2001. GrADS and DODS. *17th International Conference on Interactive Information and Processing Systems (IIPS) for Meteorology, Oceanography, and Hydrology*, Albuquerque, NM.

Williams, D.N. 2009. The planet at their fingertips: Climate modeling data heats up. *SciDAC Review Journal* 9. http://www.scidacreview.org/0902/html/esg.html (accessed October 16, 2009).

Yang, R. and M. Kafatos. 2001. Massive data sets issues in Earth observing. In *Handbook of Massive Data Sets*, eds. J. Abello, P.M. Pardalos and M.G. Resende, pp. 1093–1140. Kluwer Academic Publishers, Dordrecht.

Yang, R., X. Deng, M. Kafatos, C. Wang, and X. S. Wang. 2001. An XML-based distributed metadata server (DIMES) supporting earth science metadata. In *Proceedings of the 13th International Conference on Scientific and Statistical Database Management*, pp. 251–256. IEEE Computer Society, Fairfax, VA.

Yang, R., M. Kafatos, and X. Wang. 2002. Managing scientific metadata using XML. *IEEE Internet Computing*, 6(4):52–59.

Yang, R., X. Wang, Y. Nie, Y. Zhao, and M. Kafatos. 2003a. A web-based scientific data and information super server with a flexible XML metadata support. In *Proceedings of the 19th International Conference on Interactive Information Processing Systems (IIPS) for Meteorology, Oceanography, and Hydrology*, American Meteorological Society, Long Beach, CA.

Yang, R., M. Kafatos, B. Doty, J. Kinter, and L. Pham. 2003b. A distributed enhanced server for multi-dimensional scientific data search, access and analysis. *Computing in Science & Engineering*, 5(2):44–52.

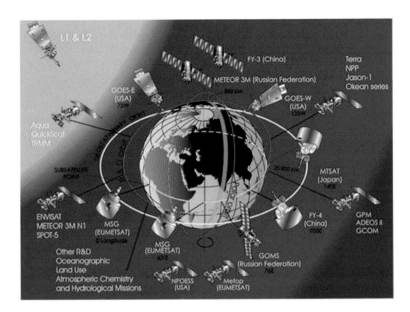

FIGURE 2.1
Schematic showing nominal orbital characteristics of Geosynchronous (GEO), low Earth orbit (LEO), and polar-orbiting (PO) satellites. (Adapted from http://www.eohandbook.com/eohb05/images/fig_03_(weather).jpg)

FIGURE 2.3
Schematic showing the Tropical Rainfall Measuring Mission (TRMM) satellite with its payloads—the TRMM Precipitation Radar (PR), the TRMM Microwave Imager (TMI), the Visible IR Spectrometer (VIRS), and the Lightning Imaging Sensor (LIS). (Adapted from http://trmm.gsfc.nasa.gov/overview_dir/trmm_instrument_large.jpg)

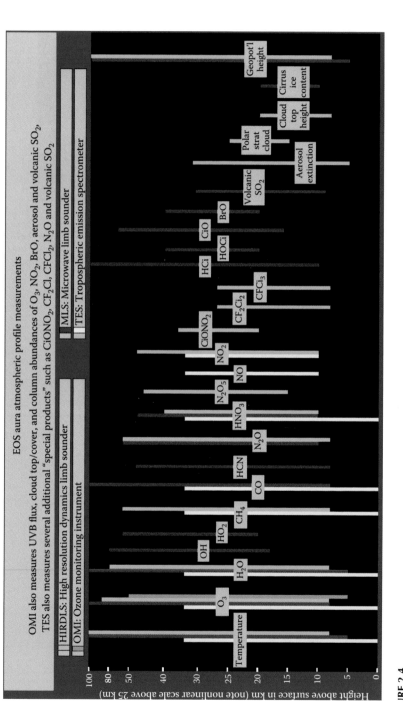

FIGURE 2.4

Table summarizing the atmospheric parameters measured by HIRDLS, MLS, OMI, and TES. The altitude range in which these parameters are measured is shown as the vertical scale. In several cases, the measurements overlap, which provides independent perspectives and cross calibration of the measurements. (Adapted from http://aura.gsfc.nasa.gov/instruments/index.html, retrieved January 15, 2010.)

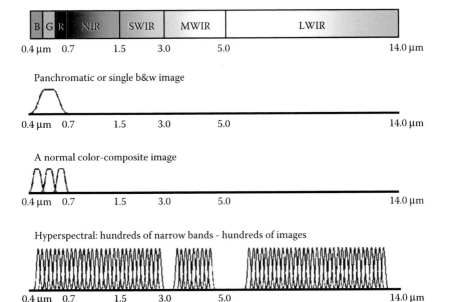

FIGURE 2.5
Sampling the electromagnetic spectrum in the 0.4 μm to 14 μm range: B = blue; G = green; R = red; NIR = near infrared; SWIR = shortwave infrared; MWIR = midwave infrared; and LWIR = longwave (or thermal) infrared. A panchromatic image is formed with one sampling function covering the visible range of the spectrum. Acquiring three images centered on the R, G, and B regions forms a normal color-composite image. The HSI data are acquired with hundreds of narrow sampling functions across the spectrum.

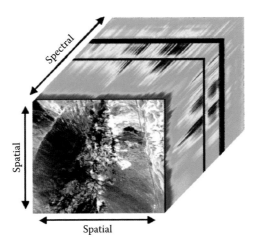

FIGURE 2.6
The image or data cube: the fundamental data structure of HSI. The cube is a three-dimensional data structure with two spatial dimensions (samples and lines of the image space) and one spectral dimension (sampling in wavelength space). The image shown is a false-color composite of NASA AVIRIS data of Cuprite, NV (R = 2.1088 μm, G = 2.2086 μm, B = 2.3381 μm).

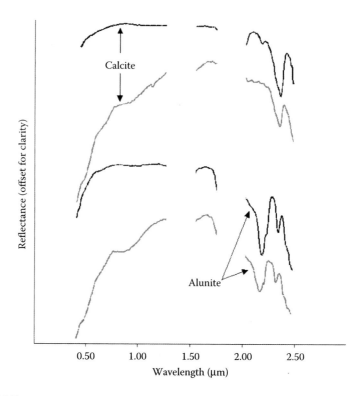

FIGURE 2.7

Spectral signatures derived from the atmospherically corrected NASA AVIRIS Cuprite, NV, HSI data. The signatures are compared with those of similar (though not identical) materials acquired with a laboratory spectrometer (Clark et al., 1993). The green spectra are from the AVIRIS data; the black spectra are the laboratory measurements. The two upper spectra are of calcite (a carbonate); the lower are of the mineral alunite. The NIR portions of the AVIRIS spectra show the presence of iron in the minerals. Diagnostic spectral features for both minerals occur in the SWIR.

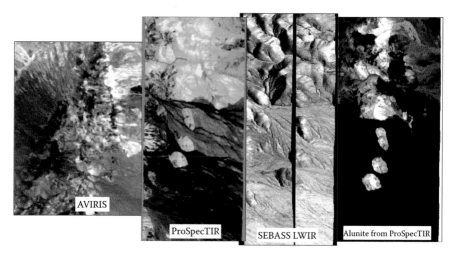

FIGURE 2.8

AVIRIS, ProSpecTIR, and SEBASS imagery of Cuprite, NV. Note the one elongate and two semicircular features common to all images. The color composite of AVIRIS data is the same as shown in Figure 2.6; the ProSpecTIR image is a PCA-derived false color composite (see text); and the SEBASS image is an LWIR "panchromatic" grayscale formed by summing all of the bands in the cube. The right-most image is a map of alunite distribution using SAM and an in-scene spectrum with a 70-band SWIR spectral subset of the ProSpecTIR HSI data.

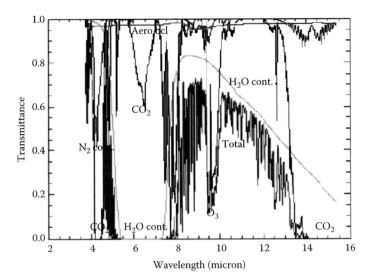

FIGURE 2.9
Atmospheric transmittance versus wavelength.

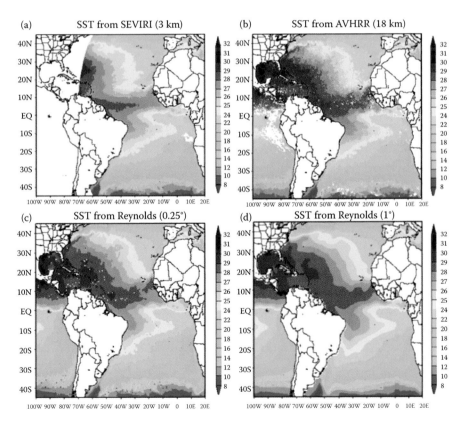

FIGURE 2.10
Comparison of monthly mean SST from (a) SEVIRI, (b) AVHRR, (c) Reynolds (0.25°), and (d) Reynolds (1°) in July 2005.

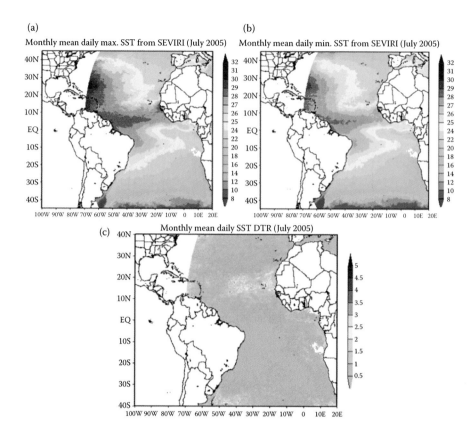

(a)
Monthly mean daily max. SST from SEVIRI (July 2005)

(b)
Monthly mean daily min. SST from SEVIRI (July 2005)

(c) Monthly mean daily SST DTR (July 2005)

FIGURE 2.11
Monthly mean daily (a) maximum, (b) minimum, and (c) DTR distribution in July 2005.

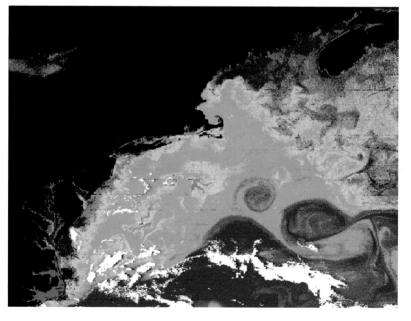

FIGURE 2.12
The CZCS ocean color scene of the Gulf Stream.

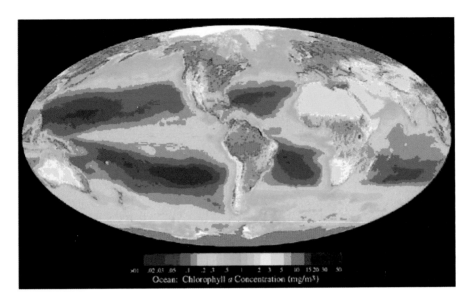

FIGURE 2.13
The SeaWiFS ocean color. (Adapted from http://earthobservatory.nasa.gov/images/imagerecords)

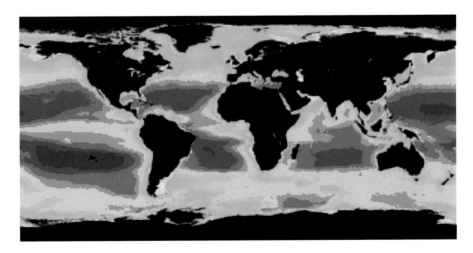

FIGURE 2.14
The MODIS ocean color product. (Adapted from http://www.nasa.gov/images/content)

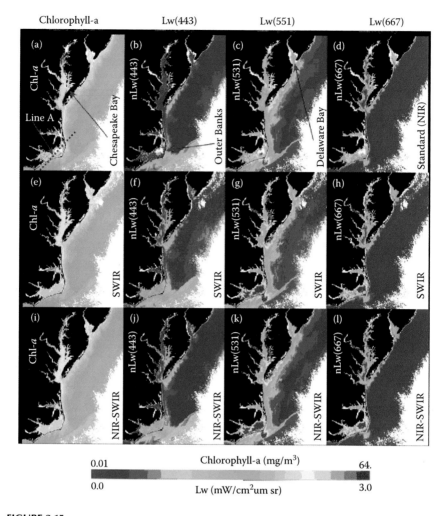

FIGURE 2.15
Comparisons of MODIS ocean color products from NIR, SWIR, and NIR-SWIR Combined Methods. (Adapted from Wang, M. and W. Shi. 2007. *Optics Express*, 15:15722–15733.)

FIGURE 3.3
A 3D global view of model cloud fields. (Adapted from Stainforth, D. et al. 2002. *Computing in Science and Engineering*, 4(3):82–89.)

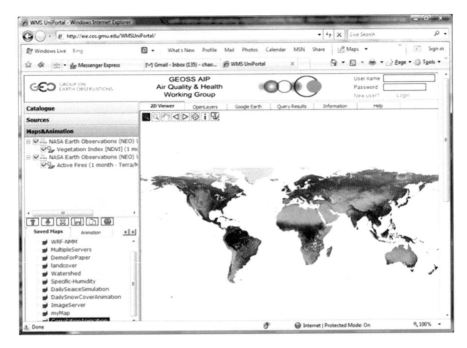

FIGURE 3.16
Land use and forest fire correlation.

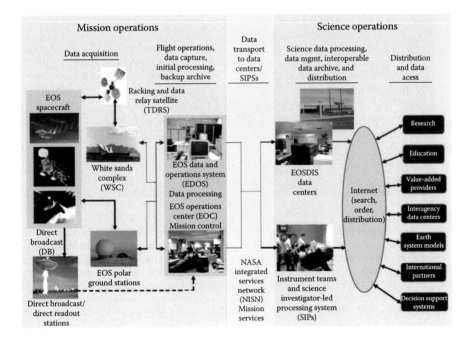

FIGURE 4.1
Context of EOSDIS.

(a)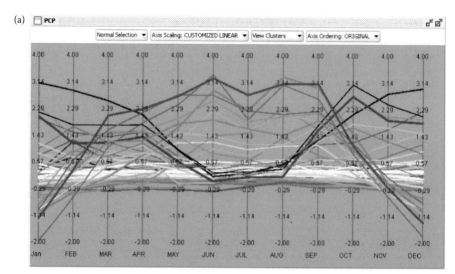

(b)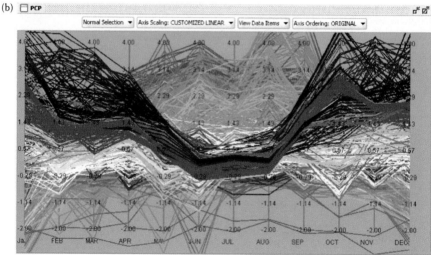

FIGURE 5.3

Seasonal and spatial patterns of surface temperature anomaly (1998–2007). (a) parallel coordinate plot of temperature change, Antarctic; (b) parallel coordinate plot of temperate change, Arctic; (c) 49 clusters derived from self-organized map (SOM); and (d) a map showing the spatial pattern of changes. (Adapted from Guo, D. 2009. *Geographic Data Mining and Knowledge Discovery*, eds. H.J. Miller and J. Han, 325–345. London and New York: Taylor & Francis; Jin, H. and D. Guo. 2009. Understanding Climate Change Patterns with Multivariate Geovisualization. In *Proceedings—IEEE International Conference on Data Mining Workshops*, 217–222. Miami, FL: IEEE Press.)

(c)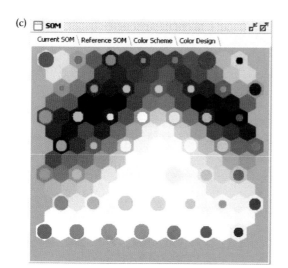

(d)

FIGURE 5.3
Continued.

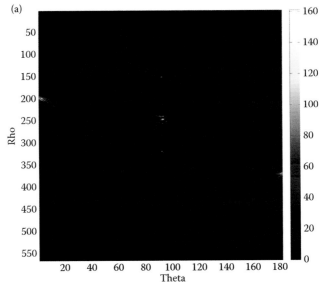

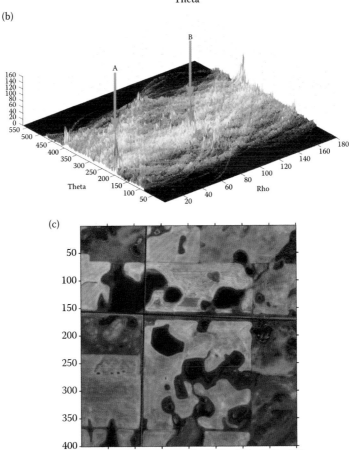

FIGURE 5.8
An example of the implementation of the Hough transform applied to the binary edge image
derived in Figure 5.3b. (a) The accumulator array (parameter space); (b) a three-dimensional
visulization of the accumulator (A and B mark the first and second highest peaks); (c) a reconstruction of the two lines corresponding to the first and second hightest peaks in the accumulator space.

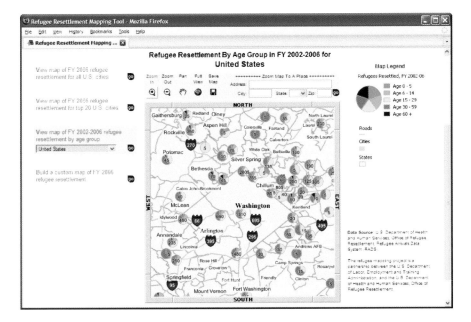

FIGURE 6.13
Single-frame dynamic loading mapping example—refugee resettlement mapping tool.

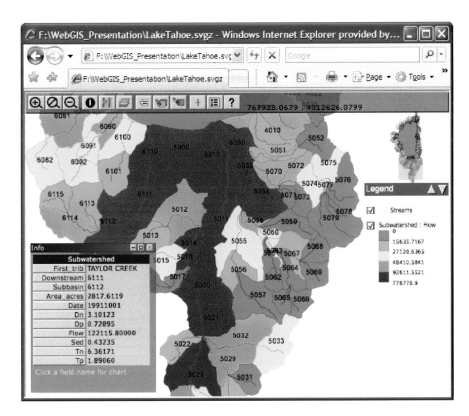

FIGURE 6.17
Scalable vector graphics-based mapping tool. (Data from GeoClient.)

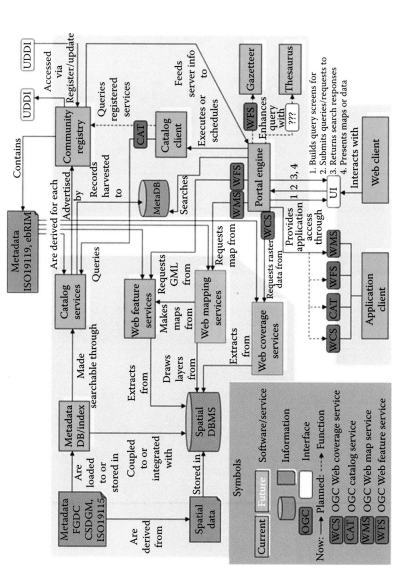

FIGURE 6.25
Data and service interaction in an SDI.

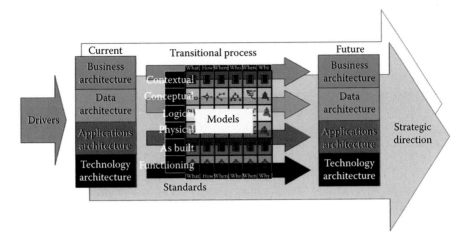

FIGURE 6.27
Conceptual process diagram for the U.S. FEA.

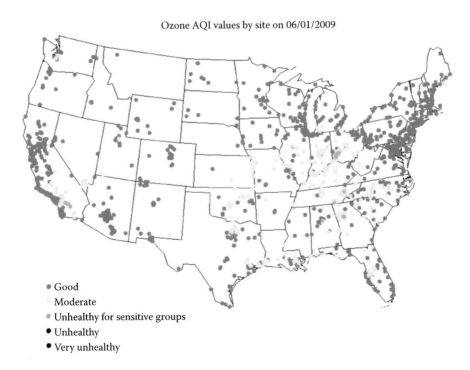

FIGURE 8.2
Ozone AQI values by sites. (Adapted from EPA's air explorer, http://www.epa.gov/airexplorer/)

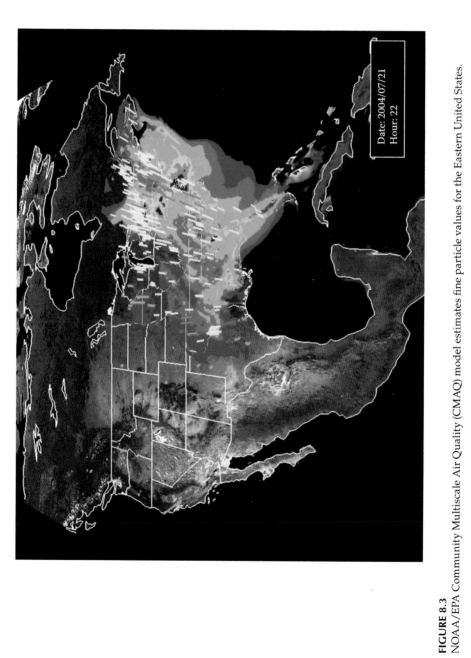

Date: 2004/07/21
Hour: 22

FIGURE 8.3
NOAA/EPA Community Multiscale Air Quality (CMAQ) model estimates fine particle values for the Eastern United States.

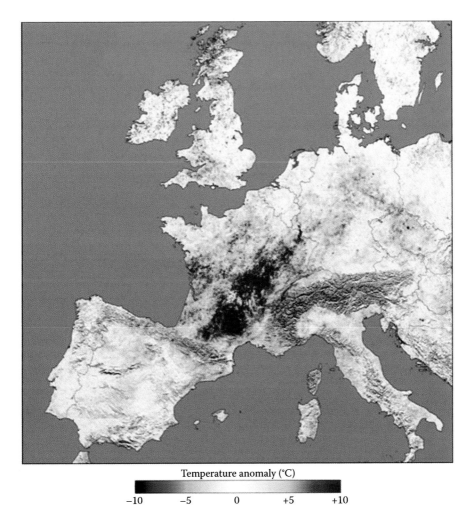

Temperature anomaly (°C)

−10   −5   0   +5   +10

FIGURE 8.8
European heat wave: surface temperature anomaly in July 2003. NASA Earth Observatory, August 16, 2003. (Adapted from http://earthobservatory.nasa.gov/IOTD/view.php?id=3714)

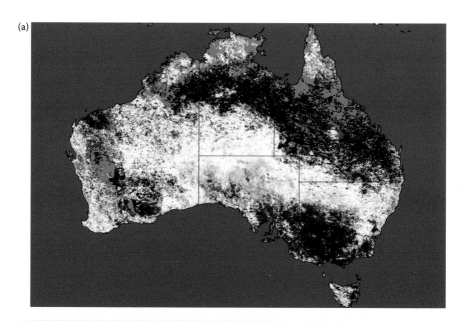

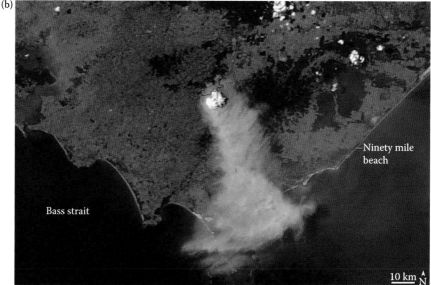

FIGURE 8.10
(a) Exceptional Australian heat wave. The image is derived from MODIS on NASA's Terra satellite. NASA Earth Observatory, February 5, 2009. (Adapted from http://earthobservatory.nasa.gov/IOTD/view.php?id=36,900); (b) Fires in southeast Australia. The image is derived from MODIS on NASA's Aqua satellite. NASA Earth Observatory, January 31, 2009. (Adapted from http://earthobservatory.nasa.gov/IOTD/view.php?id=36861.)

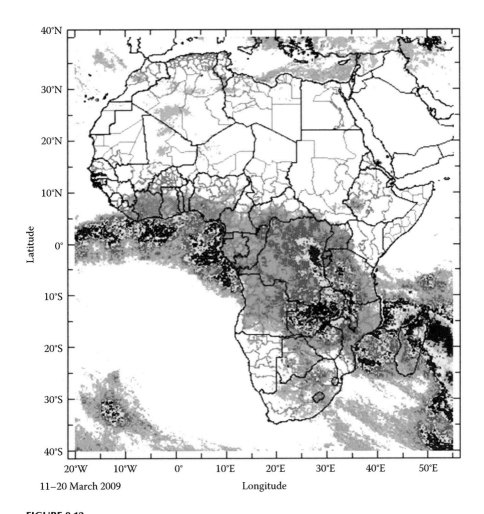

11–20 March 2009

Longitude

FIGURE 8.12

10-day precipitation estimates for malaria-endemic regions of Africa. Continuous tracking enables the identification of rainfall anomalies that signal elevated disease risk due to anticipated increase in mosquito populations. (Adapted from http://iridl.ldeo.columbia.edu/maproom/.Health/.Regional/.Africa/.Malaria/.MEWS/.)

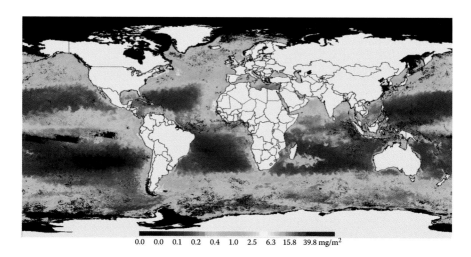

0.0   0.0   0.1   0.2   0.4   1.0   2.5   6.3   15.8   39.8 mg/m^2

FIGURE 8.15
SeaWiFS-derived chlorophyll concentration for March 1999. (Adapted from NASA DAAC.)

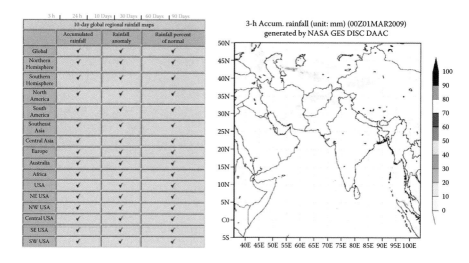

FIGURE 9.3
Left: List of maps of current conditions, providing a quick and easy access to current global and regional rainfall conditions. Rainfall anomaly and normalized anomaly can be used to identify flood or drought events. Right: A sample of the maps.

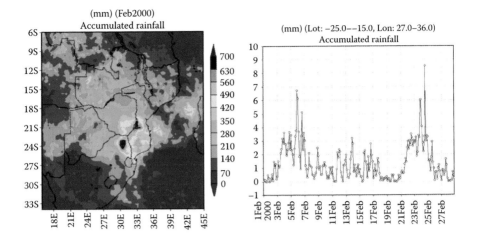

FIGURE 9.4
Left: Regional accumulated rainfall (TRMM 3B43) for February 2000. Parts of Mozambique received rainfall of over 300 mm in 1 day. Over 300 people died and 2 million had been displaced or affected according to news reports. Right: Time series of 3-hourly rainfall (TRMM 3B42) for the southern part of Mozambique (see Left). Two major rain events are identified in this plot.

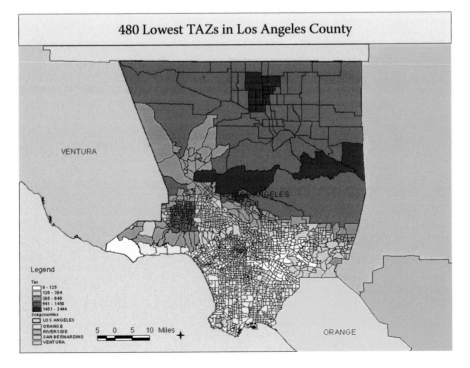

FIGURE 9.8
Identifies the locations of 480 lowest TAZs for storm water treatment plants.

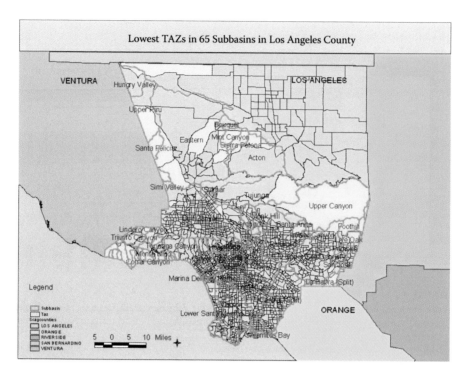

FIGURE 9.9
Identifies the locations of lowest TAZs in the 65 subbasins for storm water treatment plants.

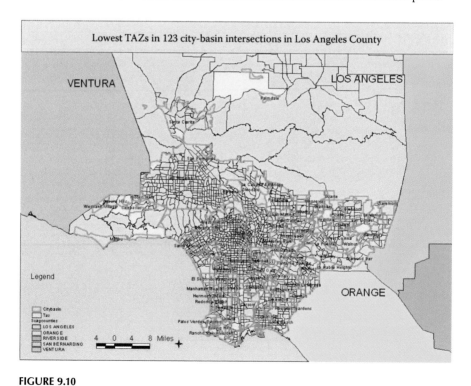

FIGURE 9.10
Identifies the locations of lowest TAZs in the 123 city–basin interactions for storm water treatment plants.

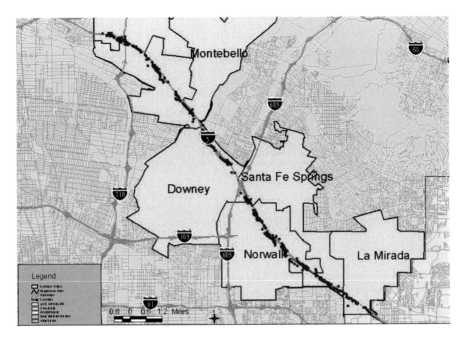

FIGURE 9.12
The identification of the displaced land parcels by geocoding.

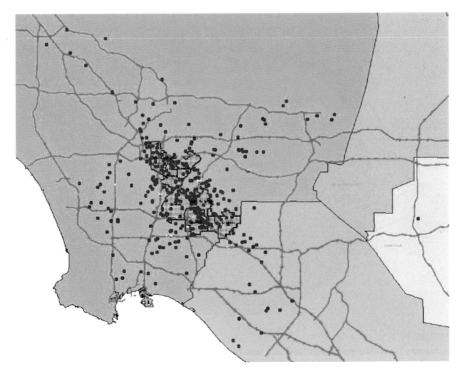

FIGURE 9.13
The distribution of the relocated (move-in) households.

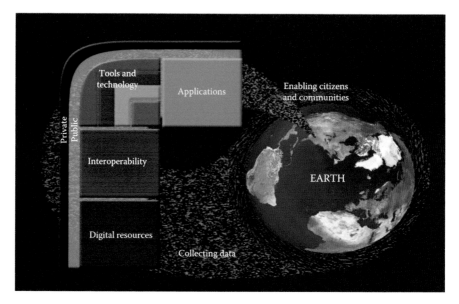

FIGURE 10.1
Components of Digital Earth.

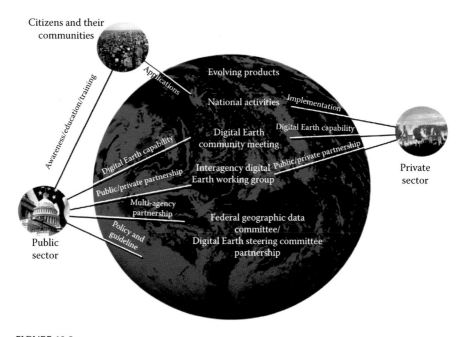

FIGURE 10.3
Roles of Public Sector, Private Section, and General Public in a Digital Earth Environment.
(Courtesy of previous NASA Digital Earth office.)

5

Spatial Data Analysis and Geoinformation Extraction

Arie Croitoru, Diansheng Guo, Fahui Wang, David Wong,
Peggy Agouris, and Anthony Stefanidis

CONTENTS

Most definitions of GIS characterize the technology as a computerized system that is designed to capture, store, manipulate, analyze, and display georeferenced data. These definitions highlight major components and functions of a traditional GIS, operating either as stand-alone systems or in a networked computing environment. Since GIS gradually transitions from desktop to distributed computing environments revolving around the Internet, this book emphasizes issues and topics more related to the latter. To some extent, the emergence of distributed computing environments has altered some of the traditional ways to capture, store, manipulate, and display spatial data, but it has not significantly changed the fundamental methodologies of analyzing and extracting information from spatial data. This chapter is one of the few places in this book to address the issue of deriving information from spatial data.

5.1 From Spatial Data to Geoinformation*

The proliferation of GIS has been accompanied by the burgeoning availability of spatial data. In fact, with the aggressive effort in developing new sensor technologies in the public and private sectors, the continuing and growing programs in launching satellite missions for Earth observing, and the building of global and national spatial data infrastructure, the amount of spatial data available has grown exponentially. These spatial data exist in various forms, from the traditional map-like data in a layer format to still images or motion videos captured by *in situ* and mobile sensors. Due to the new technologies in capturing spatial data, the massive amount of spatial data are available at various spatial scale levels, from global coverage to sub-meter pixels, and in different scopes, from covering various elements of the Earth systems to capturing the daily activity loci of individuals.

   Both the quantity and quality of these massive and heterogeneous spatial datasets create new challenges in deriving useful information from them. We have to develop new computing technologies and methodologies to handle and manipulate these data. This general challenge is being addressed by other parts of this book. In this chapter, we will focus on how we can turn data into information. Still, this general objective includes a wide variety of sub-topics, and most of them cannot be covered by this single chapter. There are many standard spatial analytical techniques already implemented in

---

* This section and the Section 5.5 are contributed by David Wong.

typical GIS software. These include overlay operations, map algebra (Tomlin, 1990), selection procedures in vector data such as buffering, and numerous functions in network analysis (Miller and Shaw, 2001; Black, 2003). Even mapping spatial data may be considered as analysis, because cartographers have to select specific information to be presented. In addition, interpreting maps may derive new information. These typical functions have been thoroughly discussed in the literature or even in GIS manuals. Therefore, we will not duplicate their efforts here.

Several areas related to analyzing spatial data will be addressed in this chapter. Analysis of spatial data requires special attention, partly because of their unique properties not found in aspatial data. One of these properties is the pervasiveness of spatial autocorrelation (another is the modifiable areal unit problem, see Wong, 2009). Tobler's first law of geography claims that all things are related, but closer things are more related (Tobler, 1970). In other words, dependence among observations in space is expected, and thus the fundamental assumption of independence among observations in classical statistics is violated (Anselin and Griffith, 1988). In this chapter, we will review some common methods to evaluate the level of spatial autocorrelation. When a significant level of spatial autocorrelation is detected, it will be necessary to incorporate spatial autocorrelation in statistical modeling. We will review several types of spatial regression models, showing how spatial autocorrelation can be captured to obtain unbiased statistical inferences.

Spatial autocorrelation may be regarded as a problem in spatial statistical analysis (Cliff and Ord, 1973), but it is clearly the nature of spatial data and the essence of geography (Gould, 1970, 443–444). Without spatial autocorrelation, we will not be able to, for instance, estimate the elevation of a point between two contour lines on a map. In general, the presence of spatial autocorrelation allows us to interpolate unknown values in locations based upon known values in other locations, because values in space are interdependent. We will review basic spatial interpolation techniques, including Kriging, which are commonly used in GIS and geosciences.

Another expected function found in GIS is spatial modeling. There are many types of models that can be potentially supported by GIS. In this chapter, we will review general categories of these models, but we will not be able to describe each type of model in detail. Readers can find more information about spatial models in Wang (2006).

Along the same line as mapping analysis is a broader technique called geovisualization (Dykes et al., 2005). Since spatial data are getting larger in size and richer in content, it is necessary to develop effective methods to tease out noises and identify systematic patterns in the data. Geovisualization, which may be regarded as a component of spatial data mining, will be addressed in section three of this chapter. There are many other topics related to analysis and modeling, but they are beyond the scope of this chapter.

Given today's state of geospatial technology, massive amounts of spatial data are in the format of still images and videos. Both types of data are derived from sensors, but they are of different nature. Image exploitation has been an important tool in analyzing high-resolution images of the Earth's surface, and these images are of abundant supply these days. On the other hand, videos captured from stationary sensors may not provide broad geographical coverage, but when individual sensors are linked to a sensor network, the collective data captured by the network can offer very rich Geoinformation (Chapter 2 of this book). Section four of this chapter will discuss the analyses of both still images and videos.

5.2 Spatial Statistics, Analysis, and Modeling*

This section first discusses various ways for defining spatial weights, then introduces some popular indices for spatial statistics, and finally examines regression models that account for spatial autocorrelation. Spatial weights capture the spatial relationship of observations and are needed in calibrating any spatial statistics. There is a wide range of spatial statistics, and this section focuses on indices for detecting spatial clusters, particularly those widely available in commercial or open-source software. Each set of indices usually includes tests for global clustering and corresponding tests for local clusters. The presence of spatial clustering (spatial autocorrelation) necessitates the usage of spatial regression models that account for its effect.

5.2.1 Measuring Spatial Autocorrelation

5.2.1.1 Defining Spatial Weights

Spatial weights define spatial relationships of observations. Defining spatial weights w_{ij} between two objects i and j can be based upon the distance or polygon contiguity between them. The following are some common methods available in some leading GIS software (e.g., ArcGIS):

1. Inverse distance ($1/d_{ij}$)
2. Inverse distance squared ($1/d_{ij}^2$)
3. Fixed distance band (= 1 within a specified distance band h and = 0 outside of the distance)
4. Zone of indifference, that is, a combination of (1) and (3) (= $1/d_{ij}$ within a distance band and = 0 outside of the distance)
5. Polygon contiguity (= 1 if two polygons are contiguous and = 0 otherwise)

---

* This section is contributed by Fahui Wang and David Wong.

Other distance-based functions, for example, $w_{ij} = \exp(-d_{ij}^2/h^2)$, can also be used to define spatial weights (Fotheringham et al., 2000, 111). Methods based on distance usually use the geometric centroids to represent areas and define distances as either Euclidean or Manhattan distances.

Spatial weights based on polygon contiguity can have two definitions: queen contiguity and rook contiguity. Borrowing the terms from a chess game, *rook contiguity* refers to two polygons sharing common boundaries, and *queen contiguity* includes polygons sharing boundaries or a point (vertex) (Cliff and Ord, 1973). Polygon contiguity may be also defined by different orders. If two polygons are immediately adjacent to each other, they are contiguous by the first order; if they are adjacent through one polygon between them, it is considered a second-order contiguity; and so on. These more complex definitions for spatial weights need to utilize specialized software such as GeoDa (http://geodacenter.asu.edu) or rely on GIS-based programming.

5.2.1.2 Indices for Global Clustering

Moran's I statistic is one of the oldest indicators that detect global clustering (Moran, 1950). It detects whether nearby areas have similar or dissimilar attributes overall, that is, positive or negative *spatial autocorrelation*, respectively. Moran's I is calculated as

$$I = \frac{N \sum_i \sum_j w_{ij}(x_i - \bar{x})(x_j - \bar{x})}{\left(\sum_i \sum_j w_{ij}\right) \sum_i (x_i - \bar{x})^2} \tag{5.1}$$

where N is the total number of units, w_{ij} are the spatial weights, x_i and x_j are the attribute values for units i and j respectively, and $\bar{x}$ is the mean of the attribute values.

It is helpful to interpret Moran's I as the correlation coefficient between a variable and its *spatial lag*. The spatial lag for variable x is the average value of x in neighboring area j defined as

$$x_{i,-1} = \frac{\sum_j w_{ij} x_j}{\sum_j w_{ij}} \tag{5.2}$$

Therefore, Moran's I varies between -1 and 1. A value near 1 indicates that similar attributes are clustered (either high values near high values or low values near low values); and a value near -1 indicates that dissimilar attributes are clustered (either high values near low values or low values near high values). If there is no significant spatial autocorrelation, the expected value of Moran's I is $-1/(N-1)$. Therefore, if N is large, the expected value will approach 0. If the observed Moran's I is close to 0, it indicates a random pattern or an absence of spatial autocorrelation.

Similar to Moran's I, *Geary's C* (Geary, 1954) detects global clustering (Geary, 1954). Unlike Moran's I, which uses the cross product of the deviations from the mean, Geary's C uses the deviations in intensities of each observation with one another. It is defined as

$$C = \frac{(N-1)\sum_i \sum_j w_{ij}(x_i - x_j)^2}{2\left(\sum_i \sum_j w_{ij}\right)\sum_i (x_i - \bar{x})^2} \tag{5.3}$$

The values of Geary's C typically vary between 0 and 2 (although 2 is not a strict upper limit), with $C = 1$ indicating that all values are spatially independent from each other. Values between 0 and 1 typically indicate positive spatial autocorrelation, whereas values between 1 and 2 indicate negative spatial autocorrelation, and thus Geary's C is inversely related to Moran's I.

In addition, Getis and Ord (1992) developed the *general G* statistic, in contrast to its local version G_i statistic. General G is a multiplicative measure of overall spatial association of values that fall within a threshold distance (d) of each other, and is defined as

$$G(d) = \frac{\sum_i \sum_{j \neq i} w_{ij}(d)x_i x_j}{\sum_i \sum_{j \neq i} x_i x_j} \tag{5.4}$$

where x_i and x_j can only be positive variables. For a general overview of significant testing for Moran's I, Geary's C, and $G(d)$, interested readers may refer to Wong and Lee (2005). For detailed theoretical explanations, readers should consult Cliff and Ord (1973) and Getis and Ord (1992).

5.2.1.3 Indices for Local Clusters

Anselin (1995) proposed the Local Indicators of Spatial Association (LISA), which include the local versions of Moran and Geary statistics to capture local pockets of instability or local clusters. Since local Geary is less well behaved, we will focus only on local Moran. Local Moran index for an area i measures the association between a value at i and values of its nearby areas, and is defined as

$$I_i = \frac{(x_i - \bar{x})}{s_x^2}\sum_j [w_{ij}(x_j - \bar{x})] \tag{5.5}$$

where $s_x^2 = \sum_j (x_j - \bar{x})^2$ is the variance. Note that the summation over j does not include the area i itself, that is, $j \neq i$. A positive I_i means either a high value is surrounded by high values (high–high) or a low value is surrounded by low values (low–low). A negative I_i means either a low value is surrounded by high values (low–high) or a high value is surrounded by low values (high–low). Therefore, local Moran (and local Geary) cannot distinguish high–high clusters from low–low clusters, unless one compares values across clusters after calculating the indices.

Similarly, Getis and Ord (1992) developed the G_i statistic, a local version of their global or general G statistic, to identify local clusters with statistically significant high or low attribute values. The G_i statistic is written as

$$G_i = \frac{\sum_j (w_{ij} x_j)}{\sum_j x_j} \qquad (5.6)$$

The summations over j may or may not include i. The index detects whether high values or low values tend to cluster in a study area. A high G_i value indicates that high values tend to be near each other, and a low G_i value indicates that low values tend to be near each other. However, G_i can distinguish high value clusters from low value clusters.

For an overview of significant testing for the local Moran's and local G_i's, one may refer to Wong and Lee (2005). For in-depth theoretical formulations of the tests, please refer to Anselin (1995) and Getis and Ord (1992).

5.2.1.4 Spatial Regression Models

Spatial cluster analysis detects spatial autocorrelation, in which values of a variable are systematically related to geographic locations. In the absence of spatial autocorrelation or spatial dependence, the ordinary least square (OLS) regression model can be used. It is expressed in matrix form as

$$y = X\beta + \varepsilon \qquad (5.7)$$

where y is a vector of n observations of the dependent variable, X is an $n \times m$ matrix for n observations of m independent variables, β is a vector of regression coefficients, and ε is a vector of random errors or residuals, which are independently distributed about a mean of zero.

When spatial dependence is present, the residuals are no longer independent from each other, and the significant testing of parameter estimates, β, in OLS regression may be biased (parameter estimates may be found to be significant but, in fact, are not due to suppressed standard errors). This section discusses two commonly used models of *maximum likelihood estimator*. The first is a *spatial lag model* (Baller et al., 2001) or *spatial autoregressive model* (Fotheringham et al.,

2000, 167). The model includes the mean of the dependent variable in "neigh-boring" areas (i.e., *spatial lag*) as an extra explanatory variable. Denoting the weights matrix by W, the spatial lag of y is written as Wy. The element of W in the i th row and j th column is defined as in Section 5.2.1.1. The model is expressed as

$$y = \rho Wy + X\beta + \varepsilon \tag{5.8}$$

where ρ is the regression coefficient for the spatial lag, and other notations are the same as in Equation 5.7. Rearranging Equation 5.7 yields

$$(I - \rho W)y = X\beta + \varepsilon$$

Assuming the matrix $(I - \rho W)$ is invertible, we have

$$y = (I - \rho W)^{-1} X\beta + (I - \rho W)^{-1} \varepsilon \tag{5.9}$$

This reduced form shows that the value of y_i at each location i is determined not only by x_i at that location (like in the OLS regression model) but also by the x_j at other locations through the spatial multiplier $(I - \rho W)^{-1}$ (not present in the OLS regression model). The model is also different from the autoregressive model in time-series analysis, where the lag is unidirectional.

The second model is the *spatial error model* (Baller et al., 2001) or *simultaneous autoregressive* (SAR) *model* (Griffith and Amrhein, 1997, 276). Instead of treating the dependent variable as autoregressive, the model considers the error term as autoregressive. The model is expressed as

$$y = X\beta + u \tag{5.10}$$

where u is related to its spatial lag such as

$$u = \lambda Wu + \varepsilon \tag{5.11}$$

where λ is a spatial autoregressive coefficient, and the second error term ε is independent.

Solving Equation 5.11 for u and substituting into Equation 5.10 yield the reduced form

$$y = X\beta + (I - \lambda W)^{-1} \varepsilon \tag{5.12}$$

This shows that the value of y_i at each location i is affected by the stochastic errors ε_j at all other locations through the spatial multiplier $(I - \lambda W)^{-1}$. Estimation of either the spatial lag model in Equation 5.9 or the spatial error model in Equation 5.12 is implemented by the maximum likelihood (ML) method, available in free software such as GeoDa. Other methods such as spatial filtering (Getis and Griffith, 2002; Griffith, 2003) have also been proposed to account for spatial autocorrelation in regression analysis.

5.2.1.5 Spatial Logit and Poisson Regression Models

There are at least two scenarios that call for a different approach from OLS regression and the aforementioned spatial regression models. Both are widely encountered. First, when the dependent variable y_i is binary (0, 1), the *logit (logistic) regression model* is used such as

$$P(y_i = 1) = \frac{\exp(\eta_i)}{1 + \exp(\eta_i)} \text{ where } \eta_i = \beta_0 + \sum_k \beta_k x_{ki} \quad (5.13)$$

The model is commonly estimated by ML. With parameters $\beta_0, \beta_1, \dots, \beta_k$ estimated from the regression, the model predicts the probability that $y_i = 1$ (Hamilton, 1992, 220–225). Similarly, the ordinary logistic regression model in Equation 5.13 performs poorly with the presence of spatial autocorrelation (e.g., a species present or absent at a given site is strongly related to its presence or absence at neighboring sites) (Wu and Huffer, 1997, 50). This calls for the use of the *autologistic model* introduced by Besag (1974), or termed as "*spatial logit regression* model," to emphasize that it is the counterpart of logit regression model controlling for spatial autocorrelation. In this model, the value y_i at i depends explicitly on its values in nearby areas. Various methods (e.g., the coding method and the maximum pseudo-likelihood method) can be used to estimate the model (Besag, 1974, 1975).

The second scenario concerns y that are discrete counts (0, 1, 2, ...). The Poisson model is effective for the analysis of such data, particularly when the mean count is low (e.g., homicide or cancer data in small geographic units). The *regular Poisson regression* is described as

$$O_i \sim \text{Poisson}(\mu_i) \quad (5.14)$$

where O_i represents the observed count in area i, and μ_i is the expected count estimated as

$$\ln \mu_i = \ln E_i + \beta_0 + \sum_k \beta_k x_{ki} \quad (5.15)$$

where E_i is the offset variable (base population). Most advanced statistical software offer a tool for estimating the model (e.g., the *PROC GENMOD* module in SAS).

The model just described is ill suited for the analysis of data with the presence of spatial autocorrelation (Griffith and Haining, 2005, 133). Some methods, such as the *Bayesian model with convolution priors* (Besag et al., 1991; Mollie, 1996), have been proposed to account for spatial autocorrelation. The model is similar to Equation 5.14 with a different formulation for the expected late-stage cancer cases, μ_i:

$$\ln \mu_i = \ln E_i + \beta_0 + \sum_k \beta_k x_{ki} + b_i + h_i \qquad (5.16)$$

where b_i can be viewed as a surrogate for unknown or unobserved variables with spatial structures, such as spatial autocorrelation between neighborhoods, which are assigned a conditional autoregressive prior; h_i captures the influence of all unknown or unobserved variables, which are assumed to be an exchangeable normal prior. The major difference between this Bayesian model in Equation 5.16 and the regular Poisson model in Equation 5.15 is the inclusion of term b_i for controlling spatial autocorrelation. Therefore, the model may be regarded as *spatial Poisson regression model*.

5.2.1.6 Applications

We have seen in the past two decades an increasing number of applications of the aforementioned spatial statistical indices, particularly after these indices became available in major GIS software such as ArcGIS. Two examples are used here to briefly illustrate the applications of these indices.

Shen (1994) used Moran's I to test two hypotheses on the impact of growth-control policies in the San Francisco area. The first hypothesis is that residents who are not able to settle in communities with growth-control policies would find the second best choice in a nearby area, and, consequently, areas of very slow population growth (i.e., negative x_i after the growth rates are standardized) would be close to areas of population growth. This leads to a negative spatial autocorrelation. The second hypothesis is related to the so-called NIMBY (Not In My BackYard) phenomenon. In this case, growth-control communities form a coalition and tend to cluster together; so do the pro-growth communities. This leads to a positive spatial autocorrelation.

Another example concerns the two types of spatial externalities in crime control. *Spatial displacement* happens when crime control measures cause crime to move away, thereby reducing crime in the target area but increasing crime in nearby areas. This would yield negative spatial autocorrelation for crime rate changes, because areas of lowering crime rates tend to be surrounded by areas of increasing crime rates. *Spatial diffusion*, in contrast, means that the benefits of crime reduction "spill over" to neighbors and lead to a decline of crime in nearby areas as well. This latter case would exhibit positive spatial autocorrelation, as trends of crime rates in nearby areas tend to be similar.

Spatial regression models have also been widely applied in various fields utilizing spatial data, particularly in biological, ecological, criminal justice, and health studies. Some examples include applications of (1) spatial lag and spatial error models in crime research (Baller et al., 2001; Mu and Wang, 2008), (2) spatial logit model in ecological research (Augustin, et al., 1996; Wu et al., 1997), and (3) spatial Poisson model in cancer research (Wang et al., 2010).

5.2.2 Spatial Interpolation and Geostatistics

In the previous section, spatial autocorrelation was regarded as a tractable problem, a perspective taken by most spatial statisticians, but some view it as a nuisance. On the other hand, Gould (1970) points out that spatial data without spatial autocorrelation will be uninteresting, similar to random numbers without meaning. The presence of spatial autocorrelation is unavoidable in most cases. In fact, the presence of spatial autocorrelation is a prerequisite condition for spatial prediction and estimation. In cartography and map reading, isolines are used to show variations of values across a region. Isolines connect points of certain values within the region. Map readers find out the value of a given location not connected by isolines by inferring from the values depicted by the nearest isolines. This inference process implicitly assumes that values in locations between two isolines are not too different from the values reflected by the two isolines. In other words, spatial autocorrelation is assumed to be present in interpreting isoline maps.

Estimating the value at a given location on an isoline map is in essence performing spatial interpolation, but in a rather subjective matter, as the process is very much performed within the map reader's head. Many spatial interpolation techniques are available, and many are already incorporated into GIS software. In this section, we will provide an overview of basic concepts and theories behind popular interpolation methods. Readers are encouraged to refer to the references provided.

There are two broad categories of spatial interpolation techniques: deterministic versus statistical. The statistical techniques often fall under the umbrella of geostatistics. Both types leverage the fact that spatial data have a certain degree of spatial autocorrelation, but they differ in how to utilize spatial autocorrelation in the interpolation. Both types attempt to address the following question: assume we know the values of a variable y in a set of sampled locations, S_i, $i = 1, 2, 3, \ldots , n$, but we want to estimate the value of y in location S_0 where we did not sample. Then the estimated value of y in S_0 is

$$\hat{y}(S_0) = \sum_{i=1}^{n} \lambda_i \, y(S_i)$$

(5.17)

where $y(S_i)$ are known values of y in respective locations, and λ_i is the weight for location S_i. In other words, we pull the known values from sampled locations, taking a fraction (λ_i) from each known value and combining them in order to estimate the y value in S_0. In statistical terms, the estimated value is a linear combination of the sampled values $y(S_i)$. Equation 5.17 is the essence of spatial interpolation and geostatistics, and the two types of techniques differ in how the weights (λ_i) are determined.

5.2.2.1 Deterministic Spatial Interpolation Methods

The deterministic approach to spatial interpolation adopts the assumption that relationship between variable values across space is strictly a function of distance separating the locations. In other words, the magnitude of spatial autocorrelation is a function of distance in this approach. This is also the typical but implicit way when map readers interpolate values on isoline maps. It is assumed that the gradient of values between isolines is linear, and thus the location in the middle of two isolines should be the averaged values of the two isolines. Based upon this assumption, the weights (λ_i) in Equation 5.17 can be determined.

In general, we expect that the strength or magnitude of a relationship decreases with increasing distance between any two given locations. In that case, weight will be inverse to distance between the two locations, or $\lambda_i = 1/d_{i0}$ where d_{i0} is the distance between the sampled location S_i and estimated location S_0. However, if sample size n is larger, the estimated value will be also be larger and vice versa. To avoid this dependence on sample size, we need to impose a condition on Equation 5.17 such that

$$\sum_{i=1}^{n} \lambda_i = 1 \qquad (5.18)$$

Due to this constraint, the weights have to been standardized or scaled in the following manner:

$$\lambda_i = \frac{1/d_{io}}{\sum_{i=1}^{n} 1/d_{i0}} \qquad (5.19)$$

Since the weights are inverses to distances, this method is often known as the inverse distance weighting (IDW) method in the geographical and GIS literature.

The above formulation is the simplest form, assuming an inverse linear relationship to distance. However, based on many empirical studies, especially in the spatial interaction modeling literature (e.g., Fotheringham and O'Kelly, 1989), the inverse to distance relationship is often nonlinear in nature, partly due to the distance-decay effect. To accommodate the nonlinear relationship to distance, a power parameter, α, is often included such that the weight is defined as

$$\lambda_i = \frac{d_{i0}^{-\alpha}}{\sum_{i=1}^{n} d_{i0}^{-\alpha}}. \tag{5.20}$$

When α is 1, the weight will be inversely proportional to distance. However, it is shown in the literature that the decline in relationship is more than proportional to distance, and a power of 2 is often applicable to many phenomena (Fotheringham and O'Kelly, 1989). Clearly, if there is evidence to support other values for the power parameter, they can easily be adopted.

The IDW method is widely used in GIS software and application studies. It is relatively easy to understand and compute. Usually, it is used in a raster environment such that a value in each grid cell will be estimated using observed values in selected locations. Thus, a value surface is generated. Theoretically, one could use all available sample observations to estimate the value of any given location. However, software often provides the option to define the number of observations to be included in the interpolation, or a distance threshold beyond which sample observations will not be included in the estimation. The IDW method does not provide information to evaluate the accuracy of interpolation. One common method to evaluate IDW results is to use a jackknifing approach to perform cross-validation: removing a sampled observation one at a time to obtain the predicted value and compare it with the original observed value.

Although IDW is easy to use and simple to understand and interpret, it clearly has several shortcomings, in addition to an absence of evaluation information. One obvious issue is the selection of the distance decay parameter, α. There is no clear criterion on how one can determine the value. Another limitation, which is also related to α, is that the same distance decay parameter is applied to the entire region. Using one parameter value for the entire study implicitly assumes that the distance decay relationship is constant in all locations, and the sampled values are equally reliable in providing estimates regardless of how sample locations are distributed. This is unlikely to be true in most cases. Therefore, an alternate approach is to vary α according to the clustering pattern of sampled locations at the local scale. Since α changes according to the local situation, this enhanced IDW method is known as the adaptive IDW technique (Lu and Wong, 2008).

5.2.2.2 Geostatistics: Kriging

The IDW method assumes that the magnitude of spatial autocorrelation is directly a function of distance with no stochastic component. This assumption can surely be challenged. Instead of assuming the nature and magnitude of spatial autocorrelation, the geostatistical approach lets the data speak for themselves. Variograms are used to evaluate the nature and magnitude of spatial autocorrelation in the data. From the variogram, a formal function describing the spatial structure of autocorrelation is derived. Using the function specification, weights in Equation 5.17 are computed. Using these weights, values in unsampled locations can be predicted. The method is generally known as Kriging, because this idea was originated from a South African miner Daniel Krige but later formalized by Georges Matheron.

There are many types of Kriging methods, depending on the assumptions and statistical characteristics of data. But the general idea is similar: use the variogram to examine the magnitude and nature of spatial autocorrelation, specify a variogram function that is best in describing the spatial structure, and use the variogram function to derive the weights in Equation 5.17 to compute the predicted values. Instead of focusing on spatial autocorrelation (similarity), geostatisticians focus on semivariance (variance or difference). Let us use h to denote the distance or spatial separation between any two locations and use S for a given location; we can compare values, $y(S)$ and $y(S + h)$. In general, when h is small (close proximity), differences among pairs of values should be relatively small. When h is large (distance locations), differences between values tend to be large. A variogram depicts how these differences change with increasing spatial separation h for all pairs of observations. An empirical variogram can be derived from

$$2\hat{\gamma}(h) = \frac{1}{n(h)} \Sigma(y_i - y_j)^2 \tag{5.21}$$

where $\hat{\gamma}(h)$ is the semivariance of a given spatial separation h between S_i and S_j, y_i and y_j are values in respective locations, and n is the total number of sampled locations. Note that in Equation 5.21, observed values in two locations are directly compared, the same concept used in Geary Ratio (Equation 5.3). Therefore, variogram analysis and spatial autocorrelation are two sides of the same coin.

In Figure 5.1a, the line represents a simple empirical variogram showing the relationship between semivariance and spatial separation. Apparently, the variogram is not smooth as it was created using a small number of samples. With a larger number of samples, it is more than likely that we will see a variogram cloud with patterns that are difficult to discern visually. The smooth and well-structured curve in Figure 5.1b is an idealized empirical variogram. Its general shape indicates that when the spatial separation is small (small h), the observed values should be similar. Nevertheless, due to stochastic processes or random errors, observations at the same location ($h = 0$) can still have different values ($\gamma(0) = C_0 > 0$). The nonzero semi-variance when $h = 0$ is called the nugget effect. The differences between values should increase when h increases, but the difference cannot be larger than a ceiling (δ^2) or the sill. The spatial separation when the semi-variance reaches its maximum is the range (r).

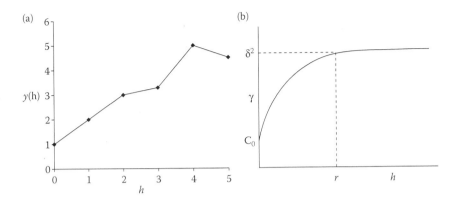

FIGURE 5.1
An empirical variogram (a) and an idealized variogram (b).

A variogram shows the spatial structure revealed by the empirical observations. Usually, it is far from smooth. The objective of constructing a variogram is to identify a mathematical or variogram function which is the best in describing the relationship between semivariance and spatial separation. Some standard functions that are popular candidates to fit a variogram include the spherical, $\gamma(h) = \sigma^2 ((3h/2r) - (h^3/2r^3))$, exponential, $\gamma(h) = \sigma^2 (1 - e^{-3h/r})$, and Gaussian, $\gamma(h) = \sigma^2 (1 - e^{-3h^2/r^2})$ models, where h is the spatial separation or distance lag, r is the range, and δ^2 is the variance or the sill. In the simplest form of Kriging, the most appropriate variogram function is used to derive the weights in Equation 5.17 using the following relationship:

$$\lambda(S_0) = \mathbf{C}^{-1} c(S_0) \tag{5.22}$$

where C is the variance–covariance matrix between observations determined by the chosen variogram function, and $c(S_0)$ is the vector of covariance between the predicted point S_0 and all sampled locations determined by the variogram function. With the weights in Equation 5.22, we can compute the estimated values for all unsampled locations.

Types of Kriging are different in the detailed specifications of how the weights are derived, but the general idea of using the variogram function to define the variance–covariance structure (C in Equation 5.22) in the study region is generally applicable. Note that both the matrix and the vector in Equation 5.22 are derived using only the variogram function and the distances between pairs of locations. The validity of this general approach rests on the two assumptions of isotropy and stationarity. Isotropy means that the relationship over space is true for all directions. Stationarity refers to the constant mean and variance over the study region. In reality, some aspects of these assumptions may be violated and special types of Kriging may be necessary. Interested readers can refer to Isaaks and Srivastava (1990) for a comprehensive discussion.

5.2.3 Spatial Models

The scope of spatial models is so broad and diverse that several overview textbooks will be required to provide a reasonable coverage. Therefore, we do not attempt to cover such breadth in this subsection. Instead, we will point to the literature covering some example models that are potentially of interest to readers. Model building has a long intellectual history in geography and geosciences. Many geographical models were developed to quantify and describe human behavior and activities. Some of them can be found in Haggett et al. (1977) and Robinson (1998). Recently, Wang (2005) offered some detailed discussions of selected models.

In geosciences, models may be domain specific (such as hydrological models), focusing on a certain aspect of geosciences, or relatively comprehensive by coupling multiple domain-specific models together (e.g., Kirtman and Shukla, 2002). These geosciences models cover a wide range of geographical scale, from global scale, such as the global circulation models (GCM, e.g., Kasahara and Washington, 1967), to regional or mesoscale scale, such as the Weather Research and Forecasting Model (WRF, Skamarock et al., 2001), and local or neighborhood scale, such as the Computational Fluid Dynamic models (CFDs, Hanna et al., 2006). Many of these models are computationally intensive, requiring advanced mathematical operations such as solving differential equations. As a result, most of these models have not been coupled with GIS, despite the fact that GIS may provide data as model inputs. An exception is hydrological models (e.g., Gurnell and Montgomery, 2000), which have been tightly integrated with GIS, partly because of its heavy reliance on digital elevation data (Mark, 1983; Band, 1986). The use of GIS in related to hydrologic research is extensive. Specific areas include the modeling of hydrodynamics such as water currents and pollutant transport in coastal regions or lakes and the impacts of storm surges in coastal areas. Literature on these applications is abundant.

Another reason that hydrological modeling is relatively well integrated with GIS is because of the capability of GIS to represent Earth surfaces. GIS is quite effective in modeling field variables or spatially continuous phenomena, such as land use changes, regardless of using the triangulated irregular network (TIN) or the raster format. As a result, many land use-land cover change (LULCC) models have been developed in GIS using various frameworks. Most of these models create simulations depicting the change in landscape. One of the approaches, cellular automata (CA), adopts the raster format to represent landscape and develops rules to program how characteristics of each cell change over time (e.g., Clark et al., 1997). This type of model is rather dynamic and flexible. Liu (2008) offers a detailed explanation of the nature of CA and how CA can be used in various simulation applications. Some applications are predictive in nature, modeling vegetation dynamics (e.g., Colasanti and Grime, 1993).

While independent of the development of CA, agent based models (ABM), which were developed under the realm of economics and social science, share some conceptual similarities with CA in terms of the framework in presenting geographical landscape in a raster format. In ABM, each agent is an autonomous entity that acts according to some predefined rules. These agents can be geographical objects, animals, or humans. The general idea is that as individuals act according to their rules, a macro-pattern or structure may emerge over time. This approach has been used to model land use changes and geographical phenomena. Parker et al. (2003) offer an overview of how ABM can play a role in spatial modeling. Benenson and Torrens (2004) offer a rather comprehensive review of geosimulation, covering both CA models and ABM models. Although geosimulation offers promises in advancing geographical research, one of its impediments is handling vector data. GIS can handle simulations in raster format reasonably well, but GIS support of simulations in the vector domain is far from adequate.

5.3 Spatial Data Mining and Spatial Knowledge Discovery*

5.3.1 Introduction

Geographers acquire new knowledge by searching for patterns, formulating theories, and testing hypotheses with observations. The beginning of this chapter has highlighted the continuing efforts through various scientific projects, government agencies, and private sectors in collecting voluminous geographic data. Modern data collection methods, such as global positioning systems (GPS), high-resolution remote sensing systems, location-aware services and surveys, and Internet-based volunteered geographic information sources, can help us obtain much more diverse, dynamic, and detailed data than ever possible. Generally speaking, geography and related spatial sciences have moved from a data-poor era to a data-rich era (Miller et al., 2009).

To analyze observational data, traditional (mostly statistical) data analysis approaches often require a priori models, measures, or hypotheses. For example, a regression model assumes that the data exhibit a linear relationship in the form of $Y = \beta_0 + \Sigma_k (\beta_k X_k) + \varepsilon$. Then the regression procedure determines the coefficients β_k and ε using the observational data. Similarly, statistical measures tend to measure the presence (or absence) of certain types of patterns. One obvious shortcoming of such confirmatory analysis is that if the hypothesis or model is mis-specified for the phenomenon being

---

* This section is contributed by Diansheng Guo.

analyzed, the analysis can at best indicate that the data do not fit, but it cannot suggest any better alternatives.

Different from the confirmatory analysis as described above, exploratory data analysis (EDA) and exploratory spatial data analysis (ESDA) (Tukey, 1977; Bailey et al., 1995; Anselin, 1999) use statistical graphics to show structural information in the data and rely on human experts to interact with data, visually identify patterns, and formulate hypotheses or models to explain the patterns. This process is often iterative, as preliminary findings may prompt the analyst to change data input, perform a new analysis, and revise the hypothesis or model.

However, traditional EDA or ESDA methods were developed in an era when data were relatively scarce and computational power was much less powerful (Miller et al., 2009). They mainly work for small datasets (e.g., dozens or hundreds of sample data points and a few variables). Consequently, they often fall short of extracting hidden information from datasets that are unprecedentedly large (e.g., millions of observations), of high dimensionality (e.g., hundreds of variables), and high degree of complexity (e.g., spatial nonstationarity, space-time dynamics, multivariate relations, and their possible interactions).

To address the above challenges, *spatial data mining* and *geographic knowledge discovery* is a new research field that focuses on the development of theory, methodology, and practice for the extraction of useful information and knowledge from spatial databases (Openshaw et al., 1987; Knorr et al., 1996; Han et al., 1997; Kulldorff, 1997; Andrienko et al., 1999; Chawla et al., 2000; Miller et al., 2001; Gahegan, 2003; Guo et al., 2003; Keim et al., 2004; Shekhar et al., 2004; Guo et al., 2006; Guo, 2008; Miller et al., 2009). Spatial data mining is closely related to both traditional spatial analysis (including spatial statistics, analytical cartography, and exploratory data analysis) and various data mining methods in statistics and computer science. A variety of methods, many of which are interactive and human-centered, have been developed in these fields to analyze large and complex spatial datasets. Very often, these new methods are placed under different umbrellas, such as spatial statistics, spatialization, geocomputation, spatial data mining, geovisualization, and visual data mining. In this section of the chapter, we label them as *spatial data mining methods* based on the following characteristics that they share:

- *Spatial:* Consider spatial properties, constraints, and processes in the analysis.
- *Exploratory:* Extract information and formulate hypotheses from data.
- *Scalable:* Can analyze large and complex datasets to discover unknown or unexpected information and knowledge.

The process of data mining and knowledge discovery is not a push-button task. Rather, it is an iterative and human-centered process that involves data

selection and preprocessing; prior domain knowledge; computational algorithms and/or visual approaches; interpretation and evaluation of results;
and the formulation or modification of hypotheses and theories (Fayyad
et al., 1996). The data mining and knowledge discovery process is exploratory and inductive in nature, wherein researchers develop and modify
theories based on the discovered information from observation data (Miller
et al., 2009, 4).

There are different types of spatial data mining tasks, and each task can be
usually completed by a number of different methods. Moreover, as new types
of geographic data become available, new methods are being developed to
search for new types of patterns. This section focuses primarily on a selected
set of spatial data mining tasks, including classification, association rule mining, clustering, and geovisualization. Each task may involve different methods, which can be computational, statistical, or visualization-based.

5.3.2 Classification

Classification is to assign observations into classes (groups) according to
their properties and similarities. Classification is also called supervised classification, because it needs a training data set to train or configure the classification model and a test data set to evaluate the performance of the trained
model. Classification methods include, but are not limited to, decision trees,
artificial neural networks (ANN), maximum likelihood estimation (MLE),
linear discriminant function (LDF), support vector machines (SVM), nearest
neighbor methods, and case-based reasoning (CBR). Among these classification methods, the decision tree produces the most understandable outputs—
a classification tree and a set of classification rules derived from the tree
(Quinlan, 1993; Mitchell, 1997).

To illustrate what constitutes a classification problem, let us use the decision tree method and a simple data set that has four attributes and a class
label (play or not play) for each record (see Table 5.1) (Quinlan, 1993). With
this training data set, a decision tree can be developed and a set of classification rules may be derived (see below). These rules can later be used to classify new records with unknown class labels.

- If outlook = sunny and humidity <= 75, then Play tennis
- If outlook = sunny and humidity > 75, then Don't play
- If outlook = overcast, then Play tennis
- If outlook = rain and windy = true, then Don't play
- If outlook = rain and windy = false, then Play tennis

Spatial classification methods consider not only attributes of the object to
be classified but also attributes of neighboring objects and their spatial relations (Ester et al., 1997; Koperski et al., 1998). A visual approach for spatial

TABLE 5.1

Demonstrative Data Set

Outlook	Temp (°F)	Humidity (%)	Windy?	Class
Sunny	75	70	True	Play tennis
Sunny	80	90	True	Don't play
Sunny	85	85	False	Don't play
Sunny	72	95	False	Don't play
Sunny	69	70	False	Play tennis
Overcast	72	90	True	Play tennis
Overcast	83	78	False	Play tennis
Overcast	64	65	True	Play tennis
Overcast	81	75	False	Play tennis
Rain	71	80	True	Don't play
Rain	65	70	True	Don't play
Rain	75	80	False	Play tennis
Rain	68	80	False	Play tennis
Rain	70	96	False	Play tennis

Source: Data from Quinlan, J. R. 1993. *C4.5: Programs for Machine Learning.* Morgan Kaufmann, San Mateo, CA.

Note: Each record has four attributes and a class label.

classification was introduced by Andrienko et al. (1999), where the decision tree derived with a traditional algorithm C4.5 (Quinlan, 1993) is combined with map visualization to reveal spatial patterns of the classification rules.

5.3.3 Association Rule Mining

Association rule mining was originally developed to discover regularities between items in large transaction databases (Agrawal et al., 1993). Let $I = \{i_1, i_2, \ldots, i_m\}$ be a set of items (e.g., items purchased in a transaction, such as a computer, milk, bike, etc.). Let D be a set of transactions, where each transaction T is a set of items such that $T \subseteq I$. Note that the quantities of items bought in a transaction are not considered, meaning that each item is a binary variable representing whether an item was present or not in a transaction. Let X be a set of items and a transaction T is said to contain X if and only if $X \subseteq T$. For example, from a supermarket transaction database, we may find such an association rule: {Milk, Bread} $\Rightarrow$ {Beef} [50%, 80%], meaning that 50 percent of the transactions in the database involve milk, bread, and beef; and out of those transactions that involve milk and bread, 80% also purchased beef. Formally, an association rule is in the form: $X \Rightarrow Y$ [$s\%$, $c\%$], where $X \subset I$, $Y \subset I$, $X \cap Y = \emptyset$, s is the *support*, and c is the *confidence* of the rule. The above example rule has a support of 50 and a confidence of 80. It is often desirable to pay attention to those rules that have reasonably large support and high confidence.

Spatial association rules can be mined in spatial databases by also considering spatial properties and relations or predicates (e.g., close to, intersect, and overlap) (Koperski et al., 1995; Han et al., 2001; Appice et al., 2003; Mennis et al., 2005). Below is an example of a spatial association rule regarding schools: close_to ("sports_center") ⇒ close_to ("park") [50%, 80%], which means that 50% of schools are close to both sports centers and parks and 80% of the schools that are close to a sports center are also close to a park. The challenge for mining spatial association rules is that there are many different spatial relationships (e.g., close_to, far_away, intersect, overlap, contain, etc.) and it is computationally expensive to consider all of them in deriving association rules from a large data set. Another potential problem with association rule mining (particularly spatial association rule mining) is that a large number of rules may be generated, but many of them are obvious or common knowledge. This means that domain knowledge is needed to filter out trivial rules and focus only on new and interesting findings.

Spatial co-location pattern mining is a unique spatial data mining task that is similar to association rule mining in principle but is very different technically (Shekhar et al., 2001). Given a data set of spatial features and their locations, the co-location pattern discovery process finds subsets of features that are frequently located together. Since a location is not a transaction and two features rarely exist at exactly the same location, a user-specified neighborhood is needed as a container to check which features co-locate in the same neighborhood. Measures and algorithms for mining spatial co-location patterns have been proposed (Shekhar et al., 2001; Huang et al., 2006).

5.3.4 Clustering

Cluster analysis is a widely used data analysis approach, which organizes a set of observations into groups (or clusters) so that observations in the same group are similar to each other and different from those in other groups (Jain et al., 1988; Gordon, 1996; Jain et al., 1999). Many different clustering methods have been developed in various research fields, such as statistics, pattern recognition, data mining, machine learning, and spatial analysis. Clustering methods may differ in many ways, including

- The definition of distance between observations (and between clusters)
- The definition of a "cluster"
- The strategy to group or divide observations into clusters
- The data type being analyzed (e.g., numerical, categorical, and/or spatial)
- Application-specific requirements and constraints

Clustering methods can be broadly classified into two groups: partitioning clustering and hierarchical clustering. Partitioning clustering methods, such as K-Means and MLE, divide a set of observations into a number of nonoverlapping clusters. A data item is assigned to the "closest" cluster based on a proximity or dissimilarity measure. Hierarchical clustering, on the other hand, organizes observations into a hierarchy with a sequence of nested partitions or groupings. Commonly used hierarchical clustering methods include the Ward's method (Ward, 1963), single-linkage clustering, average-linkage clustering, and complete-linkage clustering (Jain et al., 1988; Gordon, 1996).

Self-organizing map (SOM) (Kohonen, 2001) is a special partitioning clustering method, which not only segments data into clusters but also orders the clusters in a two-dimensional layout so that nearby clusters are similar to each other. Therefore, SOM is also considered a visualization method and a dimension reduction technique that projects multidimensional data to a 2-D space. The SOMs are widely used in various research fields and application areas (see Kaski et al., 1998; Oja et al., 2003 for comprehensive reviews). There are also numerous applications of SOM in geographic analysis, for example, the visualization of census data (Skupin et al., 2003b), spatialization of nonspatial information (Skupin et al., 2003a), and multivariate mapping and geovisualization (Guo et al., 2005; see Section 5.4.5 for more details).

5.3.4.1 *Spatially Constrained Clustering and Regionalization*

The general-purpose clustering methods introduced above do not consider geographic information or spatial constraints. Therefore, observations in a cluster are not necessarily close or contiguous in a geographic space. However, for many spatial analysis tasks, it is often desirable or required that clusters are geographically contiguous, such as the delineation of climatic regions, ecoregion analysis, map generalization, and public health analysis (Haining et al., 1994; Osnes, 1999). Regionalization is a special form of clustering that seeks to group spatial objects into spatially contiguous clusters (i.e., regions) while optimizing an objective function (Openshaw, 1977; Openshaw et al., 1995).

Existing regionalization methods that are based on a clustering concept can be classified into three groups: (1) multivariate (nonspatial) clustering followed by spatial processing to rearrange clusters into regions (Fovell et al., 1993; Haining et al., 1994); (2) clustering with a spatially weighted dissimilarity measure, which can consider spatial distance as a factor in forming clusters but cannot guarantee that each cluster is contiguous in space (Wise et al., 1997); and (3) contiguity constrained clustering, which enforces spatial contiguity during the clustering process (Guo, 2008). Figure 5.2 shows an example of regionalization, where ten regions are derived based on the percentage of population change for each county between 1990 and 2000 censuses for the United States. Although the example shown here only uses one variable for the ease of understanding, the method allows the use of multiple variables to define clusters and regions.

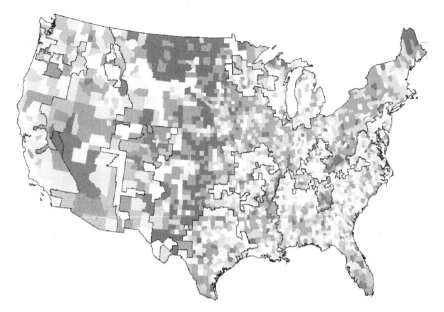

FIGURE 5.2
The map shows 10 regions derived with one variable (population change between 1990 and 2000 censuses for each county) with dark colors indicating a high population increases and decreases.

5.3.4.2 Spatial Scan Statistics

Another special class of spatial clustering methods focuses on the detection of unusual concentration of events in space, such as hot spots of crimes or geographic clusters of a disease. The general research problem is to determine whether there is an excess of observed event points (e.g., disease incidents) for a given area (e.g., within a certain distance to a location). Several scan statistics have been developed to detect such spatial clusters. A spatial scan statistic usually consists of the following steps:

- Define a strategy to enumerate a large set of local areas.
- Calculate a test statistic, which will be calculated for each local area.
- Test the significance level of the test statistic value.
- Identify areas with significant test values.

Two important examples of such spatial scan statistics are the Geographic Analysis Machine (GAM) by Openshaw (1987, 1990) and the family of space–time scan statistics by Kulldorff (1997, 2005). They both follow the above general steps but use different strategies to enumerate local areas and use different test statistics. They both use a Monte Carlo-based approach to perform significance tests for each local statistic, by constructing the distribution of the test statistics under the null hypothesis (i.e., no significant cluster).

The GAM consists of the following steps:

1. Define a two-dimensional grid covering the study area; define a minimum circle radius, a maximum radius, and a size increment.
2. Repeat the following steps until all radii are examined:
    i.   For each grid point, compute the test statistic and its significance level for the data points covered by a circle centered on the grid point.
    ii. Increase the circle radius by the given increment size.
3. Present all grid points and their circles that pass a given significance threshold.

The test statistic used in GAM is the point count (e.g., number of disease incidents) within an area (i.e., a circular region around a grid point). To determine whether the point count in an area is significant, a Monte Carlo procedure is used to generate a large number (e.g., 500) of random point sets, each representing a realization of the null hypothesis in the same area. A test statistic value is calculated for each random data set, and thus a distribution of the test statistic values under the null hypothesis is derived. By comparing the actual test statistic value (i.e., the count of points) and the derived distribution, the significance level for the test statistic in the area is obtained. A potential problem with GAM is that it is difficult to adjust for the multiple-testing problem (Rogerson et al., 2009). The multiple-testing problem refers to the fact that, when many tests are performed at the same time, the probability of incorrectly rejecting a null hypothesis is much higher than the confidence interval indicates. The computational workload for GAM is also a disadvantage, but more or less all scan statistics need considerable computational power to search for and test local clusters.

The spatial scan statistics developed by Kulldorff (1997, 2005) calculates a likelihood ratio for each local area. To overcome the multiple-testing problem, the scan statistic uses the ML ratio (which is the ML ratio among all local areas) as the test statistic. Therefore, the scan statistic method reports the most likely cluster, although a set of secondary clusters is also provided. It first calculates the likelihood ratio for each local neighborhood (defined by a circle for example) and finds the maximum of all likelihood ratios. To derive the significance level, replications of the data set are generated under the null hypothesis, conditioning on the total number of points. For each replication, the test statistic value is calculated again (i.e., the ML ratio is found over all enumerated local areas). Then the actual test statistic value is compared with the test values of all replications to derive the significance level for the most likely cluster (and the secondary clusters).

5.3.5 Geovisualization

Geovisualization concerns the development of theories and methods to facilitate knowledge construction through visual exploration and analysis of spatial data and the implementation of visual tools for subsequent knowledge retrieval, synthesis, communication, and use (MacEachren et al., 2001). As an emerging domain, geovisualization has drawn interests from various cognate fields and evolved along a diverse set of research directions (Dykes et al. 2005; Taylor, 2005). The main difference between traditional cartography and geovisualization is that the former focuses on the design and use of maps for the communication of known information, whereas the latter emphasizes the discovery of unknown information with the development of highly interactive maps and associated tools for data exploration, hypothesis generation, and knowledge construction (MacEachren, 1994; MacEachren et al., 1997).

To cope with today's large and diverse spatial datasets and facilitate the discovery and understanding of complex information from data, geovisualization needs to address several major challenges: (1) process very large datasets efficiently and effectively; (2) handle multiple perspectives and many variables simultaneously to discover complex patterns; and (3) the design of effective user interfaces and interactive strategies to facilitate the discovery process. Research efforts for the third challenge have emerged as an active subfield called *visual analytics* (Thomas et al., 2005). We briefly introduce the first two challenges in the next paragraph.

Large data volumes cause serious problems for most existing visualization techniques. First, a large number of observations often lead to a cluttered visual display (e.g., points overlapping in a scatter plot) and thus make it very difficult (if possible at all) for the analyst to visually perceive patterns. Second, a large data set requires considerable and often prohibitive amount of time to process if the visualization involves complex statistical computing (such as the calculation of a semi-variogram for spatial autocorrelation or geostatistical analysis). A type of solution is to rely on the user to dynamically filter, select, zoom, and adjust detail levels in the visualization, meaning that only a small subset of data is visualized at one time. Another type of solution is to combine efficient computational methods (such as clustering, classification, and association rule mining) with geovisualization, with the former finding patterns quickly and the latter helping users explore and understand the patterns (Andrienko and Andrienko, 1999; Ward, 2004; Guo et al., 2005).

Multivariate mapping has long been an interesting and challenging research problem. Generally, multivariate mapping methods can be classified into three types. The first type, such as the Chernoff face (Chernoff et al., 1975), depicts each variable with a certain visual attribute (such as color, shape, size, or orientation) and then integrates all variable depictions into one symbol to show on the map or a figure (Grinstein et al., 1992; DiBiase et al., 1994; Wittenbrink et al., 1995; Gahegan, 1998; Zhang et al., 2004). The second type uses multiple linked views (or maps) that show one (or more) variables per view (Monmonier, 1989; Dykes, 1998; MacEachren et al., 1999). For example, Carr et al. (2005) proposed the conditioned choropleth maps (CCmaps), which use a two-way layout of maps (arranged by two potential explanatory variables) to facilitate the exploration of potential associations between a dependent variable (as represented in colors) and two explanatory variables. The third type of solution is to project data to a lower-dimensional (normally 1D or 2D) space through clustering and then map the clusters (Guo et al., 2003; Guo et al., 2005).

Figure 5.3 shows an example of multivariate clustering and geovisualization (Guo et al., 2005; Guo et al., 2006) adopting an integrated approach that couples an SOM, multidimensional visualization (a parallel coordinate plot-PCP), and a map component. The SOM derived 49 clusters colored with a systematic coloring scheme. The PCP and the map can help understand these clusters. Red or reddish clusters, primarily located in the Arctic area, have high positive values for winter months (November, December, January, and February) and around zero (i.e., no change) for summer months (June, July, and August). This means that during the past decade, winter in the Arctic area was much warmer than before but was relatively stable during summer. In contrast, the green or light-green clusters mainly located in the Antarctic area have high positive values for winter months (April–September) and negative values for summer months (January, February, November, and December). This means that during the past decade, the Antarctic area was much warmer in winter but cooler during summer.

5.3.6 Summary

It is worth re-emphasizing that spatial data mining is not a push-button task. The data cannot tell stories unless we formulate appropriate questions to ask and use appropriate methods to solicit the answers from the data. Data mining is not only data-driven but also, more importantly, human-centered, with the user controlling the selection and integration of data, cleaning and transformation of the data, choosing analysis methods, and interpreting the results.

The abundance of spatial data provides exciting opportunities for new research directions but also demands caution in using these data. The data are often from different sources and collected for different purposes under various conditions, such as measurement uncertainty, biased sampling, varying area units, and confidentiality constraints. It is important to understand the data quality and characteristics and assess their suitability for the targeted tasks. Careful selection, preprocessing, and transformation of the data are needed to ensure a meaningful analysis and result.

The choice of analysis methods is also critical. First, for a complex problem, a single method is often not enough to complete the task. A suite of methods may be needed to examine the data from different perspectives and collectively address a large question. Second, each method has its own

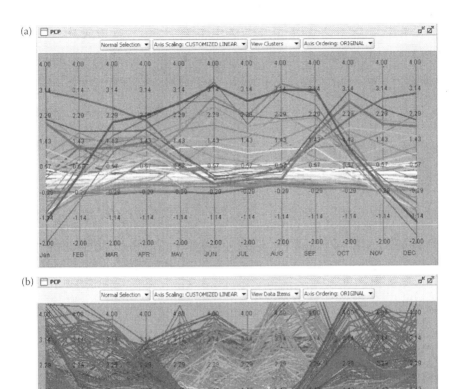

FIGURE 5.3
(See color insert following page 144.) Seasonal and spatial patterns of surface temperature anomaly (1998–2007). (a) parallel coordinate plot of temperature change, Antarctic; (b) parallel coordinate plot of temperate change, Arctic; (c) 49 clusters derived from self-organized map (SOM); and (d) a map showing the spatial pattern of changes. (Adapted from Guo, D. 2009. *Geographic Data Mining and Knowledge Discovery*, eds. H.J. Miller and J. Han, 325–345. London and New York: Taylor & Francis; Jin, H. and D. Guo. 2009. Understanding Climate Change Patterns with Multivariate Geovisualization. In *Proceedings—IEEE International Conference on Data Mining Workshops*, 217–222. Miami, FL: IEEE Press.)

limitations and assumptions. It is important to understand what a method can do and more importantly, what it cannot do. Third, new data types and applications often require the development of new data mining methods and the discovery of new types of patterns.

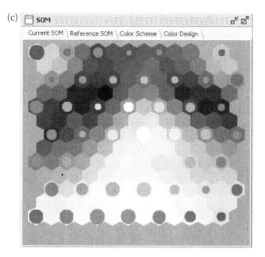

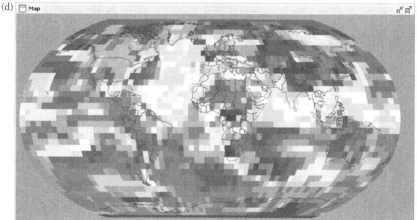

FIGURE 5.3
Continued.

5.4 Still and Motion Image Analysis*

The advancement of digital camera technology and the emergence of novel camera deployment strategies (e.g. onboard commercial remote sensing platforms, onboard unmanned aerial vehicles, or distributed in a network) are making digital imagery the main source of Geoinformation. As a consequence, the volume of digital imagery collected today for Geoinformation applications in a single day far exceeds the amount of data collected over

---

* This section is contributed by Arie Croitoru, Peggy Agouris, and Anthony Stefanidis.

multiple years in the past. For example, GeoEye-1, launched in 2008, is able to collect approximately 700,000 square kilometers daily, an area comparable to the State of Texas, at sub-meter resolution. We need efficient algorithms that allow us to automatically identify features within this imagery for this imagery to be processed in a timely manner. The types of features that can be identified include

- Points; for example, the corner of a rooftop or a manhole on the street
- Lines; for example, the outline of a road segment
- Areas; for example, the area covered by a specific type of vegetation

In this section, we will present representative techniques to extract points and lines from digital imagery. Area extraction is typically the subject of remote sensing classification techniques, and the reader is referred to standard remote sensing textbooks (e.g. Lillesand et al., 2007).

Further, as video sensors are becoming increasingly more reliable, the geoinformatics community is slowly starting to exploit the opportunities offered by the availability of spatiotemporal information (e.g. in the form of trajectories). Thus, we include a brief segment at the end of this section, outlining major issues related to motion imagery data processing.

5.4.1 Edge and Junction Detection

The detection of edge and junction pixels is essential to the processing of spatial and spatiotemporal imagery. For example, junction points are pivotal to the ability to orient images in a stereo imagery pair or determine the georeferencing parameters of an image. Similarly, edges in imagery are essential for automated feature extraction (e.g., roads, lakes, buildings, etc.) and scene understanding. The detection of edge and junction pixels is based on the application of an image operator designed to produce a strong local response in locations where the image function values correspond to sharp changes. Ultimately, such operators should be designed to provide (Agouris et al., 1989)

a. *Good detection:* the ability to minimize occurrences of missing or falsely detected edges and junctions
b. *Good localization:* the ability to detect edges and junction points as close as possible to their true location
c. *Robustness to noise:* the ability to mitigate the effects of image noise on the operator estimation
d. *Computational efficiency:* the ability to minimize the number of computations required while marinating all previous requirements

Given these requirements, commonly used edge and junction detection techniques employ three processing steps, namely *noise reduction, edge detection*, and *localization*. The primary purpose of noise reduction is lowering the level of noise in the image while preserving edge information. Once noise has been suppressed, edges are detected by convolving the image with an operator (kernel) that produces a strong response to edges (i.e., local maxima) in the image. Here, local derivatives of the image function are typically used under the assumption that the image can be described as an analytic function. Finally, based on the magnitude and directional information derived from the edge operator response, a selection process of the most significant edges is carried out. In its simplest form, the selection process is carried out by a threshold operation, selecting stronger edges with specific direction. These three processing steps are often based on a set of assumptions and a priori knowledge with respect to the image data, such as the nature of the image noise, the type of the edges (e.g., a step function, a ramp, etc.), and the type of objects to which the edges correspond.

In light of these requirements and challenges, considerable effort has been invested in the development of various edge and junction detection operators over the last several decades. Rather than being a comprehensive review of these efforts, the purpose of this section is to explore some of the key principles that guided the development of such operators. The interested reader may find a comprehensive comparison of edge and junction detection techniques in Ziou and Tabbone (1998), Heath et al. (1998), Mehrotra et al. (1990), and Bouchara and Ramdani (2006).

5.4.1.1 Edge Pixel Detection

The existence of abrupt changes in pixel intensity levels is an important cue in the detection and delineation of *edges*, which are often associated with boundaries of features and transition zones between features (e.g., buildings, lakes, roads, etc.) in the imaged scene. Considering the image pixel values as a continuous function $f(x, y)$, an edge (i.e., a sharp change) can be modeled as a step function, for which the gradient vector ∇f can be estimated at a given image pixel location (x, y):

$$\nabla \vec{f}(x,y) = \begin{bmatrix} \dfrac{\partial f(x,y)}{\partial x} \\ \dfrac{\partial f(x,y)}{\partial y} \end{bmatrix} = \begin{bmatrix} f_x(x,y) \\ f_y(x,y) \end{bmatrix} \tag{5.23}$$

The magnitude of the gradient can be approximated by

$$\left| \nabla \vec{f}(x,y) \right| = \sqrt{\left(f_x(x,y) \right)^2 + \left(f_y(x,y) \right)^2} \approx \left| f_x(x,y) \right| + \left| f_y(x,y) \right| \tag{5.24}$$

and its direction (perpendicular to the edge orientation) is given by

$$\theta = \tan^{-1}\left(\frac{f_y(x,y)}{f_x(x,y)}\right) \tag{5.25}$$

Note that the approximation in Equation 5.24 is used to reduce the overall computational load required for the estimation of the gradient. Since sharp changes in the value of pixels will lead to high derivative response, the magnitude of the gradient can be used directly to detect pixels that are associated with edges, whereas the gradient direction can be used to estimate the edge direction. The problem of detecting edge pixels is therefore closely associated with the estimation of image derivatives.

Derivatives in images can be estimated *directly*, based on the image pixel values, or *indirectly*, as an optimization problem based on the neighborhood of each pixel. To examine the direct estimation scheme, let us examine the first-order discrete forward or backward derivative approximation f_x in the x direction:

$$\begin{cases} f_x^{\text{forward}}(x,y) = f(x+1,y) - f(x,y) \\ f_x^{\text{backward}}(x,y) = f(x,y) - f(x-1,y) \end{cases} \tag{5.26}$$

Note that we assume that in a discrete image $|\Delta x| = |\Delta y| = 1$. Averaging both approximations yields

$$\hat{f}_x(x,y) = \frac{1}{2}\left(f(x+1,y) - f(x-1,y)\right) \tag{5.27}$$

A similar expression can be derived for the y derivative:

$$\hat{f}_y(x,y) = \frac{1}{2}\left(f(x,y+1) - f(x,y-1)\right) \tag{5.28}$$

These expressions can be represented by the following three-by-three kernels, which correspond to the Roberts operator:

$$\hat{f}_x = \frac{1}{2}\begin{bmatrix} 0 & 0 & 0 \\ -1 & 0 & 1 \\ 0 & 0 & 0 \end{bmatrix}, \quad \hat{f}_y = \frac{1}{2}\begin{bmatrix} 0 & -1 & 0 \\ 0 & 0 & 0 \\ 0 & 1 & 0 \end{bmatrix} \tag{5.29}$$

(a)

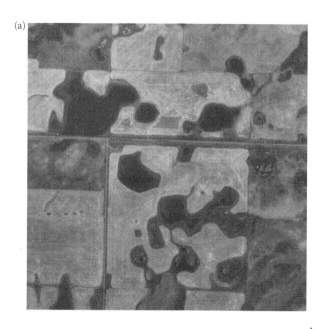

(b)

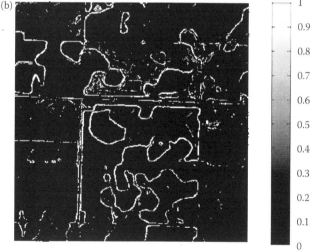

FIGURE 5.4

An example of the Roberts operator. (a) a gray level image; (b) detected edge pixels, after magnitude thresholding.

An example of the application of these kernels to a sample image can be seen in Figure 5.4.

The indirect estimation of the derivatives in the x and y direction is based on approximating the function $f(x, y)$ using the neighborhood of a pixel, hence resulting in an estimation less susceptible to highly localized variation and noise. Consider an indirect estimation scheme in which a 3×3 neighborhood of a pixel is approximated as a plane f^P:

$$f^P(x,y) = a + bx + cy \tag{5.30}$$

where a, b, and c are the plane coefficients, and the derivatives are estimated as $f_x = b$ and $f_y = c$. Assuming that the location for which we wish to estimate the derivatives is the origin of a local coordinate system $((x,y) = (0,0))$, it is possible to write 9-plane equations for this 3×3 neighborhood:

$$\underbrace{\begin{bmatrix} 1 & -1 & -1 \\ 1 & -1 & 0 \\ 1 & -1 & 1 \\ 1 & 0 & 1 \\ 1 & 1 & 1 \\ 1 & 1 & 0 \\ 1 & 1 & -1 \\ 1 & 0 & -1 \\ 1 & 0 & 0 \end{bmatrix}}_{A} \begin{bmatrix} a \\ b \\ c \end{bmatrix}_{X} = \underbrace{\begin{bmatrix} f(-1,-1) \\ f(-1,0) \\ f(-1,1) \\ f(0,1) \\ f(1,1) \\ f(1,0) \\ f(1,-1) \\ f(0,1) \\ f(0,0) \end{bmatrix}}_{L} \Rightarrow AX = L \tag{5.31}$$

A solution to this set of equations can be obtained by least squares from which an estimation for f_x (i.e., b) and f_y (i.e., c) is, therefore, given by

$$b = \frac{1}{6}\left(f(1,1) + f(1,0) + f(1,-1)\right) - \frac{1}{6}\left(f(-1,-1) + f(-1,0) + f(-1,1)\right)$$

$$c = \frac{1}{6}\left(f(-1,1) + f(0,1) + f(1,1)\right) - \frac{1}{6}\left(f(-1,-1) + f(1,-1) + f(0,1)\right) \tag{5.32}$$

which is the well-known Prewitt edge detection operator. A comparison of the Roberts and Prewitt operators is shown in Figures 5.5a and b. It can be seen that the estimation of the derivatives using a 3×3 neighborhood results in a stronger gradient response and thicker edge regions. Similarly, the Sobel operator can be derived by assigning higher weights to the pixels along the x and y directions, respectively:

$$b = \frac{1}{8}\left(f(1,1) + 2f(1,0) + f(1,-1)\right) - \frac{1}{8}\left(f(-1,-1) + 2f(-1,0) + f(-1,1)\right)$$

$$c = \frac{1}{8}\left(f(-1,1) + 2f(0,1) + f(1,1)\right) - \frac{1}{8}\left(f(-1,-1) + 2f(1,-1) + f(0,1)\right) \tag{5.33}$$

(a)

(b)

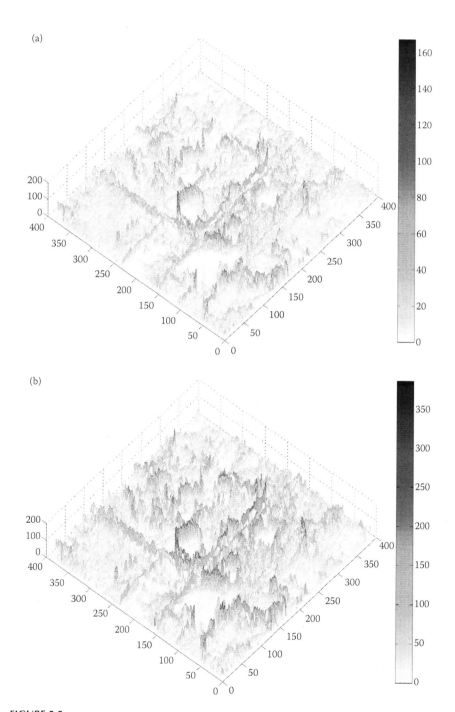

FIGURE 5.5
A three-dimensional visualization of the gradient estimation using the Roberts (a) and Sobel (b) operators (derived from the image in Figure 5.3a).

The second derivative of an image function may also be used for detecting edge pixels. In this approach, the second derivative of the image function is used to estimate the Laplacian, which can be approximated by a central difference numerical differentiation approximation (Agouris et al., 1989):

$$\nabla^2 f = f_{xx} + f_{yy}$$

$$= \frac{f(x+1,y) + f(x-1,y) - 2f(x,y)}{\Delta x^2} + \frac{f(x,y+1) + f(x,y-1) - 2f(x,y)}{\Delta y^2}$$

$$= f(x+1,y) + f(x-1,y) + f(x,y+1) + f(x,y-1) - 4f(x,y)$$

(5.34)

Note that the Laplacian is a direction-independent (isotropic) operator. The second derivative produces a double response with opposite signs for step functions, thus enabling to detect edges by identifying the *zero-crossing* pixels, that is, pixels for which the sign of the Laplacian is reversed. Since the second derivative used in the Laplacian operator (Equation 5.34 is particularly susceptible to noise, a Gaussian smoothing operator with a user-defined standard deviation (σ) is first applied to the image before the Laplacian is computed. The combination of these two operators results in the Laplacian of Gaussian (LoG) operator. Although the Gaussian operator is essential for the reliable estimation of the second derivative of the image function, the level of smoothing used (σ) may affect the localization accuracy of the edges that will be detected. An example of the application of this approach is given in Figure 5.6, which shows the effect of increasing smoothing on the detected edges. As can be seen from a comparison of Figure 5.6a with Figures 5.5a and b, the application of the Gaussian smoothing contributes significantly to the reduction of noise. However, these examples also demonstrate that increased smoothing in the LoG operator may result in the elimination of weak edges, smoothing of the remaining edges, and the potential drift of the remaining edges (e.g, compare Figures 5.5a and c).

A more advanced approach to the edge detection problem has been developed by Canny (1986). Similar to the LoG, the Canny operator is based on image smoothing and on analyzing the magnitude of the gradient. However, unlike the LoG, the edge detection localization process is based on two sequential steps, namely *nonmaxima suppression* and *hysteresis thresholding*, which incorporate both the orientation and magnitude of the gradient. In the nonmaxima suppression step, pixels are coded based on their gradient orientation angle. Then, for each pixel a search is conducted for neighboring pixels with a similar gradient direction (up to ±180°) and a smaller gradient magnitude. These pixels are suppressed and their value is set to zero, otherwise the pixel is preserved. This process ensures that the resulting edges have a thickness of

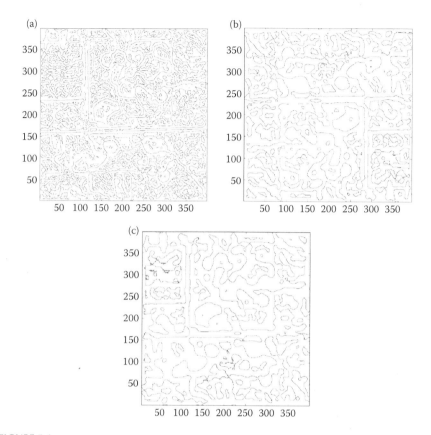

FIGURE 5.6
The LoG operator. (a) the zero-crossing of the LoG operator, $\sigma = 2$; (b) the zero-crossing of the LoG operator, $\sigma = 4$; (c) the zero-crossing of the LoG operator, $\sigma = 5$.

one pixel. In the final hysteresis step, the resulting edge pixels are thresholded with a user-defined value.

5.4.1.2 Junction Detection

Image junction points (often referred to as *junctions corners*, or *interest* points) are formed as the result of the intersection of two or more straight image edges, indicating the intersection of *physical* features in the imaged scene (Ziou and Tabbone, 1998). Based on this definition, junctions are distinct points along image edges for which a significant change in the edge direction occurs (Mehrotra et al., 1990). Consequently, the basic requirements that were outlined for edge detection should also be applied to junction detection, with an additional requirement for enabling the estimation of the corner angle and orientation. Generally, three primary strategies can be applied for junction detection (Schmid et al., 2000):

a. *Contour-based methods* are based on the detection of junction points in contours that have been already extracted from the image by identifying the location of maxima in the curvature of the image curve. Given a two-dimensional curve $f(x,y)$, the curvature κ can be estimated as

$$\kappa(x,y) = \frac{f_x f_{yy} - f_y f_{xx}}{\left(f_x^2 + f_y^2\right)^{3/2}} \tag{5.35}$$

where f_{xx}, f_{yy} are the second derivatives along the x and y directions, respectively.

b. *Intensity-based methods* are based on the detection of junction points using image intensity values. Typically, this method is based on some measure of the change in the gradient direction, such as the well-known *cornerness* measure, $C(x,y)$, suggested by Kitchen and Rosenfeld (1982):

$$C(x,y) = \frac{f_{xx} f_y^2 - 2 f_{xy} f_x f_y + f_{yy} f_x^2}{f_x^2 + f_y^2} \tag{5.36}$$

where f_{xy} is the corresponding second derivative.

c. *Parametric methods* are based on fitting a parametric junction model to an image neighborhood. For example, the SUSAN operator (Smith and Brady, 1995) is based on an analysis of the immediate neighborhood of each pixel to analyze local gray value variations. Uniform neighborhoods are eliminated as potential locations and, conversely, neighborhoods characterized by substantial variations of gray values are merged to form longer edge regions.

In order to demonstrate some of the basic ideas behind junction detection, this section will explore an intensity-based method, the Harris and Stephens (1988) junction detection operator. The guiding principle of this operator is that by considering the behavior of the gradient in a small image neighborhood, three different cases may be identified: (a) constant intensity values that indicate an absence of edges; (b) strong intensity changes in one direction that indicate an image edge; and (c) strong intensity changes in two different directions that indicate a junction point. Given a small image neighborhood (u,v) and a corresponding two-dimensional weight function

w, our goal is, therefore, to estimate the change in the image intensity values as a result of a small shift Δx, Δy. This can be expressed by

$$E(\Delta x, \Delta y) = \sum_u \sum_v w(u,v)\left[f(u + \Delta x, v + \Delta y) - f(u,v)\right]^2$$

$$= \sum_u \sum_v w(u,v)\left[f_x(u,v)\Delta x + f_y(u,v)\Delta y\right]^2 \tag{5.37}$$

Since Δx and Δy are small, this equation may be rewritten as

$$E(\Delta x, \Delta y) \cong \begin{bmatrix} u & v \end{bmatrix} M \begin{bmatrix} u \\ v \end{bmatrix} \quad \text{where } M = \sum_u \sum_v w(u,v) \begin{bmatrix} f_x^2 & f_x f_y \\ f_x f_y & f_y^2 \end{bmatrix} \tag{5.38}$$

Note that this form of M can also be seen as a *local autocorrelation function* (see Schmid et al., 2000 for the full proof). To detect corners, Harris and Stephens suggested the following function for measuring the *response* of the detector:

$$R = \det(M) - k\left(tr(M)^2\right) \tag{5.39}$$

where $det(M) = \lambda_2$, $tr(M) = \lambda_1 + \lambda_2$, ($\lambda_1$, λ_2 are the eigenvalues of M) and k is an empirically derived constant (typically 0.035–0.065). It is important to note that the value of R depends only on the eigenvalues of M and that it is invariant to rotation. Consequently, given a user-defined threshold T, it is possible to detect corners and edges using Equation 5.39, as follows: (a) if $R > 0$ and $|R| > T$, a corner is detected; (b) if $R < 0$ and $|R| > T$, an edge is detected; and (c) $|R| < T$, an area of constant intensity. An example of the results of the Harris corner detector is shown in Figure 5.7. As can be seen from Figure 5.7a, areas that are characterized by small intensity variations result in an R value close to zero, whereas areas that are associated with potential corners appear as peaks. The detected corners after the application of a user-defined threshold are shown in Figure 5.7b.

5.4.2 Linear Feature Extraction

Although the detection of edge and junction pixels are of great importance in many applications, often the overarching goal for geoprocessing imagery is the extraction of linear (or curve-linear) features. Roads, man-made boundaries, shorelines, and buildings are but a few examples of spatial features that are often represented by linear features. Generally, the extraction of linear features can be carried out using either a *bottom-up* or a *top-down* strategy.

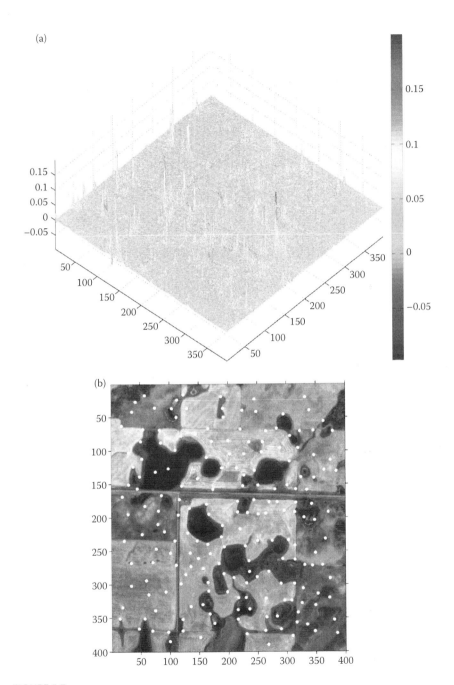

FIGURE 5.7

An example of the Harris corner detector. (a) the R values (Equation 5.40) for the image in Figure 5.3a; (b) the detected corners after applying a threshold of $T = 0.075$.

In the bottom-up strategy low-level image cues, such as edge pixels and junction points, are grouped together in support of the construction of a feature. In this process, both geometrical (e.g., orientation and proximity) and radiometric (e.g., average gray level and gradient magnitude) can be used. To demonstrate this approach, we review in this section the *Hough transform*, a voting-based algorithm for detecting linear features such as buildings (e.g., Croitoru and Doytsher, 2004). In the top-down strategy, an initial linear feature is created in the image space, and its final (optimal) configuration is determined using low-level image cues. This allows incorporating external knowledge into the feature extraction process. This approach is demonstrated here through *active contours* (snakes). Recent work (Agouris et al., 2001; Gyftakis et al., 2005) has shown that active contours can also be used for effective change detection through the incorporation of prior spatial vector data and imagery.

5.4.2.1 Hough Transform

Hough transform evolved from the Radon transform, which was introduced by Johann Radon in 1917. The basic idea behind the approach presented by Radon is the transformation, g, of a two-dimensional function $f(x, y)$ into a two-dimensional parametric space (a, b) by integrating the function values along a slanted line, where the line is defined by its slope (a) and offset (b) (Durrani and Bisset, 1984):

$$g(a,b) = \int_{-\infty}^{+\infty} f(x, ax + b) dx \qquad (5.40)$$

By representing the slant line through a delta function (δ) and using a polar form for the line, this transformation can be rewritten as

$$g(\rho,\theta) = \int_{-\infty}^{+\infty} f(x, y) \delta(\rho - x\cos\theta - y\sin\theta) dx\, dy \qquad (5.41)$$

where ρ and θ are the polar line parameters (range and angle, respectively). When the Radon transform is applied to a digital image, a discrete form of the transformation is used, preceded by summarizing pixels along a one pixel-wide strip (Toft, 1996).

By systematically applying the Radon transformation for all ρ and θ values, a *parameter-space representation* of $f(x, y)$ is constructed. Hough transform can be seen as a special case of the Radon transform in which an additional constraint is implemented on the integration (or summation in the discrete case) along each line: Instead of summing all pixels, in the Hough transform it is assumed that a binary image is provided, and the summation is applied only to *nonzero* pixels. As the input function $f(x, y)$ is binary and since the transformation is carried out through a summation in the discrete case,

Hough transform can be seen as a process of *accumulating evidence* ('votes') on the existence of a line for a given set of parameter values. The process of accumulating evidence is carried out in a discrete (ρ, θ) parameter space (Hough space) that represents all possible line parameters in image space. Let us now focus on how accumulation of evidence occurs in Hough space. Consider a discrete Hough space, a two-dimensional array in which each cell is $\Delta\rho$ by $\Delta\theta$. In addition, consider two nonzero pixels, p and q, with coordinates (x_p, y_p) and (x_q, y_q). For each of these pixels, it is possible to write an equation of the form:

$$\rho(\theta_i) = x\cos\theta_i + y\sin\theta_i \tag{5.42}$$

By applying all possible θ_i values in the range $[0,\pi]$ (with an increment of $\Delta\theta$) to this equation and computing the corresponding $\rho(\theta)$, a sinusoidal curve will be formed (note that for a given nonzero pixel, x and y are constants). The application of Equation 5.42 to p and q will, therefore, result in two sinusoidal curves, which will intersect at *one point* in Hough space—the point corresponding to the ρ and θ values of the line passing through *both* pixels. The application of the Hough transform begins with the initialization of an accumulator space with a parameter resolution of $\Delta\rho$ by $\Delta\theta$. Then, the transformation is applied to all nonzero pixels in the (edge) binary image, where for each pixel Equation 5.42 is applied, and for each θ_i the accumulator array closest to $[\theta_i, \rho(\theta_i)]$ is incremented by one. At the end of the process, local peaks are detected in the accumulator space, and the corresponding line parameters are retrieved.

An example of the application of the Hough transform is shown in Figure 5.8, where Figures 5.8a and b depict the accumulator resulting from the Hough transform (note the different sinusoidal curves in the accumulator space). Figure 5.8c shows the lines (in red) that were reconstructed from the two most dominant peaks in the accumulator.

A key advantage of the Hough transform is its robustness to noise and occlusions. However, the Hough transform may also suffer from several limitations. First, although the Hough transform can indicate whether a set of pixels are co-linear, it cannot indicate whether a set of edge pixels *should* be linked together. For example, consider the vertical line in Figure 5.8c, where pixels at the top of the image and pixels at the bottom of the image belong to the same line even though they belong to the boundaries of different fields. Second, since a discrete accumulator is used, the line parameters can only be resolved up to the precision of the accumulator. Although it is possible to assign smaller $\Delta\rho$ and $\Delta\theta$ values to increase the precision, this may in fact result in a degradation of the accumulation effect and the inability to detect peaks. Finally, it should be noted that as the magnitude of the peaks in the accumulator is directly related to the number of edge pixels that contribute to a given line parameter set, shorter lines in the image (e.g., closer to the boundary of the image) are less likely to be detected.

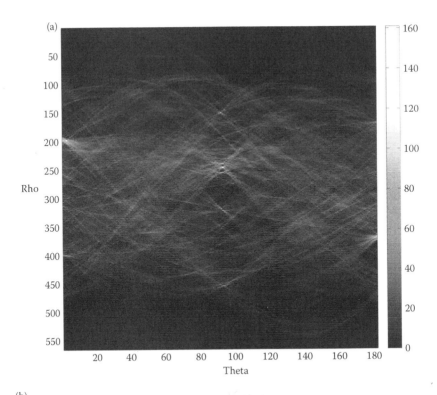

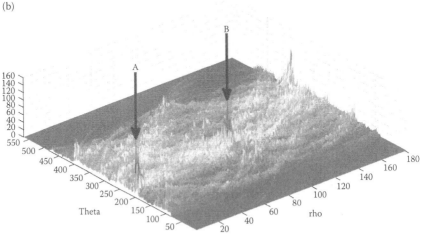

FIGURE 5.8

(See color insert following page 144.) An example of the implementation of the Hough transform applied to the binary edge image derived in Figure 5.3b. (a) The accumulator array (parameter space); (b) a three-dimensional visulization of the accumulator (A and B mark the first and second highest peaks); (c) a reconstruction of the two lines corresponding to the first and second hightest peaks in the accumulator space.

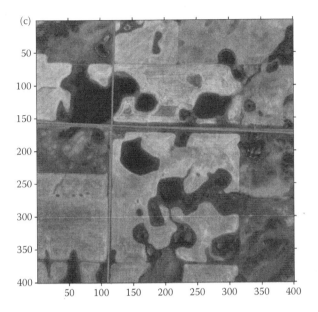

FIGURE 5.8
Continued.

5.4.2.2 Active Contours (Snakes)

Active contours ("snakes"*) are a class of curves capable of deforming (i.e., changing their location and shape, hence "active") based on a set of internal and external forces that are applied in image space. Given such forces, the active contour is initialized by creating an initial curve in image space, and then an iterative process of evolving the contour is carried out until the active contour reaches a configuration in image space that *minimizes* the applied forces. Generally, two types of active contours have been introduced in recent research (Xu and Price, 1997): *parametric active contours* (e.g., Kass et al., 1987), and *geometric active contours* (e.g., Malladi et al., 1995). In this section, we will focus on the classic parametric active contour model as introduced in the seminal work of Kass et al. (1987). In this model, an active contour is a planar curve $l(s) = (x(s), (y(s))$ in image space, where s is the normalized curve parameter ($s \in [0,1]$). For a given imbedding of l in image space, its energy may be expressed through the following function:

$$E_{snake} = \int_0^1 E_{int}\left(l(s)\right) + E_{ext}\left(l(s)\right) + E_{con}\left(l(s)\right) ds \tag{5.43}$$

---

* In this section, we will use the terms "active contour" and "snakes" interchangeably.

where E_{int} is the internal energy, E_{ext} is the external energy, and E_{con} is the constraints energy. When high-level external knowledge about the desired location of the active contour is available, such knowledge can be formulated as constraints and integrated in E_{con}. For simplicity, we will focus our discussion on the first two energy terms and assume that E_{con} is zero (the interested reader may find more details on this topic in Kass et al., 1987). The internal energy describes the tension and rigidity of l, which prevents the curve from tearing and controls the amount of bending, and is given by

$$E_{int} = \frac{1}{2}\left(\alpha \left| \frac{\partial l(s)}{\partial s} \right|^2 + \beta \left| \frac{\partial^2 l(s)}{\partial s^2} \right|^2 \right)$$ (5.44)

where α and β are user-defined parameters that balance between the tension and rigidity in the active contour. For example, setting β to zero will allow the active contour to "bend," that is, become second-order discontinuous, thus creating a corner. The external energy describes the transition from the feature of interest to its surroundings (e.g., the local edge magnitude). For a gray-level image, the external force can be estimated using one of these equations (Xu and Prince, 1997):

$$E_{ext}^a = - \left| \nabla f(x,y) \right|^2$$

$$E_{ext}^b = - \left| \nabla \left(G_\sigma(x,y) * f(x,y) \right) \right|^2$$ (5.45)

where G_σ is a two-dimensional Gaussian, ∇ is the gradient operator, and $*$ is the convolution operator. The objective of the evolution process of an active contour in image space is to *minimize* its energy, as given in Equation 5.43, that is, to move toward a location at which the internal and external forces balance out. To achieve this, the active contour must fulfill the Euler equation (Kass et al., 1987):

$$\alpha \frac{\partial^2 l}{\partial s^2} + \beta \frac{\partial^4 l}{\partial s^4} + \nabla E_{ext} = 0$$ (5.46)

This equation is iteratively solved through a time parameter t by gradually updating s so that the total energy is reduced between consecutive updates. A solution is achieved when the energy function cannot be further reduced.

The parameters in Equation 5.46 should, therefore, be treated as time-dependent. In discrete form, Equation 5.43 can be rewritten as

$$E_{snake} = \sum_{i=1}^{n} E_{int}(i) + E_{ext}(i) \tag{5.47}$$

where n is the number of points along the active contour. Let us now focus on point i in the active contour. Using a discrete approximation of the second and fourth derivatives, the energy at that point is, therefore, given by

$$\begin{cases} E_x(i,t) = \alpha \cdot \begin{bmatrix} 1 & -2 & 1 \end{bmatrix} \begin{bmatrix} x_{i-1} \\ x_i \\ x_{i+1} \end{bmatrix} + \beta \cdot \begin{bmatrix} -1 & 4 & -6 & 4 & -1 \end{bmatrix} \begin{bmatrix} x_{i-2} \\ x_{i-1} \\ x_i \\ x_{i+1} \\ x_{i+2} \end{bmatrix} + \gamma \cdot \left[E_{ext}(x_i, y_i) \right]_x \\[3em] E_y(i,t) = \alpha \cdot \begin{bmatrix} 1 & -2 & 1 \end{bmatrix} \begin{bmatrix} y_{i-1} \\ y_i \\ y_{i+1} \end{bmatrix} + \beta \cdot \begin{bmatrix} -1 & 4 & -6 & 4 & -1 \end{bmatrix} \begin{bmatrix} y_{i-2} \\ y_{i-1} \\ y_i \\ y_{i+1} \\ y_{i+2} \end{bmatrix} + \gamma \cdot \left[E_{ext}(x_i, y_i) \right]_y \end{cases} \tag{5.48}$$

where $[E_{ext}]_x$ and $[E_{ext}]_y$ are the x and y components of the external energy field, respectively, and α, β, and γ are constants. Based on this, the evolution equations for point i can be written as

$$\begin{cases} x_i(t+1) = x_i(t) + \mu E_x(i,t) \\ y_i(t+1) = y_i(t) + \mu E_y(i,t) \end{cases} \tag{5.49}$$

where μ is the time step size. The implementation of these evolution equations is carried out through an iterative process. The process begins with an initialization of the active contour points $p_i = (x_i, y_i)$, $(i = 1 \dots n)$ in image space by the application of *a priori* knowledge (i.e., predicted location of the contour) or by a manual digitization by a human operator. In addition, the maximum number of updates, N, and the minimum thresholds for the location updating, T_x and T_y, are set. Here, the external energy is calculated using one of the terms in Equation 5.45. The active contour evolution process then begins by incrementally updating the initial contour position using Equation 5.49. The evolution process terminates when either the maximum number of iterations is reached or when the updates for all the contour points are below the user-defined thresholds. An example of this process is shown in Figure 5.9, which depicts the process of extracting the boundary of a lake through the evolution of an active contour using the external force E^b. The smoothness of the final boundary in this example is partially due to the Gaussian smoothing in E^b before the active contour evolution.

In practice, the implementation of the active contour scheme described here requires some additional consideration to ensure satisfactory results. First, the determination of the energy constants (α, β, and γ) may require some empirical evaluation and tuning. Second, the evolution scheme described earlier may be sensitive to the distribution and density of points along the initial active contour. Finally, it has been noted that the classic active contour model, as introduced by Kass et al. (1987), does not perform well in the presence of concave shape elements. To alleviate some of these limitations, several improvements of the classic model have been suggested. For example, Xu and Prince (1997) have suggested using Gradient Vector Flow (GVF) as an external force to accommodate shape concavities. Cohen (1991) suggested the use of a "balloon" force to enable active contours to expand, thus preventing the contour from collapsing on itself and permitting it to reach edges that are not close to the initial curve position. Finally, Ravinda and Rajapakse (2003) introduced active contours based on nonuniform rational b-splines (NURBS), which provide more control over the shape of the curve without the introduction of additional points. In addition, active contours have been extensively used for object tracking (Blake and Isard, 1998).

5.4.3 Motion Image Analysis

Motion imagery analysis is supporting spatial data collection in rapid monitoring applications. The term *motion imagery* is typically used to refer to video (30 frames per second) or quasi-video rate (e.g., 1–2 frames per second) sequences of digital imagery. The type of information that is typically extracted from such datasets is trajectories of objects, information that provides critical support to a variety of applications. Sample applications may range from classic surveillance and human motion analysis using static cameras (Lim et al., 2003) to complex traffic monitoring (Kastrinaki et al., 2003) and object tracking using unmanned aerial vehicles (Dobrokhodov et al., 2006).

From an image analysis point of view, the major challenge related to motion image analysis is to track objects in sequences of images. This entails first detecting moving objects by comparing frames and then linking those objects across long frame sequences to establish long trajectories.

Assuming for example that a static camera is observing a scene in which one or more objects are moving, change can be detected by subtracting

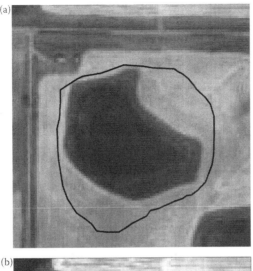

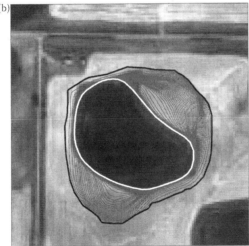

FIGURE 5.9
An example of active contour evolution. (a) the active contour initialization (manual digitization); (b) the evolution process (gray) and final contour position (white).

successive image frames $f_{t-1}(x, y)$ and $f_t(x, y)$ to generate a *difference image* $d_t(x, y)$:

$$d_t(x,y) = \left| f_t(x,y) - f_{t-1}(x,y) \right| \qquad (5.50)$$

In this difference image, pixels are marked as *moving* or *stationary*, according to a comparison of their values to a threshold value k. If $d_t(x, y) \geq k$, a pixel is marked as moving, whereas if $d_t(x, y) < k$, the pixel is marked as stationary.

Threshold selection is typically based on a statistical analysis of image content, to ensure that the radiometric difference between the compared instances is sufficiently large. Rosin (2002) and Rittscher et al. (2000) offer an overview of popular thresholding techniques, whereas Luthon et al. (2002) present an entropy-based approach. In addition to threshold selection, temporal analysis may also be used to minimize the effects of noise in video tracking, exploiting the high temporal redundancy of motion imagery. Accumulative frame differencing records difference information over a sequence of images. By eliminating objects with a brief temporal occurrence, we can eliminate noise and other illumination-induced artifacts.

Equation 5.50 indicates that successive frames are compared with each other in order to detect motion; however, comparison is typically taking place between an incoming frame and an image $f_{back}(x, y)$ of the background scene for the processed feed.

$$d_t(x,y) = \left| f_t(x,y) - f_{back}(x,y) \right| \tag{5.51}$$

The background scene image may be generated *a priori* (e.g., by capturing a frame before objects start moving in the scene) or by averaging the actual frames. If the number of objects moving in a scene is sufficiently low, a background may be generated by averaging (or median filtering) a sequence of frames taken with a static sensor (see Figure 5.10).

The individual pixels marked as moving pixels through a differential analysis can be clustered together to form complete moving objects using

FIGURE 5.10
Sample video frames (top) and the resulting background composite image (bottom). Notice the removal of the person walking on the snow in the background composite image.

FIGURE 5.11
An example of object tracking. A frame from the video feed is shown on the left, whereas a tracked vehicle extracted from this feed is shown on the right.

standard morphological processes like dilation and erosion or variations thereof (Figure 5.11).

As tracking proceeds from two to numerous frames within a feed, linking the same object across these frames becomes a challenging task. As the object moves over time in a sequence of frames, its image follows a path in the image coordinate space.

$$f_{t+1}(x + d_x, y + d_y) = f_t(x, y) + q(x, y) \Rightarrow$$

$$f_{t+1}(x + d_x, y + d_y) = f_t(x, y) + \frac{\partial f}{\partial x} d_x + \frac{\partial f}{\partial y} d_y + \frac{\partial f}{\partial t} d_t \tag{5.52}$$

This leads to the optical flow equation, describing this path:

$$\frac{\partial f}{\partial x} v_x + \frac{\partial f}{\partial y} v_y + \frac{\partial f}{\partial t} = 0 \tag{5.53}$$

Although background subtraction models tend to work well in situations where the camera is stationary, they are not suitable for situations where the camera itself may also be moving. To overcome this problem, kernel-based tracking approaches often proceed by representing tracked objects in a feature space (e.g., the color probability density function). The tracking problem is then reduced to match a moving target object to matching candidates, by comparing their corresponding feature space representations. In the popular mean-shift approaches (Comaniciu et al., 2000), the similarity between the target and candidate objects is estimated using the Bhattacharyya coefficient, and the mean-shift procedure handles the optimization process. Mean-shift techniques have proved to be computationally efficient and very robust in the presence of clutter, partial occlusions, camera motion, and target scale variations.

FIGURE 5.12
The trajectory of the moving person detected in the video feed of Figure 5.10, and overlaid on the background frame.

Contour tracking approaches offer an alternative to kernel-based solutions. They track objects by tracking their outline variations in a sequence of images. The snakes model presented above in this section is highly suitable for such applications, and it has been extended by modeling visual features (e.g., color, texture) and shape information to handle partial occlusions in video tracking applications (Yilmaz and Shah, 2004).

Using the techniques outlined above to detect and link moving objects over a complete sequence of frames allows us to generate long trajectories (Figure 5.12). These trajectories represent a novel type of Geoinformation, deviating from traditional measurements in terms of content and applications, as they describe information that extends beyond the traditional boundaries of the geoinformatics community. A trajectory, for example, not only describes the start and end point of the monitored individual but also allows the labeling of his or her activities through the identification of patterns of change in it, the application of reasoning techniques, and the prediction of future events. This brings forward both opportunities and challenges for our field. The challenges primarily relate to the fact that spatiotemporal trajectory information is highly redundant in both space and time, thus necessitating novel concise descriptions of these trajectories (Agouris and Stefanidis, 2003) and the development of novel reasoning approaches (Cohn et al., 2003; Gabelaia et al., 2005). Substantial challenges also relate to the potential large-scale configurations of data capturing, especially as sensors become mobile. For example, UAV-captured motion imagery is still processed visually, by trained operators, as the available automated solutions fail to capture information at adequate accuracy and rate. However, the opportunities far outweigh the challenges. For example, motion imagery analysis is a necessary step toward *smart spaces*, spaces that sense human activities and react to them (Gottfried et al., 2006; Bernandin and Stiefelhagen, 2007; Jakkula and Cook, 2007; Menon et al., 2008). Similarly, motion imagery analysis is critical for large-scale persistent monitoring and surveillance applications

(Valera and Velastin, 2005), especially in the context of geosensor networks (Nittel et al., 2008).

5.5 Summary

The focus of this chapter is on extracting information from spatial data of various types. We discuss how spatial autocorrelation can be evaluated and modeled, using both the spatial statistical-regression approach and geostatistical-Kriging approach. We also review some common techniques in mining and visualizing patterns buried in spatial data. Spatial data exist in many forms, and the supply of still and motion images is abundant. This chapter also provides a review of some techniques to extract information from both types of images. Apparently, the set of techniques and models covered in this chapter is limited. For related topics not covered in this chapter, we point readers to relevant literature.

Despite the advances in geocomputational environments, the development of cyberinfrastructure, and the migration of GIS from desktop computing platform to the Internet, basic principles and techniques in extracting information from spatial data have not changed substantially, although implementation of certain methods and techniques may have to be adjusted to the new environments and settings. Particularly, the distributed computing environment and massive amount of spatial data can pose new challenges in the implementations of spatial analytical techniques. These issues will be partly addressed by other chapters in this book volume.

References

Agouris, P., A.F. Schenk, and A. Stefanidis. 1989. Zero-crossings for edge detection. *Proceedings of the 1989 ASPRS-ACSM Fall Convention*, pp. 91–99. Baltimore, MD.

Agouris, P., A. Stefanidis, and S. Gyftakis. 2001. Differential snakes for change detection in road segments. *Photogrammetric Engineering & Remote Sensing*, 67(12):1391–1399.

Agouris, P. and A. Stefanidis. 2003. Efficient summarization of spatiotemporal events. *Communications of the ACM*, 46(1):65–66.

Agrawal, R., T. Imielinski, and A. Swami. 1993. Mining association rules between sets of items in large databases. In *ACM SIGMOD International Conference on Management of Data*, pp. 207–216.

Andrews, D.F. 1972. Plots of high-dimensional data. *Biometrics* 29:125–136.

Andrienko, G. and N. Andrienko. 1999. Data mining with C4.5 and interactive cartographic visualization. In *User Interfaces to Data Intensive Systems*, eds. N.W. Paton and T. Griffiths, pp. 162–165. IEEE Computer Society, Los Alamitos, CA.

Anselin, L. 1995. Local indicators of spatial association—LISA. *Geographical Analysis*, 27:93–115.

Anselin, L. 1999. Interactive techniques and exploratory spatial data analysis. In *Geographical Information Systems—Principles and Technical Issues*, eds. P.A. Longley, M.F. Goodchild, D.J. Maguire, and D.W. Rhind, pp. 253–266. John Wiley & Sons, Inc., New York, NY.

Anselin, L. and D.A. Griffith. 1988. Do spatial effects really matter in regression? *Papers in Regional Science* 65(1):11–34.

Appice, A., M. Ceci, A. Lanza, F.A. Lisi, and D. Malerba. 2003. Discovery of spatial association rules in geo-referenced census data: A relational mining approach. *Intelligent Data Analysis*, 7:541–566.

Augustin, N.H., M.A. Mugglestone, and S.T. Buckland. 1996. An autologistic model for the spatial distribution of wildlife. *Journal of Applied Ecology*, 33:339–347.

Bailey, T.C. and A.C. Gatrell. 1995. *Interactive Spatial Data Analysis*. John Wiley & Sons, Inc., New York, NY.

Baller, R.D., L. Anselin, S.F. Messner, G. Deane, and D.F. Hawkins. 2001. Structural covariates of U.S. county homicide rates: Incorporating spatial effects. *Criminology*, 39:561–590.

Band, L.E. 1986. Topographic partition of watersheds with digital elevation models. *Water Resource Research*, 22:15–24.

Benenson, I. and P. Torrens. 2004. *Geosimulation:Automata-Based Modeling of Urban Phenomena*. John Wiley & Sons, West Sussex, England.

Bernardin, K. and R. Stiefelhagen. 2007. Audio-visual multiperson tracking and identification for smart environments. *Proceedingsof the 15th International Conference on Multimedia*, pp. 661–670, Augsburg, Germany.

Besag, J. 1974. Spatial interaction and the statistical analysis of lattice systems (with Discussion). *Journal of the Royal Statistical Society, Series B*, 36:192–236.

Besag, J. 1975. Statistical analysis of non-lattice data. *The Statistician*, 24:179–195.

Besag J., J. York, and A. Mollie. 1991. Bayesian image restoration, with two applications in spatial statistics. *Annals of the Institute of Statistical Mathematics*, 1:1–59.

Black, W.R. 2003. *Transportation: A Geographical Analysis*. The Guilford Press, New York.

Blake, A. and M. Isard. 1998. *Active Contours*. London, Springer-Verlag.

Bouchara, F. and S. Ramdani. 2007. Statistical behavior of edge detectors. *Signal, Image and Video Processing*, 1(3):273–285.

Canny, J., 1986. A computational approach to edge detection. *IEEE Transactions on Pattern Analysis and Machine Intelligence*, 8(6):679–698.

Carr, D.B., D. White, and A.M. MacEachren. 2005. Conditioned choropleth maps and hypothesis generation. *Annals of the Association of American Geographers*, 95:32–53.

Chawla, S., S. Shekhar, W. Wu, and U. Ozesmi. 2000. Extending data mining for spatial applications: A case study in predicting nest locations. In *ACM SIGMOD Workshop on Research Issues in Data Mining and Knowledge Discovery (DMKD 2000)*. Dallas, TX.

Chernoff, H. and M.H. Rizvi. 1975. Effect on classification error of random permutations of features in representing multivariate data by faces. *Journal of American Statistical Association*, 70:548–554.

Clarke, K.C., S. Hoppen, and L.J. Gaydos. 1997. A self-modifying cellular automation model of historical urbanization in the San Francisco Bay area. *Environment and Planning B*, 24:247–261.

Cliff, A.D. and J.K. Ord. 1973. *Spatial Autocorrelation*. Pion, London.

Cohen, L.D. 1991. On active contour models and balloons. *Computer Vision, Graphics, and Image Processing. Image Understanding*, 53(2):211–218.

Cohn, A., D. Magee, A. Galata, D. Hogg, and S. Hazarika, 2003. Towards an architecture for cognitive vision using qualitative spatiotemporal representations and abduction. *Spatial Cognition III*, LNCS, 2685:1034–1045.

Colasanti, R.L. and J.P. Grime. 1993. Resource dynamics and vegetation processes: A deterministic model using two-dimensional cellular automata. *Functional Ecology*, 7:169–176.

Comaniciu, D., V. Ramesh, and P. Meer. 2000. Real-time tracking of non-rigid objects using mean shift. *Proceedings IEEE CVPR'00*, pp. 142–149.

Cook, D., A. Buja, J. Cabrera, and C. Hurley. 1995. Grand tour and projection pursuit. *Journal of Computational and Graphical Statistics*, 4:155–172.

Croitoru, A. and Y. Doytsher. 2004. Right-angle rooftop polygon extraction in regularized urban areas: Cutting the corners. *The Photogrammetric Record*, 19(108): 311–341.

DiBiase, D., C. Reeves, J. Krygier, A.M. MacEachren, M.V. Weiss, J. Sloan, and M. Detweiller. 1994. Multivariate display of geographic data: Applications in earth system science. In *Visualization in Modern Cartography*, eds. A.M. MacEachren and D.R.F. Taylor, pp. 287–312. Pergamon, Oxford, UK.

Dobrokhodov, V., I. Kaminer, K. Jones, and R. Ghabcheloo. 2006. Vision-based tracking and motion estimation for moving targets using small UAVs. *Proceedings of the IEEE American Control Conference*, Minneapolis, MN, pp. 1428–1433.

Durrani, T.S. and D. Bisset. 1984. The radon transform and its properties. *Geophysics*, 49(8):1180–1187.

Dykes, J. 1998. Cartographic visualization: Exploratory spatial data analysis with local indicators of spatial association using Tcl/Tk and cdv'. *The Statistician*, 47:485–497.

Dykes, J., A.M. MacEachren, and M.-J. Kraak. 2005. *Exploring Geovisualization*. Elsevier, Amsterdam.

Ester, M., H.P. Kriegel, and J. Sander. 1997. Spatial data mining: A database approach. In *Advances in Spatial Databases*, Vol. 33, pp. 47–66. Springer-Verlag, Berlin.

Fayyad, U., G. Piatetsky-Shapiro, and P. Smyth. 1996. From data mining to knowledge discovery—an review. In *Advances in Knowledge Discovery*, eds. U. Fayyad, G. Piatetsky-Shapiro, P. Smyth, and R. Uthurusay, pp. 1–33. AAAI Press/The MIT Press, Cambridge, MA.

Fischer, M.M. 1998. Computational neural networks: A new paradigm for spatial analysis. *Environment and Planning A*, 30:1873–1891.

Fischer, M.M., M. Reismann, and K. Hlavackova-Schindler. 2003. Neural network modeling of constrained spatial interaction flows: Design, estimation, and performance issues. *Journal of Regional Science*, 43:35–61.

Fotheringham, A.S., C. Brunsdon, and M. Charlton. 2000. *Quantitative Geography: Perspectives on Spatial Data Analysis*. Sage, London.

Fotheringham, A.S. and M.E. O'Kelly. 1989. *Spatial Interaction Models: Formulations and Applications*. Kluwer Academic Publishers, Dordrecht.

Fovell, R.G. and M.-Y.C. Fovell. 1993. Climate zones of the conterminous United States defined using cluster analysis. *Journal of Climate*, 6:2103–2135.

Gabelaia, D., R. Kontchakov, A. Kurucz, F. Wolter, and M. Zakharyaschev, 2005. Combining spatial and temporal logics: Expressiveness vs. complexity. *Journal of Artificial Intelligence Research*, 23:167–243.

Gahegan, M. 1998. Scatterplots and scenes: Visualization techniques for exploratory spatial analysis. *Computers, Environment and Urban Systems*, 22:43–56.

Gahegan, M. 2003. Is inductive machine learning just another wild goose (or might it lay the golden egg)? *International Journal of Geographical Information Science*, 17:69–92.

Geary, R. 1954. The contiguity ratio and statistical mapping. *The Incorporated Statistician*, 5:115–145.

Getis, A. and D.A. Griffith. 2002. Comparative spatial filtering in regression analysis. *Geographical Analysis*, 34:130–140.

Getis, A. and J.K. Ord. 1992. The analysis of spatial association by use of distance statistics. *Geographical Analysis*, 24:189–206.

Gopal, S., W. Liu, and C. Woodcock. 2001. Visualization based on the fuzzy ARTMAP neural network for mining remotely sensed data. In *Geographic Data Mining and Knowledge Discovery*, eds. H.J. Miller and J. Han, pp. 315–336. Taylor & Francis, London and New York.

Gordon, A.D. 1996. Hierarchical classification. In *Clustering and Classification*, eds. P. Arabie, L. J. Hubert, and G. D. Soete, pp. 65–122. World Scientific Publisher, River Edge, NJ.

Gottfried, B., H. Guesgen, and S. Huebner, 2006. Spatiotemporal reasoning for Smart Homes. *Designing Smart Homes*, LNCS, 4008:16–34.

Gould, P.R. 1970. Is Statistic Inference the geographical name for a wild goose? *Economic Geography*, 46:439–448.

Griffith. D.A. 2003. *Spatial Autocorrelation and Spatial Filtering*. Springer, Berlin.

Griffith, D.A. and C.G. Amrhein. 1997. *Multivariate Statistical Analysis for Geographers*. Prentice–Hall, Upper Saddle River, NJ.

Griffith D.A. and R. Haining. 2005. Beyond mule kicks: The Poisson distribution in geographical analysis. *Geographical Analysis*, 38:123–139.

Grinstein, G., J.C.J. Sieg, S. Smith, and M.G. Williams. 1992. Visualization for knowledge discovery. *International Journal of Intelligent Systems*, 7:637–648.

Guo, D. 2008. Regionalization with dynamically constrained agglomerative clustering and partitioning (REDCAP). *International Journal of Geographical Information Science*, 22:801–823.

Guo, D. 2009. Multivariate spatial clustering and geovisualization. In *Geographic Data Mining and Knowledge Discovery*, eds. H.J. Miller and J. Han, pp. 325–345. Taylor & Francis, London and New York.

Guo, D., J. Chen, A.M. MacEachren, and K. Liao. 2006. A visualization system for space–time and multivariate patterns (VIS-STAMP). *IEEE Transactions on Visualization and Computer Graphics*, 12:1461–1474.

Guo, D., M. Gahegan, A.M. MacEachren, and B. Zhou. 2005. Multivariate analysis and geovisualization with an integrated geographic knowledge discovery approach. *Cartography and Geographic Information Science*, 32:113–132.

Guo, D., D. Peuquet, and M. Gahegan. 2003. ICEAGE: Interactive clustering and exploration of large and high-dimensional geodata. *GeoInformatica*, 7:229–253.

Gurnell, A.M. and D.R. Montgomery. (eds.) 2000. *Hydrological Applications of GIS*. John Wiley & Sons, West Sussex, England.

Gyftakis, S., P. Agouris, and A. Stefanidis, 2005. Image-based change detection of aerial objects using differential snakes. *Proceedings of ACM-GIS'05*, pp. 135–142. Bremen, Germany.

Haggett, P., A.D. Cliff, and A. Frey. 1977. *Locational Methods*. John Wiley & Sons, New York.

Haining, R.P., S.M. Wise, and M. Blake. 1994. Constructing regions for small area analysis: Material deprivation and colorectal cancer. *Journal of Public Health Medicine*, 16:429–438.

Hamilton, L.C. 1992. *Regression with Graphics*. Duxbury, Belmont, CA.

Han, J. and M. Kamber. 2001. *Data Mining: Concepts and Techniques*. Morgan Kaufmann Publishers, San Francisco, CA.

Han, J., K. Koperski, and N. Stefanovic. 1997. GeoMiner: A system prototype for spatial data mining. In *ACM SIGMOD International Conference on Management of Data*, pp. 553–556. Tucson, AZ.

Hanna, S.R., M.J. Brown, F. Camelli, S.T. Chan, W.J. Coirier, O.R. Hansen, A.H. Huber, S. Kim, and R.M. Reynolds. 2006. Detailed simulations of atmospheric flow and dispersion in downtown Manhattan: An Application of five computational fluid dynamics models. *Bulletin of the American Meteorological Society*, 87:1713–1726.

Harris, C. and M. Stephens, 1988. A combined corner and edge detector. *Proceedings of The Fourth Alvey Vision Conference*, pp 147–151. Manchester.

Harris, R.L. 1999. *Information Graphics: A Comprehensive Illustrated Reference*. Oxford Press, Oxford, UK.

Huang, Y., J. Pei, and H. Xiong. 2006. Mining co-location patterns with rare events from spatial data sets. *Geoinformatica* 10:239–260.

Inselberg, A. 1985. The plane with parallel coordinates. *The Visual Computer* 1:69–97.

Isaaks, E.H. and R.M. Srivastava. 1990. *An Introduction to Applied Geostatistics*. Oxford University Press, New York.

Jain, A.K. and R.C. Dubes. 1988. *Algorithms for Clustering Data*. Prentice-Hall, Englewood Cliffs, NJ.

Jain, A.K., M.N. Murty, and P.J. Flynn. 1999. Data clustering: A review. *ACM Computing Surveys (CSUR)*, 31:264–323.

Jakkula, V. and D. Cook, 2007. Using temporal relations in smart environment data for activity prediction, *Proceedings of the 24th International Conference on Machine Learning*, Corvallis, OR.

Jin, H. and D. Guo. 2009. Understanding climate change patterns with multivariate geovisualization. In *Proceedings—IEEE International Conference on Data Mining Workshops*, pp. 217–222. IEEE Press, Miami, FL.

Kasahara, A. and W.M. Washington. 1967. NCAR global general circulation model of the atmosphere. *Monthly Weather Review*, 95(7):389–402.

Kaski, S., J. Kangas, and T. Kohonen. 1998. Bibliography of self-organizing map (SOM) papers: 1981–1997. *Neural Computing Surveys*, 1:102–350.

Kass, M., A. Witkin, and D. Terzopoulos, 1988. Snakes: Active contour models. *International Journal of Computer Vision*, 1(4):321–331.

Kastrinaki, V., M. Zervakis, and K. Kalaitzakis, 2003. A survey of video processing techniques for traffic applications. *Image and Vision Computing*, 21:359–381.

Keim, D.A., C. Panse, M. Sips, and S.C. North. 2004. Visual data mining in large geospatial point sets. *IEEE Computer Graphics and Applications*, 24:36–44.

Kirtman, B.P. and J. Shukla. 2002. Interactive coupled ensemble: A new coupling strategy for CGCMs. *Geophysical Research Letters*, 29(10):5.1–5.4

Knorr, E.M. and R.T. Ng. 1996. Finding aggregate proximity relationships and commonalities in spatial data mining. *IEEE Transactions Knowledge and Data Engineering*, 8:884–897.

Kohonen, T. 2001. *Self-Organizing Maps.* Springer, Berlin, New York.

Koperski, K. and J. Han. 1995. Discovery of spatial association rules in geographic information databases. In *The 4th International Symposium on Large Spatial Databases (SSD95)*, pp. 47–66. Maine, USA.

Koperski, K., J. Han, and N. Stefanovic. 1998. An efficient two-step method for classification of spatial data. In *1998 International Symposium on Spatial Data Handling SDH'98*, pp. 45–54. Canada, Vancouver, BC.

Kulldorff, M. 1997. A spatial scan statistic. *Communications in Statistics—Theory and Methods,* 26:1481–1496.

Kulldorff, M., R. Heffernan, J. Hartman, R.M. Assunção, and F. Mostashari. 2005. A space–time permutation scan statistic for the early detection of disease outbreaks. *PLoS Medicine,* 2:216–224.

Lillesand, T., R. Kiefer, and J. Chipman. 2007. *Remote Sensing and Image Interpretation.* Wiley, New York.

Lim, S.N., L. Davis, and A. Elgammal, 2003. A scalable image-based multi-camera visual surveillance system. *IEEE Conference on Advanced Video & Signal Based Surveillance (AVSS)*, pp. 205–212. Miami.

Liu, Y. 2008. *Modelling Urban Development with Geographical Information Systems and Cellular Automata.* CRC Press, Boca Raton, FL.

Lu, G.Y. and D.W. Wong. 2008. An adaptive inverse-distance weighting spatial interpolation technique. *Computers and Geosciences,* 34(9):1044–1055.

Luthon, F., M. Lievin, and F. Faux, 2004. On the use of entropy power for threshold selection, *Signal Processing,* 84(10):1789–1804.

MacEachren, A. 1994. Visualization in modern cartography: Setting the agenda. In *Visualization in Modern Cartography,* eds. A.M. MacEachren and D.R.F. Taylor, pp. 1–12. Pergamon, Oxford, UK.

MacEachren, A.M. and M.J. Kraak. 1997. Exploratory cartographic visualization: Advancing the agenda. *Computers and Geosciences,* 23:335–343.

MacEachren, A. and M.J. Kraak. 2001. Research challenges in geovisualization. *Cartography and Geographic Information Science,* 28:3–12.

MacEachren, A.M., M. Wachowicz, R. Edsall, D. Haug, and R. Masters. 1999. Constructing knowledge from multivariate spatiotemporal data: Integrating geographical visualization with knowledge discovery in database methods. *International Journal of Geographical Information Science,* 13:311–334.

Malladi, R., J.A. Sethian, and B.C. Vemuri, 1995. Shape modeling with front propagation: A level set approach. *IEEE Transactions on Pattern Analysis and Machine Intelligence* (PAMI), 17(2):158–175.

Mark, D.M. 1983. Automatic detection of drainage networks from digital elevation models. *Cartographica,* 21:168–178.

Meegama, R.G.N. and J.C. Rajapakse, 2003. NURBS Snakes. *Image and Vision Computing,* 21(2003):551–562.

Mehrotra, R., S. Nichani, and N. Ranganathan, 1990. Corner detection. *Pattern Recognition,* 23(11):1223–1233.

Mennis, J. and J.W. Liu. 2005. Mining association rules in spatio-temporal data: An analysis of urban socioeconomic and land cover change. *Transactions in GIS,* 9:5–17.

Menon, V., B. Jayaraman, and V. Govindaraju, 2008. Integrating recognition and reasoning in smart environments. *IET International Conference on Intelligent Environments,* pp. 1–8. Seattle, WA.

Miller, H. and J. Han. 2009. Geographic data mining and knowledge discovery: An overview. In *Geographic Data Mining and Knowledge Discovery*, eds. H. Miller and J. Han, pp. 1–26. CRC Press, Taylor & Francis Group, Boca Raton, FL.

Miller, H.J. and J. Han. 2001. Geographic Data Mining and Knowledge Discovery: An overview. In *Geographic Data Mining and Knowledge Discovery*, eds. H.J. Miller and J. Han, pp. 3–32. Taylor & Francis, London and New York.

Miller, H.J. and S.L. Shaw. 2001. *Geographic Information Systems for Transportation: Principles and Applications*. Oxford University Press, New York.

Mitchell, T.M. 1997. *Machine Learning*. McGraw-Hill, New York.

Mollie A.1996. Bayesian mapping of disease. In *Markov Chain Monte Carlo in Practice*, eds. W.R. Gilks, S. Richardson, and D.J. Spiegelhalter, pp. 359–379. Chapman & Hall, New York.

Monmonier, M. 1989. Geographic brushing: Enhancing exploratory analysis of the scatterplot matrix. *Geographical Analysis*, 21:81–84.

Mu, L. and F. Wang. 2008. A scale-space clustering method: Mitigating the effect of scale in the analysis of zone-based data. *Annals of the Association of American Geographers*, 98:85–101.

Nittel, S., A. Labrinidis, and A. Stefanidis, 2008. *Advances in GeoSensor Networks*, Lecture Notes in Computer Science, Vol. 4540, Springer, Berlin.

Openshaw, S., M. Charlton, C. Wymer, and A. Craft. 1987. A mark 1 geographical analysis machine for the automated analysis of point data sets. *International Journal of Geographical Information Science*, 1:335–358.

Openshaw, S., A. Cross, and M. Charlton. 1990. Building a prototype geographical correlates exploration machine. *International Journal of Geographical Information Science*, 4:297–311.

Osnes, K. 1999. Iterative random aggregation of small units using regional measures of spatial autocorrelation for cluster localization. *Statistics in Medicine*, 18:707–725.

Parker, D.C., S.M. Mason, M.A. Janssen, M.J. Hoffmann, and P. Deadman. 2003. Multi-agent systems for the simulation of land-use and land-cover change: A review. *Annals of the Association of American Geographers*, 93:314–337.

Pickett, R.M., G. Grinstein, H. Levkowitz and S. Smith. 1995. Harnessing preattentive perceptual processes in visualization. In *Perceptual Issues in Visualization*, eds. G. Grinstein and H. Levkowitz, pp. 33–45. Springer, New York.

Quinlan, J.R. 1993. *C4.5: Programs for Machine Learning*. Morgan Kaufmann, San Mateo, CA.

Rittscher, J., J. Kato, S. Joga, and A. Blake, 2000. A probabilistic background model for tracking, *European Conference on Computer Vision (ECCV'00)*, Lecture Notes in Computer Science, Vol. 1843, pp. 336–350, Springer-Verlag, Berlin.

Robinson, G.M. 1998. *Methods and Techniques in Human Geography*. John Wiley and Sons, West Sussex, UK.

Rogerson, P. and I. Yamada. 2009. *Statistical Detection and Surveillance of Geographic Clusters*. Taylor & Francis Group, London and New York.

Rosin, P., 2002. Thresholding for change detection, *Computer Vision and Image Understanding*, 86(2):79–95.

Schmid, C., R. Mohr, and C. Buckhage, 2000. Evaluation of interest point detectors. *International Journal of Computer Vision*, 37(2):151–172.

Skamarock, W.C., J.B. Klemp, and J. Dudhia, 2001. Prototypes for the WRF (Weather Research and Forecasting) model. Preprints, *Ninth Conference on Mesoscale Processes*, pp. J11–J15. Ft. American Meteorological Society, Ft. Lauderdale, FL.

Shekhar, S. and Y. Huang. 2001. Discovering spatial co-location patterns: A summary of results. In *Advances In Spatial And Temporal Databases, Proceedings, Lecture Notes in Computer Science*, eds. C. Jensen, M. Schneider, B. Seeger, and V. Tsotras, pp. 236–256. Springer-Verlag, Berlin.

Shekhar, S., P. Zhang, Y. Huang, and R. Vatsavai. 2004. Trends in spatial data mining. In *Data Mining: Next Generation Challenges and Future Directions*, eds. H. Kargupta, A. Joshi, K. Sivakumar, and Y. Yesha, pp. 357–381. AAAI/MIT Press, Cambridge, MA.

Shen, Q. 1994. An application of GIS to the measurement of spatial autocorrelation. *Computer, Environment and Urban Systems*, 18:167–191.

Skupin, A. and R. Hagelman. 2003. Attribute space visualization of demographic change. In *Proceedings of the Eleventh ACM International Symposium on Advances in Geographic Information Systems*, pp. 56–62. ACM Press, New Orleans, LA.

Smith, S.M. and J. Brady, 1995. SUSAN—A new approach to low level image processing. *International Journal of Computer Vision*, 23(1):45–78.

Taylor, D.R.F., 2005. *Cybercartography: Theory and Practice*. Elsevier Science, Amsterdam.

Thomas, J.J. and K.A. Cook. 2005. *Illuminating the Path: The Research and Development Agenda for Visual Analytics*. IEEE Computer Society, Los Alametos, CA.

Tobler, W.R. 1970. A computer movie simulating urban growth in the Detroit region. *Economic Geography*, 46:234–240.

Toft, P., 1996. *The radon transform—Theory and implementation*. PhD thesis, Department of Mathematical Modeling, Section for Digital Signal Processing, University of Denmark.

Tomlin, D. 1990. *Geographic Information Systems and Cartographic Modeling*. Prentice-Hall, Englewood Cliffs, NJ.

Tukey, J. 1977. *Exploratory Data Analysis*. Addison-Wesley, Reading, MA.

Valera, M. and S. Velastin, 2005. Intelligent distributed surveillance systems: A review. *IEEE Vision, Image and Signal Processing*, 152(2):192–204.

Wang, F. 2006. *Quantitative Methods and Applications in GIS*. CRC Press/Taylor & Francis, Boca Raton, FL.

Wang, F., L. Luo, and S. McLafferty. 2010. Healthcare access, socioeconomic factors and late-stage cancer diagnosis: An exploratory spatial analysis and public policy implication. *International Journal of Public Policy*, 5:237–258.

Ward, J.H. 1963. Hierarchical grouping to optimise an objective function. *Journal of the American Statistic Association*, 58:236–244.

Ward, M.O. 2004. Finding needles in large-scale multivariate data haystacks. *Computer Graphics and Applications*, 24:16–19.

Wise, S.M., R.P. Haining, and J. Ma. 1997. Regionalization tools for the exploratory spatial analysis of health data. In *Recent Developments in Spatial Analysis: Spatial Statistics, Behavioural Modelling and Neuro-Computing*, eds. M. Fischer and A. Getis. Springer-Verlag, Berlin.

Wittenbrink, C.M., E. Saxon, J.J. Furman, A. Pang, and S. Lodha. 1995. Glyphs for visualizing uncertainty in environmental vector fields. *IEEE Transactions on Visualization and Computer Graphics*, 2:266–279.

Wong, D. 2009. The Modifiable Areal Unit Problem (MAUP). In *The SAGE Handbook of Spatial Analysis*, eds. A.S. Fotheringham and P.A. Rogerson, pp. 95–112. Sage, London.

Wong, D.W.S. and J. Lee. 2005. *Statistical Analysis and Modeling of Geographic Information*. John Wiley & Sons, Hoboken, NJ.

Wu, H. and F.W. Huffer. 1997. Modelling the distribution of plant species using the autologistic regression model. *Environmental and Ecological Statistics*, 4:49–64.

Xu, C. and J.L. Prince, 1997. Gradient Vector flow: A new external force for snakes. *Proceedings of IEEE Conference on Computer Vision and Pattern Recognition*, pp. 66–71. San Juan, Puerto Rico.

Yilmaz, A. and M. Shah, 2004. Contour-based object tracking with occlusion handling in video acquired using mobile cameras, *IEEE Transactions on Pattern Analysis and Machine Intelligence*, 26(11):1531–1536.

Zhang, X. and M. Pazner. 2004. The icon image map technique for multivariate geospatial data visualization: Approach and software system. *Cartography and Geographic Information Science*, 31:29–41.

Ziou, D. and S. Tabbone, 1998. Edge detection techniques—An overview. *International Journal of Pattern Recognition and Image Analysis*, 1998(8):537–559.

6

Geoinformation Infrastructure (GII)*

Xuan Shi, Doug Nebert, Chuanrong Zhang, Haihong Yang, Huayi Wu,
Peisheng Zhao, Zhenlong Li, Liping Di, Qunying Huang, Jing Li,
Wenwen Li, Min Sun, and Genong Yu

CONTENTS

* From Webopedia. http://www.webopedia.com/DidYouKnow/Hardware_Software/2006/
thin_client_applications. asp (accessed December 22, 2009). With permission.

Geoinformation is collected, archived, and utilized in a geographically dispersed fashion. Heterogeneous problems are prevalent in different data, information, processing, algorithm, models, and applications. Geoinformation, however, once collected can be shared in many different applications and at different geographic scopes. An information infrastructure, therefore, is needed to provide interoperating platforms for discovery and access and to share the Geoinformation collected. This chapter provides an overview of the origin and different aspects of information infrastructure needed for Geoinformation sharing, and the U.S. NSDI is used as an example.

6.1 Need and Emergence of GII*

The concept of a "National Information Infrastructure" (NII) was initiated in the mid-1990s by U.S. Vice President Al Gore (1993). The vision of NII includes not only physical facilities and equipment but also information, computer hardware and software, network transmission mechanisms and standards, and the people. In 2003, when the National Science Foundation defined the term "Cyberinfrastructure" (CI), CI was defined as consisting of "... hardware, software, personnel, services and organizations" (Atkins 2003). The emergence of information infrastructure (II) is a natural response to the needs of a human society; reducing the duplication of efforts and resources is a key promotional strength.

In the Geoinformation domain, Executive Order 12906 (1994) defines the NSDI as "the technology, policies, standards, and human resources necessary to acquire, process, store, distribute, and improve utilization of spatial data." The FGDC is designated as the coordinating entity responsible for developing and implementing national strategies to advance the goals of the NSDI and provides a contemporary interpretation and vision of its strategic plan for the NSDI in 2010 as follows:

> Federal, state, and local government agencies, the private sector, academia, and others engaged in an unprecedented national project to design, build, and maintain a highly accurate, distributed, and consistent spatial framework during the first decade of the twenty-first century. This revolutionary partnership continues to leverage resources, technologies, and investments to create the spatial framework that supports the national information infrastructure and enables informed decision-making at all levels of society (FGDC 2004).

The NSDI is a critical component in the NII and CI. This section will introduce the changing Geoinformation technology as an evolutionary information infrastructure and the main challenges in sharing and integrating Geoinformation data and computational resources within the information infrastructure.

6.1.1 Changing Geoinformation Technology and Infrastructure

In the 1960s and 1970s, GIS originated in the electronic data processing pursued by cartographers and mapping groups that primarily operated in separated and isolated computing environments. Such early systems used mainframe and minicomputers and workstations running in batch mode or using DOS commands. Spatial analysis was limited, and no explicit graphical

---

* This section is contributed by Xuan Shi.

facilities were available to provide a user-friendly interface (Coppock et al. 1991). Researchers and organizations had limited computer communication, and considerable efforts were expended in getting spatial data into machine-readable format.

The GIS made substantial progress in the 1980s with advances in computer technology and the theory of spatial data processing and analysis. In particular, the theoretical complexity of spatial data structures was refined and this was fundamental to the varied methods of data collection, generation, quality, standards, analysis, and database organization (Batty 1999; Lo et al. 2002). The geometric operations and algorithms performed on the spatial features of points, lines, and polygons were the core of spatial data analysis. To this end, GIS enabled both computer cartography and spatial data analysis to be integrated within one computational framework. Advances in computer hardware also enabled better graphics to be achieved, aided by improved graphical user interfaces and improved image processing (Batty 1999). Further, microcomputers were becoming more widely available throughout the late 1970s and 1980s, and this enabled desktop GIS to be available to a large number of users. Today, a desktop computer is a typical setup for GIS professionals and includes a computer monitor, user-friendly interface, GIS software, and data installed within the microcomputer.

During the 1990s, the power of personal computers was enhanced radically; advances in computer hardware have increasingly blurred the distinction between personal computers and multi-user platforms, such as mainframes and workstations. The GIS technology also matured as significant advances in computer technology were made. The GIS as a stand-alone application could now be implemented on a variety of platforms and operational systems. Both computer graphics and graphical user interfaces were greatly improved, and spatial database management systems were more robust and stable. New GIS tools and functions were developed to manipulate increasingly available digital spatial data. The price of computer hardware continued to fall, and this trend provided even better personal and institutional access to GIS technology. For a number of reasons, however, it is argued that historically GIS has splintered into "islands" as specialized and independent portions of mainstream information technology (Bishr 1998), and proprietary GIS has contributed to the isolation of the GIS technology (Alameh 2001).

The initiation of Internet and Web technology in the 1990s also changed the landscape of GIS development. The Internet, as a computer network, allows for the interaction and communication among distributed computer systems, and transfers the traditional stand-alone GIS toward a network of systems. Historically, GIS data was generated by different proprietary software systems with monolithic data formats. The availability of different GIS data formats generated the incompatibility issue, such that even if spatial data could have been exchanged and delivered over the Internet, such data was problematic to use in different GIS systems. The GIS data interoperability, and especially semantic data interoperability, became a significant theme in GIScience research (Sondheim et al. 1999).

Internet and Web technology allowed users to access information and resources through a Web browser starting in the 1990s. How to enable Web users to derive geographic information and knowledge from spatial data posed a major research challenge to GIS professionals. Early Web mapping has largely focused on the exchange of static maps. More recently, focus has shifted toward interactive mapping applications. The latest GIS Web portals now provide users with integrated information and mapping products through multiple distributed GIS servers nationwide.

Although the Internet enables sharing of data and computing resources, how to enable distributed GIS to interact seamlessly across computer networks has been a major question. The core of such a challenge is data incompatibility between GIS systems. Moreover, when GIS systems were connected through computer networks, seamless system integration and interaction became problematic. As a result, the functions of one GISystem could not be used and shared by other GISystems. Interoperable GIS, as distinct from interoperable data, was raised as a research initiative by the National Center for Geographic Information Analysis (NCGIA) between 1996 and 1997: "In principle, interoperability offers one possible way of making GIS more useful and accessible to scientific research, by making the processes of interaction with GIS easier, and obviating the need for complex techniques to overcome incompatibilities between *software systems* and *data sets*" (Egenhofer and Goodchild 1997). Such a goal could not be achieved until 2000, when the new Web service technology was developed within mainstream IT as a solution for broad software interoperability.

How Web service technology enables software interoperability is discussed in detail in Section 6.4. *Web service*, unfortunately, is an ambiguous terminology, because such a new technology may not have any relationship with the *Web* at all. Each *service* is actually an independent functional module or component of a software package, which can provide data, information, and processing services, such as a spatial query function in a GIS software package. Through Web services, functional modules in different GIS configurations can be accessed and invoked individually through the Internet by users or computer agents using either a Web browser or some other types of programming tools. To this end, such functional modules or components are a kind of service provided over the Internet. In essence, Web services refer to individual functional modules that can be accessed and used over computer networks.

Desktop GIS provides user-friendly interfaces through the stand-alone software, whereas Internet GIS provides user-friendly interfaces through a Web browser. Usually, for desktop applications, spatial data is uploaded and saved inside the same computer in which the GIS software is installed. For current Internet GIS applications, the GIS server retrieves spatial data that can be located on the same server machine or through a spatial data engine that is connected to the server. In a Web service enabled environment, however, the fundamental difference is that the new technology does not provide any user interface. Instead, it offers application programming interfaces (API) to access and use varied kinds of data and computing resources over the network through dynamic and seamless data and system integration. Further, machines that host the GIS software may or may not be connected with GIS data.

The new Web service technology separates data from software function and its interface definition, but it enables the dynamic integration of the data and computing resources at run time over the computer network through the standardized programming interface definition. It is the first time that GIS and geographers have had the same opportunity as domain experts and developers to integrate the knowledge of geography and GIS with others under a common computing architecture. Under such a new radical architecture and information infrastructure, GIS methods and tools can be shared not only among GIS professionals but also among researchers in other disciplines. Web service technology, as a solution for software interoperability, could have a significant impact on the integration of GIS into mainstream IT. Through Web services, GIS functions can be accessed, deployed, and integrated by researchers and professionals with different academic and technology backgrounds without using any proprietary GIS software. Geographers and GIS professionals can benefit from such a framework by directly accessing and deploying the data and computing resources developed in other specialized knowledge domains. Web services have already begun to have a major impact on GIS development and geographical research.

The challenges arising from the use of Web services in GIS applications are presented in at least two important aspects. First of all, the new technology only provides APIs for communication and integration and not a general user interface. Thus, programming is a requirement whenever users wish to share or use the online distributed data and computing resources through Web services APIs. For example, programming may be needed to use ESRI's ArcGIS online Web services, which provide address geocoding and reverse geocoding functions that are useful in many application modes. Meanwhile, Web service technology itself is not mature yet and has suffered from semantic problems in that programmers and computer agents cannot understand the meaning of the services.

6.1.2 Spatial Data Distribution and Mapping over the Internet

The need for spatial data sharing contributed significantly to the formation of the National GIS Data Clearinghouse under the auspices of the FGDC in the United States. The clearinghouse concept pursued by the FGDC represents a decentralized system of servers located on the Internet. The goal of the GIS data clearinghouse is to provide access to digital spatial data through metadata-based searches. The clearinghouse functions as a detailed catalog service that supports links to distributed spatial data and maps. Today, many state agencies provide such data clearinghouse services for the local and regional user communities, such as the West Virginia GIS data clearinghouse (http://wvgis.wvu.edu/data/data.php). The clearinghouse provides hypertext linkages within metadata entries that enable users to directly download digital data in one or more available formats.

Accessing and transferring spatial data through a clearinghouse over the Internet was an important first step to deploy the power of information infrastructure. The next step will be to derive understandable Geoinformation and knowledge from the accessed spatial data. Normally, the accessed spatial data is processed by GIS professionals to present the results in the form of maps, reports, or analysis. Increasingly, the Internet provides a timely, easy, and economic way to visualize the accessed spatial data through maps. However, deriving understandable maps over the Web remains a significant challenge. One problem is that such maps are often static and cannot be easily manipulated by the end users. Further, GIS software that is required to create and publish Internet maps is typically expensive and is not readily available to those without access to GIS resources.

Web mapping started by providing static maps as images on the Web page. To enhance the interaction between users and a GIS application over the Internet, dynamic or interactive Web mapping techniques have been developed and deployed for both client-side and server-side applications since the late 1990s. The client-side solution includes client-side Web scripting such as Dynamic Hypertext Markup Language (DHTML), JavaScript, VBScript, Document Object Model (DOM), and Cascading Style Sheet (CSS) and client applications such as plug-ins, Java Applets, and ActiveX. The server-side solution focuses mainly on enhancing the Web server applications to improve the server performance in handling client requests through Servlet, Active Server Page (ASP), and ColdFusion (Peng and Tsou 2003).

Although Web mapping applications have certain GIS functionalities such as buffer creation and geocoding, they are created by the application providers rather than by the users to primarily deploy the spatial data resources contained within their server sites. The users are thus restricted from selecting data or combining data from other Internet sources. The vision of the recent FGDC deployment of the Geospatial One-Stop (GOS) (GOS 2009) Portal is to provide an online access point to a collection of spatial data that allows users to obtain and view the data they require. The capability of the GOS Portal to dynamically generate GIS maps from distributed data servers is an important distinction and advance on the earlier National GIS Data Clearinghouse. The portal is based on the originating objectives of the NSDI to create "a physical, organizational, and virtual network designed to enable the development and sharing of this nation's digital geographic information resources" and to enhance interoperability among government activities by providing a common user interface for spatial data and services distributed throughout all levels of government. The portal does not store or maintain the data that is distributed nationwide across many servers. Each server is maintained by the agency or organization that is responsible for the data. The critical element in establishing a Web portal is to have open standards and specifications that can accommodate the underlying proprietary nature of GIS and spatial data formats.

One feature that is common to the above-mentioned clearinghouses, Web mapping, and Web portals is that all such applications depend on spatial data that is located on the server side. Such data-centric applications, thus, focus on data transfer and displays from the server to the client. In this framework, it is difficult, if not impossible, for clients to utilize computing resources that accept client-side spatial data for geoprocessing, spatial analysis, and integrated mapping over the network. The Web services technology, as introduced in Section 6.4, may provide a solution to such a problem.

6.1.3 Interoperability Challenge to Information Infrastructure

Contemporary GIS are now essentially distributed systems, because both data and software are often located on different computers. Achieving interoperable GIS has been a challenge due to the need to share interoperable GIS data and GIS methods seamlessly over computer networks. Figure 6.1 describes the main themes and issues in interoperable GIS, and this can help identify the context and position of the needs of interoperability in the II or the CI.

The GIS interoperability can be realized through three levels of interoperation to explore technical and semantic solutions as well as institutional collaboration. At the technical level, interoperable data means that data generated in heterogeneous formats can be exchanged, integrated, and deployed by different GIS software products that can process and analyze the data without the concern of incompatibility. Interoperable software means that functional modules in different software products can be seamlessly

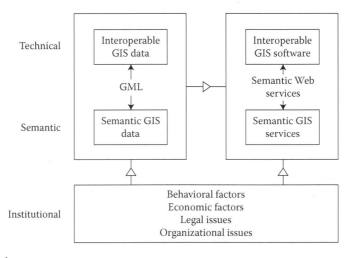

FIGURE 6.1
Levels of the GIS interoperability in information infrastructure.

deployed by others who can make a composite workflow to integrate software components to accomplish a specific task. At the semantic level, interoperable GIS requires that the concept and interpretation about the data and computational modules are meaningful to both the providers and the consumers of such digital resources. At the institutional level, interoperable GIS requires the collaboration between domain knowledge communities that share their geospatial data and computational resources. Such collaboration targets both technical and semantic levels of interoperability to generate and adopt standards and specifications on both data and software components. Section 6.2 mainly targets such issues and approaches pursued to resolve partial problems in order to enable data interoperability. Interoperable software involves the possible sharing of logical software components that can work together seamlessly, regardless of the platform upon which they were deployed or the development tools that were used to build them (Snell and Glover 2003). Section 6.3 discusses the evolving Web services technology as a partial technical solution for software interoperability.

6.2 Interoperability*

6.2.1 Heterogeneity Problems

As mentioned in Section 6.1, one challenge for achieving interoperable GIS is the heterogeneity problems in GIS data. Many geodatabases have been developed over different periods and for different purposes based on different GIS software, such as ESRI ArcInfo and ArcView, Smallworld GIS, Intergraph GeoMedia, MapInfo professional, and Clark Lab's Idrisi. Since geodatabases built on these different GIS software have their own proprietary data models and database storage structures, they cannot communicate and share information without data conversion. In order to exchange information and share heterogeneous geodatabases, conversion tools have to be developed to transfer data from one format into another.

However, data conversion is costly and time consuming and may, therefore, be unsuitable for many time-critical applications such as emergency response, location-based services, and real time traffic management, all of which need real-time access to diverse data to make quick decisions and take instantaneous actions (Zhang and Li 2005). Further, as the number of applications grows, GISystems are also presented with an ever-expanding range of resource types with regard to the types of devices, platforms, computer technologies, and languages. This causes more difficulty when communicating

---

* This section is contributed by Chuanrong Zhang, who holds a joint appointment at the Department of Geography & Center for Environmental Sciences and Engineering, University of Connecticut.

and exchanging spatial information. Although the development of the World Wide Web (WWW) and many Internet GIS programs provides proprietary ways to allow users to quickly access, display, and query spatial data over the Web (Peng and Zhang 2004), these Internet GIS programs also have the problems of proprietary software designs, data models, and database storage structures. Mapping and geoprocessing resources distributed over the Web by these Internet GIS programs cannot be shared and interoperated in real time. Data sharing facilitated by the advances of network technologies is hampered by the incompatibility of the variety of data models and formats used at different sites (Choicki 1999). Data that are in different formats are often remotely accessible only through simple protocols (e.g., FTP) that do not allow queries and filtering. Enabling two or more systems to communicate and exchange information becomes a costly and time-consuming process. Determining how to allow disparate heterogeneous geographic systems to share and integrate spatial data and Geoinformation in a cost-effective way is one of the most difficult problems in the geospatial world. Therefore, building truly interoperable distributed geographic information systems to share spatial data is imperative (OGC 1998).

Further, heterogeneity of GISystems and the difficulty of sharing Geoinformation cause a spatial data duplication problem. Redundant efforts are commonplace in the legacy of GISystems. Although there has been a massive increase in the number of GIS applications over the past couple of decades, it is often the case that these applications were built with little knowledge of other applications with which they could share information. As a result, many agencies and companies are trying to maintain spatial databases that coexist but that are not integrated. There are many duplicate data and processes that occur in separate departments and applications. The costs attributed to redundant spatial data and duplicated efforts are huge, because spatial data set creation is expensive and complex. Thus, it is necessary to reduce overlaps of spatial data. The last decade in particular has seen an enormous investment in spatial data integration initiatives. However, early solutions mainly focus on data translation and offline data replication. Although this has undoubtedly improved the sharing and synchronization of information across the diverse resources of typical organizations, it has led to intensive data redundancy and numerous updating or synchronization problems. In spite of the obvious benefits in terms of efficiency and effectiveness to be derived from sharing spatial data and information both within and between organizations, duplications of data collection and storage lead to inefficiencies in many GIS applications (Nedovic-Budic and Pinto 2000).

Although the geospatial community realizes the costs of spatial data duplication and the costs associated with heterogeneity of existing GISystems and applications that seriously limit sharing spatial data and thus inhibit the synchronization between GIS and businesses, they cannot simply discard these GISystems and start new systems from scratch, because it takes too long to develop a new system. A solution that builds on initial expenditures rather than on starting a new one from scratch will allow agile integration of these spatial data in a cost-effective way; such a solution is required.

6.2.2 Interoperability

Interoperability is the ability of a system, or components of a system, to provide information portability and interapplication cooperative process control (Brodie 1992). Two kinds of interoperability can be distinguished. For a program, data interoperability means the ability to utilize a range of data formats. For a data set, program interoperability means that it can be used by different types of programs (Laurini 1998). An interoperable database refers to the data-level interoperability. It can be used by different types of programs and applications. With interoperable databases, users can request and integrate data easily irrespective of whether the databases are stored locally or remotely. The interoperability of data from heterogeneous sources is extremely important in the context of geographical applications, because there are large amounts of spatial data with different geographical formats and because there are increased demands for reuse of existing spatial data.

How can we realize the goal of data interoperability? There are two approaches to data interoperability—database integration and standardization (Devogele et al. 1998). Database integration is the most sophisticated approach. A very basic method is to provide users with a global catalog of accessible information sources, where each source is described by associated metadata, including representation mode, scale, last update date, data quality level, and so on (Uitermark 1996). Current database integration has evident drawbacks related to a lack of scalability, consistency, and duplication (Devogele et al. 1998). The second approach to data interoperability is through standardization. The definition of standard data modeling and manipulation features provides a reference point that facilitates data exchange among heterogeneous systems (Devogele et al. 1998).

In the past, several useful standards have been developed to facilitate data exchange. Among them, Geographic Data File (GDF) and the Spatial Data Transfer Standard (SDTS) have been used and accepted. The GDF is specifically designed for spatial data exchange for Intelligent Transportation Systems (ITS). It defines a set of spatial features, attributes, and relationships that are particularly relevant to ITS applications and specifies a set of useful data structures and data formats. This makes it readily usable for off-line data exchange. The SDTS is a general purpose standard that is flexible and adaptive (NIST 1994). With anticipated extensions and refinements, SDTS was expected to become an important data format for ITS spatial data transfer or a neutral format for data archiving (Arctur et al. 1998). However, several barriers block the popularity of SDTS. These barriers include the complexity of SDTS, slowness in the development of practical SDTS profiles, restriction of each SDTS data set to a single profile, lack of a clear definition of spatial features in SDTS, and ambiguity in the means of specifying cardinality of relationships in a data model (Arctur et al. 1998). Currently, both GDF and SDTS are not as widely used as originally anticipated. The creation of a new standard data exchange format, Geography Markup Language (GML), represents another important step taken by the geospatial community toward data interoperability. The emergence of OGC Web services and services-oriented architecture provides a way to overcome the heterogeneous problems of geodatabases (Zhang et al. 2003). Significant progress has been made in terms of the implementation of global, regional, and local SDIs based on OGC Web services and services-oriented architecture (Yang et al. 2006; 2007). Since OGC Web services are based on open standards, they can provide data interoperability in network environments and permit easy access and effective exploitation of distributed spatial data and information from heterogeneous sources and formats without requiring complex preprocessing or time-consuming translations. The OGC Web services make it possible to access, visualize, share, and disseminate spatial data "on-the-fly" across the Internet. In addition to providing interoperability, OGC Web services can be created by using any platform, operating system, programming language, and object model.

While Web services and services-oriented architecture are helpful in dealing with access to heterogeneous spatial data and services, they cannot resolve semantic heterogeneity issues in spatial data and services. The OGC Web services and GML only provide a good mechanism to organize and standardize spatial data; they cannot define the data in terms of semantics by themselves. They alone do not contain rules and formal programming logic that can perform certain types of runtime automated reasoning. Thus, OGC Web services and GML data can facilitate human understanding of structures and meanings of spatial data, whereas it is difficult for computers to automatically understand and fuse structures and meanings of diverse information sources. In order to share data from different sources, data sources have to exactly follow standard models and terminology. This is time consuming and demotes the motivation of data sharing, because applications have to redo their data model and adjust data elements. Thus, although OGC Web services provide standard syntactic interfaces and protocols for invoking heterogeneous geodatabases, they do not specify the intended meaning of their terms in machine-readable form. The OGC Web service descriptions only allow for syntax specifications of basic service contents and provide no semantic descriptions of the meaning of these data contents. However, differences in semantics used in different data sources are one of the major problems in spatial data and Geoinformation interoperability (Bishr 1998). Two identical GML/XML descriptions may mean very different things depending on the context of their uses. To achieve semantic interoperability, spatial data must be put in a machine-understandable representation. A more formal structure must be developed and the semantics of the words must be made explicit for a computer to understand the meaning of the information.

One possible approach to overcome the problem of semantic heterogeneity and to achieve semantic interoperability among disparate and dispersed

GISystems is through ontology (Smith and Mark 1998). The concept of a Geospatial Semantic Web (GSW) was recently suggested to address the semantic challenges and achieve automation in service discovery and execution. The GSW can be seen as an extension of the current Web but where Geoinformation is given a well-defined meaning by ontology, thus allowing users to query spatial data based on semantics (Duke et al. 2005). Based on the work in spatial ontology, some recent research explores the concept and feasibility of developing the GSW (e.g., Egenhofer 2002; Zhang et al. 2007). Many of these studies, however, are still at the initial stage of proposing frameworks for raising queries in GSW, using the concept of global ontology and local ontology (Cruz et al. 2002), or ontology integration (Fonseca et al. 2002). Figure 6.2 illustrates a proposed framework of GSW to achieve data and information interoperability at a semantic level based on OGC Web services, services-oriented architecture, and ontology.

The framework is composed of four elements: service provider, service broker, service client, and ontology server. The service provider uses OGC data

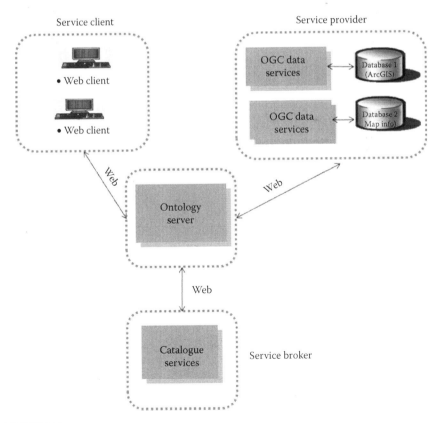

FIGURE 6.2
A framework for Geospatial Semantic Web for data and information interoperability.

services such as WFS, WMS, and WCS to publish heterogeneous spatial data that are connected to legacy GISystems. The service broker provides a registry for service advertisement and discovery. The service client provides a user interface for users to query and visualize results. The ontology server ensures semantic interoperability of service clients and service providers. Web services are connected via OWL-based Web service Ontology (OWL-S)—ontology services built on Web Ontology Language (OWL)—to facilitate Web service discovery, invocation, and composition. The main advantage of the proposed framework is that it not only resolves technical interoperability via Web services and services-oriented architecture but also resolves the semantic heterogeneity problem. By combining ontology with Web services, it can deal with differences in data semantics in diverse data sources. The proposed GSW provides a way of defining the semantics of spatial data available on the Web and, thus, offers efficient and flexible ways to conduct complex queries involving semantic knowledge and locating geoinformation and services on the Web.

6.2.3 Benefits and Challenges

Data and information interoperability ensures organizations to maximize opportunities for exchange and reuse of information whether internally or externally. It can avoid having information islands among which information cannot flow effectively. Data and information interoperability provides significant benefits, including (1) reuse of existing heterogeneous data as well as updates and maintenance of data remotely across the Web, thus providing a potential way to alleviate the duplication problem and reduce related costs; (2) being useful for organizations with scare resources such as time, expertise, and the funds to implement geodatabases; (3) making it easier to distribute spatial data and applications across platforms, operating systems, and various computer languages as well as allowing users to find, access, and use needed information over the Web; (4) allowing GIS applications to adapt to changing environments quickly, thus increasing their flexibility and agility; (5) reusing existing infrastructures. Thus, the vast majority of existing legacy GISystems and applications that are typically closed and proprietary can interact easily with each other, and the tremendous effort spent over many years in developing and enhancing legacy GIS applications does not need to be lost; and (6) providing timely, accurate, and consistent information to allow users and key decision makers to access the most up-to-date spatial data over the Web.

Interoperability has potential benefits, whereas the realization of interoperability has challenges. Although the technologies of GSW have been proposed to address the semantic interoperability problem and many projects have been funded in this area nationally and internationally, for example, the SPIRIT project (Jones et al. 2003) and the METEOR-S project (Rajasekaran et al. 2004), GSW is still at the initial research stage, and there are still many issues waiting for further studies to make it a practical technology. For example, how can geospatial Web services be discovered automatically? With currently implemented geospatial Web services such as catalog services, it is only possible to search and access spatial data by keywords in metadata. Despite the efforts that the geospatial community has made by using ontologies and GSW to discover geospatial Web services, content-based searching at a semantic level still remains a challenging problem. Approaches or algorithms that can automatically search and access spatial data or services by their contents are needed. For another example, how can heterogeneous spatial ontologies be integrated? Although a single domain-specific ontology might be a desirable goal, it is only realistic to expect that the currently evolving GSW will be highly heterogeneous in terms of knowledge represented. It is impractical to develop a single global ontology for all applications that support the tasks envisaged by a distributed environment like GSW. In environments with multiple independent systems, each system might have its own ontology. Thus, possibilities of conflicts and mismatches may exist. Understanding how to integrate heterogeneous spatial ontologies needs further research. In general, although OGC Web services, services-oriented architecture, and GSW have made progress with regard to spatial data and Geoinformation interoperability, we still have a long way from seamless interoperability.

6.3 Web Services and Service-Oriented Computing (SOA)*

Since interoperating spatial data require a seamless interaction between systems, GIS software interoperability also requires that the logical programming components work together regardless of the platform on which they are deployed or the development tools that were used to build them (Snell and Glover 2003). Software interoperability problems arise due to different computer platforms, operating systems, programming languages, data types, and the semantic meaning of the terminologies used; whereas semantic and institutional interoperability issues have likewise been mixed into software interoperability as well. Current GIS software packages have limited communication capabilities for sharing and exchanging system functionality and components. Further, software interoperability is also contingent on resolving incompatible data formats, data structures, and spatial data models that exist in the legacy heterogeneous GIS software systems.

Previously, Distributed Component Object Model (DCOM) from Microsoft and Common Object Request Broker Architecture (CORBA) were occasionally used to enable communication between software systems. Remote Procedure Call (RPC) is a common concept to Unix systems, but Unix RPC could never be widely deployed by other vendors. Similarly, Sun's Java Remote Method Invocation (RMI) and the Enterprise Java Bean (EJB) may

---

* This section is contributed by Xuan Shi.

enable successful communication among Java packages, but they could hardly receive complete industry support. Obviously, communication outside of these vendor-specific object models was not possible, and thus they provided a partial solution to the problem of GIS system interoperability. Web service is a recent solution to resolving the technical issues associated with software interoperability though it does not intend to resolve the semantic system interoperability problems, as discussed later.

According to the WWW Consortium, there are many things that might be called *Web services* in the world at large. However, for the purpose of this working group and this architecture, and without prejudice toward other definitions, we will use the following definition: "A Web service is a software system designed to support interoperable machine-to-machine interaction over a network. It has an interface described in a machine-processable format (specifically Web Services Description Language [WSDL])" (W3C, 2004b). Other systems interact with the Web service in a manner prescribed by its description using Simple Object Access Protocol (SOAP) messages, typically conveyed using HTTP with an XML serialization in conjunction with other Web-related standards (W3C 2004a).

As a solution to software interoperability, it is interesting to note that Web services may or may not have any relationship with the Web at all, because interoperable computing resources can be connected through WSDL by exchanging SOAP messages. The W3C's definition, as quoted above, indicates that there can be many different approaches to interoperable machine-to-machine interaction over a network. For example, the so-called Representational State Transfer (REST) approach also provides Web based services implemented via a Web browser over the Internet. The latter is more familiar to GIS professionals, as the OGC has generated many standards and specifications, such as Web map services, which are based on the REST approach.

6.3.1 SOAP-Based Web Services

At the core, Web service technology is designed to improve the interoperability among many diverse application development platforms that exist today (Snell 2002). The fundamental goal of interoperability in Web services is essentially to remove the incompatibility that exists among the various development environments used to implement the service so that developers using those services do not have to consider which programming language or operating system the services are hosted on. Considerable literature on this topic can be found on the Internet (Ballinger 2001; Tost 2003).

For GIS interoperability, Web services promise a new level of interoperability among applications. Each service encapsulates specific business functionalities that can be invoked through SOAP over HTTP. The most useful functionality offered by SOAP is the ability to convert data between platform-specific or program-specific formats and the XML format used in SOAP messages: a process called *serialization* and *de-serialization*, as described in Figure 6.3. It is

FIGURE 6.3
Communication between heterogeneous systems through SOAP.

this functionality that allows an application running on one platform to convey information to an application running on another platform (Deem 2002).

In the past, communication among distributed GIS and other commercial software packages was object specific, as evidenced by the object products CORBA and DCOM. The CORBA is a standard specified by the Object Management Group (OMG) to enable software components, written in different computer languages and running on varied computers as individual modules or groups of functional modules, to interoperate. However, it is impossible for CORBA and DCOM objects to interact, because CORBA and DCOM are competitive and non-interoperable (Peng and Tsou 2003). Since such object-specific service requests cannot be passed through Internet firewalls, there is no mechanism established to register and discover such distributed components over the Web. Web service technology simplifies this process.

The fundamental differences between Web services and other previous technologies such as CORBA and DCOM are that Web services are XML (eXtensible Markup Language) based. The XML is a standard protocol for data and system communication. Like Hyper Text Markup Language (HTML), it is designed to describe data and presentation. In contrast to HTML, however, XML cannot only describe the data but can also be a format for data exchange. The XML defines the content of the data and separates data content from its presentation. The HTML consists of a set of predefined tags that predefine the representation of data, such as the text style, color, and font. Unlike HTML, XML tags are not predefined but can be defined by different domain terminologies, thereby making the data more meaningful. In addition, XML is case sensitive whereas HTML is not, and XML also has stricter rules that follow Document Type Definition (DTD) or an XML Schema to define the data syntactically. Importantly, XML is both machine understandable and human readable and thus provides a foundation for developing intelligent systems. For these reasons, XML provides a neutral, interoperable, and meaningful standard for both data formatting and computer communication. To date, despite some shortcomings such as large file sizes, XML has demonstrated its superiority for data exchange over traditional file-based, vendor-specific, and binary geospatial data.

Web services have fundamentally changed contemporary distributed GIS development and applications over the Internet. Such a technology can include not just client-server architecture but also peer-to-peer relationships in which all parties provide the dual role of both server and client for the sharing of data and GIS computation functions and resources. Currently,

Internet GIS applications are end-user oriented, and GIS data and functional modules are delivered and presented directly to end users over a computer network. However, applications such as Java Applets or plug-ins on the Web browser are not easily deployed or integrated by third-party developers. Web service technology, on the other hand, provides a foundation for the dynamic binding of distributed computing systems and enables the exchange and interaction of data and functions among heterogeneous computer networks. A Web service is a developer-oriented software component that can be accessed and integrated by application developers through standard APIs over a computer network. Developing Web services is a process that entails decomposing the whole system into functional components and modules (componentization or modularization). Utilizing Web services enables application developers to find and assemble the necessary components to build specific integrated applications. As a result, functions for spatial data analysis in any proprietary GIS can be developed into individual functional modules or components that can then be accessed through a Web service interface.

6.3.2 Representational State Transfer Web Services

Originally, the REST, as proposed by Roy Fielding (2000) in his PhD dissertation research, sought to describe an architectural style of networked systems. Fielding explained that "Representational State Transfer is intended to evoke an image of how a well-designed Web application behaves: a network of Web pages (a virtual state-machine), where the user progresses through an application by selecting links (state transitions), resulting in the next page (representing the next state of the application) being transferred to the user and rendered for their use." Thus, in Web-mapping applications, any map service URL is a representation of the Web resource that is placed in a certain state. When a user browses a Web site to retrieve a map, the URL of the Web site represents the current state of the resource the user is accessing, for example, a map service URL to access TerraServer: http://terraserver.microsoft.com/OgcWms.aspx?version=1.1.1&request=GetMap&Layers=DOQ&SRS=EPSG:26917&BBOX=549600,4183900,551200,4184900&width=800&height=500&format=image/jpeg

This URL defines multiple variables that are specified in the OGC WMS specifications. The above GetMap request retrieves a map image of a Digital Orthophoto Quarterquad (DOQ) as the layer name. The Spatial Reference System (SRS) is defined by the European Petroleum Survey Group (EPSG) code 26917, which refers to the north seventeenth zone of the Universal Transverse Mercator (UTM) projection with the North American Datum of 1983. The WMS GetMap request also defines the map width, height, image format, and the Bounding Box (BBOX) of the map extent, in terms of MinX, MinY, MaxX, and MaxY. When the user zooms in, or zooms out, or pans to another area, a new representation places the client application into another state. This new state can be examined by changing the definition of the BBOX variables. Thus, the client application state is transferred with a different resource representation, for example, another map service URL to access TerraServer: http://terraserver.microsoft.com/OgcWms.aspx?version=1.1.1& request=GetMap&Layers=DOQ&SRS=EPSG:26917&BBOX=576000,4152400, 678400,4216400&width=800&height=500&format=image/jpeg

The REST is a simple and easy way for users to access online resources. The REST implementations leverage existing standard HTTP methods, such as GET, POST, PUT, and DELETE, to identify and deploy the resources available over the Internet based on the traditional client–server architecture. The REST is a so-called Web service, because it is based on the Web and provides service to the end user. However, SOAP-based Web services target software interoperability that cannot be covered by REST, which can only operate on the four HTTP methods specified above. The SOAP is a more powerful protocol for interoperable computing, because it can define any number of methods, whereas the REST is gaining popularity for its simplicity in developing mashup applications.

6.3.3 Web Services Architecture or SOA

The three fundamental components of Web services technology are the service provider, service requester, and service registry. Figure 6.4 describes the role of these three components in developing and utilizing Web services in an SOA. In this process, the service provider traditionally has the responsibility for developing, describing, and publishing the services and for listening for and responding to service requests. The service requester, on the other hand, looks for the appropriate services through the service registry, constructs a service composition, aggregates and mediates service invocation

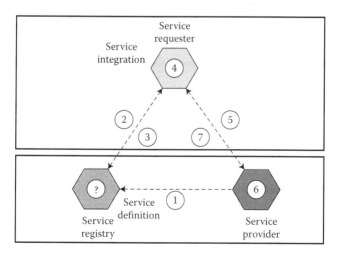

FIGURE 6.4

Fundamental Web service processes. (1) Service publication. (2) Service discovery. (3) Service matchmaking. (4) Service composition. (5) Service request. (6) Service invocation. (7) Service response. (?) No real service registry. (Data from Shi, X. 2007. *IEEE IT Professional*, 9(4):42–45.)

processes, creates applications to invoke services in a specified order, and, finally, monitors the process.

The service registry, known as the Universal Description, Discovery, and Integration (UDDI), accepts service registrations and publishes them to the service requesters. To enable service requesters to find the necessary services, the service discovery could be implemented by a service registry such as the UDDI, a geoportal, or an index system like Google or the Peer-to-Peer (P2P) architecture. Within the P2P Web architecture, a Web service is a node in a network of peers who query neighbors over the network in search of a suitable Web service (W3C 2004b). All interactions among these three components are currently expected to work through the WSDL interface. The question mark in Figure 6.4 indicates the questionable role and functionality of UDDI that is currently under redesign and reconstruction.

6.3.4 SOA Design Principles and Concerns

Service-oriented computing (SOC) and SOA are not new ideas but the natural evolution in developing distributed systems, as evidenced by the previous models and developments like CORBA, DCOM, RPC, RMI, and EJB. Before SOC or SOA became popularized in 2000s, the dominant development models were object-oriented programming (OOP) in 1980s followed by component-oriented programming (COP) in 1990s (Szyperski 2002). It is common in all these approaches to construct complex distributed systems from simple components. For Web services, while noting that Web services standards received industrial support, at least two prominent distinctions have to be highlighted in SOC; these are loose coupling and message exchange.

In software engineering, *coupling* refers to the degree of whether and how one component is dependent upon the other components in an application or system architecture. Tightly coupled systems and architectures are dependent on each other, thus changes made in one component may lead to changes in many other components. As a result, it increases the difficulty in testing and changing components, especially in cases to bind components at the compile time or runtime in a synchronous state. Data and system integration is more difficult in the communication between distributed objects that were created by different programming languages on heterogeneous platforms, because many such objects are RPC-based and connected by proprietary interfaces, whereas vendors only compete on varied specific Object Request Brokers without reaching a solution for software interoperability.

Loosely coupled architecture demonstrated certain advantages in constructing the service-oriented systems to enable effective and efficient communication between heterogeneous systems. In a loosely coupled architecture, changing one component may not reflectively require replacing other components in the system. In contrast to the tightly coupled system, it increases the agility and neutrality for system management and implementation, because developers do not have to depend on certain technology or platforms to realize the system goals. Thus, it benefits cross-organizational and cross-disciplinary development and system integration, because services can be deployed through published interfaces without knowing the implementation details behind the services.

Message orientation is another distinction in SOC and SOA. One common misconception about Web services and SOA might be related to the SOAP protocol. The name SOAP seems to imply the remote procedure call (RPC) to distributed and remote objects. However, Web service is not the same concept as a distributed object. Such an RPC-related feature was revitalized by focusing on message exchange in the form of an XML document (Vogels 2003). Instead of deploying method invocation on a remotely referenced object, service communication is formalized by exchanging messages between the service providers and service requesters. As the successor of component oriented programming, service oriented technology incorporates COP concepts such as self-description and encapsulation so that requesters can deploy the services without knowing the details about how the service is constructed.

There are still many challenges and concerns with regard to accepting and promoting the development of Web services technology and SOA. Service discovery problems have not been ideally solved. Unfortunately, varied services have been discussed for almost a decade, but we are still far from discovering the right services (Brodie 2007; Shi 2008), considering that in the English language, service means the "work done for others." If others cannot find services, then there are no services; what we have are only the tools for interoperable software engineering and application development.

Although Web services technology enables the sharing of data and computational resources to avoid the duplication of labor and resources, reliability has been a big concern in the user community. Most commonly, a service host server may stop service temporarily for maintenance for a few hours or a day. This may not be a big deal in many cases. Service requesters may find alternative services offered by different providers. However, in case of emergency situations when information infrastructure might be destroyed or broken due to natural or human factors, distributed systems might not even be connected. Related issues about quality of services will be discussed in Section 6.8.

6.4 SOA and OGC Web Services*

6.4.1 Service-Orientated Architecture

By following the standards ranging from data to orchestration, as shown in Figure 6.5, SOA and Web services are emerging as the basis for

---

* This section is contributed by Peisheng Zhao, Liping Di, and Genong Yu.

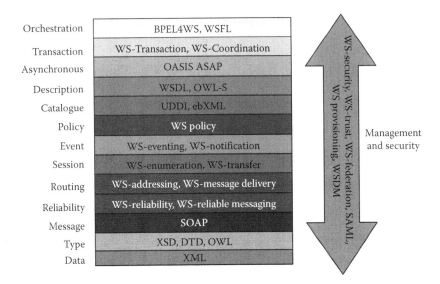

FIGURE 6.5
Web service protocol stack.

distributed computing. Large networks of collaborating applications are changing the way computer software is designed, developed, and deployed. More and more spatial content and capabilities are available online as Web services in recent years. This increase significantly enhances the ability of users to collect, analyze, and derive spatial data, information, and knowledge.

The SOA is a framework in terms of services and service-based application development (Harding 2006). Within the context of SOA, a system consists of a collection of loosely coupled services that communicate with each other by passing data from one service to another to coordinate an activity. New systems can be created dynamically by combining new application-specific services with existing services (Erl 2005; Newcomer and Lomow 2005). The basic SOA paradigm is shown in Figure 6.6:

- *Consumers:* Entities that make use of the functions offered by services
- *Applications:* Specific process logic implemented by service orchestrations
- *Service Support:* A set of facilities or tools that provide background support functions for the SOA, such as service catalog and workflow engine
- *Services:* A set of distributed units that perform specific tasks when invoked through explicit interfaces

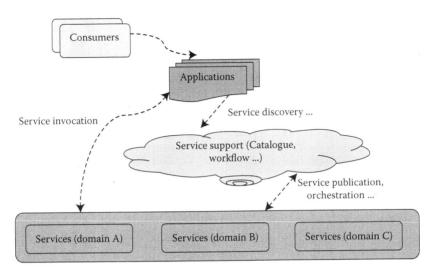

FIGURE 6.6
SOA basic paradigm.

The SOA enables systems with the following distinguishing characteristics:

- *Consistency:* All components are implemented as services with explicit interfaces
- *Interoperability:* Self-described services are machine-to-machine discoverable and executable through standard protocols
- *Orchestration:* Services can be assembled into a service chain to solve a more complicated problem
- *Efficiency:* Reusable services can be easily integrated into a new system
- *Flexibility:* Services can be distributed over the network, run on different platforms, and implemented in different programming languages

Moreover, SOA and Web services are becoming the basis for several other recent distributed computing technologies, such as grid services, cloud computing, and semantic Web services

6.4.2 OGC Web Services

In parallel with the development of general-purpose Web services, many geospatial Web services designed to be interoperable with distributed spatial data have emerged. The OGC has successfully addressed geospatial interoperability requirements and standards to enhance the discovery, retrieval, and use of geospatial Web services. A series of OGC services, such as WCS (Whiteside and Evans 2008), WFS (Vretanos 2005), WMS (Beaujardiere 2004), and Catalogue Service for Web (CS-W) (Nebert et al. 2007), have been used and deployed by government agencies, universities, and companies worldwide. NASA Earth Science Gateway (ISO 2005) allows users access to scientific and research products, including data, models, and visualizations provided by a variety of national and international organizations, through open standard Web protocols. The GeoBrain Processing Web Services (CSISS 2007) based on the Geographic Resource Analysis Support System (GRASS) functionality modules provide the capabilities of spatial data management, raster image processing, spatial modeling and analysis, graphic map generation, and data visualization over the Internet. Adam Web Services leverages the Algorithm Development and Mining Toolkit (ITSC) to enable mining to be remotely sensed and other scientific data dynamically over the network for pattern recognition, image processing and optimization, and association rule exploration. The term *geospatial Web service* is straightforward insofar as it refers to using Web service technologies to manage, analyze, and distribute spatial data, information, and knowledge. A geospatial Web service can be sorted and searched through its spatial characteristics, such as domain concepts and spatial features. Moreover, this term further involves the standards, architecture, and semantics that make geospatial SOA feasible (Di et al. 2005; Zhao et al. 2006).

Interoperability of geospatial Web services is achieved by using given standards, mainly from the FGDC, the International Organization for Standardization (ISO), and the OGC. Geospatial Web service standard stack can be partitioned into four parts—data (message encoding), interface (transport protocol), metadata, and process, as shown in Figure 6.7. At the data level, the standards specify the data formats used for communicating between Web services and applications, such as OGC GML and FGDC SDTS. At the interface level, the standards define common interfaces and message encodings for both applications or Web services and human users, such as OGC WCS and OGC WFS. At the metadata level, a set of consensus data types and descriptions are associated with each Web service or data, such as ISO 19119 and ISO 19115. At the process level, standards specify a set of spatial queries and analysis interfaces, such as OGC Web Processing Service (WPS) (Schut 2007).

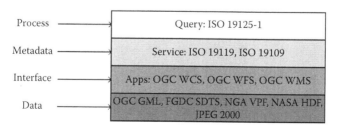

FIGURE 6.7
Geospatial Web service standard stack.

Theoretically, any programming language can be used to develop Web services on any platform. A typical process for implementing a geospatial Web service from scratch may include (1) analysis of the spatial data and processes to determine the proper interfaces to be exposed, preparation of the relevant WSDL to generate the program skeleton, or writing of the program interfaces to generate the relevant WSDL; (2) development of internal modular functions with specific message styles (RPC or document); and (3) testing and publication of the implemented service. Another approach is to adapt existing programs and expose some of their functions as Web services. For example, the GRASS is a comprehensive system of more than 350 programs for spatial analysis, map generation, data visualization, data modeling, database management, and data manipulation (GRASS 2005). In order to convert GRASS functions to Web services, the first problem is to transfer these functions from a desktop environment to a Web-accessible environment. The second problem is to determine which functions are appropriately deployed as Web services and what is the basic level of operations to be exposed—command level or combined script (Zhao et al. 2006). A flexible layered architecture for geospatial Web service development (Dadi and Di 2007) integrates the low level (command level) interfaces and conceptual level (combined script) interfaces.

A catalog service acts as a directory in SOA: service providers advertise the service availability by using meta-information and service, thereby allowing consumers to discover the desired services by querying meta-information. There are two prominent general models for catalog services: the Electronic Business Registry Information Model (ebRIM) (OASIS 2002) and the UDDI (OASIS 2004). The ebRIM is adopted as the geospatial registration information model of OGC catalog service, because it is more general and extensible. In the ebRIM, as shown in Figure 6.8, the class *ClassificationScheme* defines a tree structure made up of *ClassificationNodes* to describe a structured way for classifying or categorizing *RegistryObjects*. For instance, the OGC WCS is one *ClassificationNode* belonging to the *ClassificationSchema*, which are OGC Services. The class *Association* uses an *associationType* attribute to identify the relationship between objects. For example, *Service* is associated with *Data set* by using *Association*, where *associationType* is *operateOn*, *sourceObject* is *Service*, and *targetObject* is *Data set*. The class *CSWExtrinsicObject* derived from the class *ExtrinsicObject* represents all of the metadata objects that describe those not intrinsic to the catalog. The class *Data set* is derived from *CSWExtrinsicObject* to describe geographic data sets. Many metadata from the ISO 19115 and its draft part 2 are added to *Data set*. As a plug-in, *Slot* is used to add the new service attributes derived from the ISO 19119 into the *Service* class.

However, mismatching in service discovery may arise due to semantic ambiguity. The lack of semantics in Web services makes it impossible to implement reliable and large-scale interoperation by computer programs or agents (McIlraith et al. 2001). One solution to the above problems is to

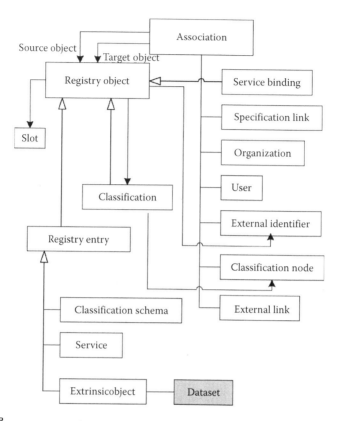

FIGURE 6.8
A simplified illustration of ebXML in OGC CS-W (Adapted from OASIS. 2002. OASIS/ ebXML Registry Information Model v2.0. http://www.oasis-open.org/committees/regrep/ documents/2.0/specs/ebrim.pdf (accessed December 22, 2009); Martell, R. 2005. OGC™ Catalogue Services—ebRIM (ISO/TS 15000-3) profile of CSW. Open Geospatial Consortium, Inc., 62pp.)

annotate geospatial Web services with semantics using proper ontologies. An ontology is a specific set of vocabularies and relationships that have explicit assumptions about the intended meanings and relationships of the vocabulary in the set (Sivashanmugam et al. 2003). The GSW Service is a technology that enables the computer-understandable description of Web services. Emerging standards for the Semantic Web are OWL, OWL-S, and Semantic Web Rule Language (SWRL) (Horrocks et al. 2003; Dean and Schreiber 2004; Martin et al. 2004). These standards evolve the description of Web services from a syntactical description to a semantic description, mainly based on description logic. Recently, several projects have developed ontologies across different geospatial domains. The ontologies within the Semantic Web for Earth and Environmental Terminology (SWEET) (Raskin 2008) contain several thousand terms spanning a broad extent of Earth system science and related concepts, such as NASA's Global Change Master Directory (GCMD), Earth System Modeling Framework (ESMF), grid computing, and those of the OGC. The SWEET provides a high-level semantic description of Earth system science. The ontologies for geographic information metadata, defined in ISO 19115 (Zhao 2004), add semantic meanings to the standard metadata by which data sets are explicitly associated with providers, instruments, sensors, and disciplines and the relationships across these concepts. The spatial ontologies can be used to describe the semantics of spatial data and geospatial Web services and define their relationships with one another. A correct service match can further be made by matching both data semantics for service inputs and outputs and functional semantics for the Web services.

Discovery and use of a geospatial service is the initial step of geospatial SOA implementation. Geospatial Web service orchestration is the assembly of individual geospatial Web services into a service chain for representing a more complicated geospatial model and process flow. It is essential for complex geospatial applications and knowledge discovery. A geospatial Web service orchestration can be completed in three approaches, as stated in the definition of a service chain in ISO 19119 (ISO/TC211 2005):

- *Transparent:* The user knows everything about the intended work and plays a central role in finding and composing all of the required services and data. The composite service chain can be invoked either in a user-controlled sequence or in a system-controlled process.
- *Translucent:* The user queries the system for each orchestration step, and then the system assists the user to select and configure the most suitable services and data for building a service chain.
- *Opaque:* The user presents a problem, and then the system uses its embedded knowledge to automatically build a service chain using the best services and data, without the user's intervention.

6.4.3 GeOnAS: SOA-Based Geospatial Online Analysis System

Geospatial SOA and Web services are changing the way the geospatial applications are designed, developed, and deployed. The GeOnAS (2007), a fully extensible online analysis system designed to use geospatial Web services to discover, retrieve, analyze, and visualize distributed spatial and other network data, is introduced here to illustrate the geospatial SOA paradigm. It is an open data system by which users can get the spatial data from any OGC-compliant data service in the world. It is also an open application system in which all functions are provided through interoperable Web services allowing users to integrate Web services easily to carry out a specific analysis. Further, it is a collaboration system allowing different users to contribute geospatial processes and data products for sharing, exchanging, and reusing. Thus, new solutions can be created dynamically by composing together

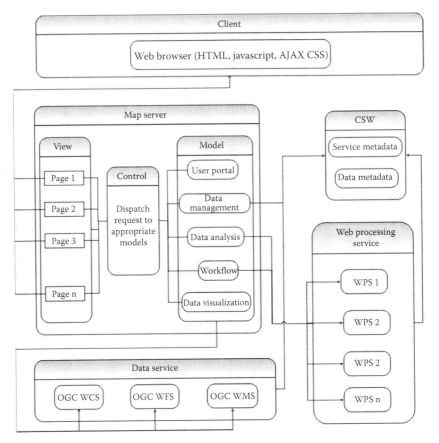

FIGURE 6.9
GeOnAS architecture. (Data from Di, L. et al. 2007. GeoBrain Web Service-based Online Analysis System (GeOnAS). In *NASA Earth Science Technology conference* 2007.)

new application-specific services and existing services. The GeOnAS architecture is shown in Figure 6.9.

Map Server follows the Model-View-Controller (MVC) design pattern to provide the following models:

- *User Portal:* Stores the current usage state in an OGC WMC (Sonnet 2005) document that can be imported again later to restore the portal's state.

- *Data Management:* Retrieves spatial data from a remote service and temporarily stores them with the related session ID on the map server in a network-accessible location.

- *Data Analysis:* Integrates and invokes preferred processing services to perform data analysis.

- *Workflow:* Enables the user to build a chain of services to perform a complicated task.
- *Data Visualization:* Allows the user to set up data display preferences, such as map sequence, data subset, and image palette. A set of different rendering services are deployed for the different purposes of data visualization.

Data services provide users with a common data environment in which they can work with data in an interoperable way. The OGC WCS provides intact multidimensional and multitemporal spatial data as a *coverage* to meet the requirements of client-side rendering, scientific model inputs, and other clients beyond simple viewers. The OGC WFS supports the networked exchange of geographical vector data as *features* encoded in GML. The OGC WMS provides spatial data as a *map* dynamically generated from real geographical data, which is generally rendered in spatially referenced pictorial image formats (such as PNG, GIF, or JPEG).

The OGC CS-W is used to enable registry, discovery, and retrieval of a wide variety of distributed resources (for instance, spatial data, applications, and services). This CS-W can access other distributed catalog services, such as the Group on Earth Observations' GEOSS Clearinghouse and NASA's ECHO, to search for more data. It has some additional extensions to accommodate the ISO model. The data set class has been added to the ebRIM to provide a flexible way to describe network-accessible data. This CS-W also supports a variety of classification methods, including the definitions from OGC specifications, ISO 19119, and NASA GCMD, for service publishers to indicate the domain to which a service belongs.

The OGC WPS provides a domain-specific computational model, which might be a simple spatial calculation or a complex global climate change model, to enable users to do data analysis over the network (Schut 2007). For manipulating and analyzing vector and raster spatial data, the GeOnAS provides more than 20 built-in WPSs and more than 50 relevant operations that are developed based on the GRASS. The WPSs can also be chained together to perform more complex analysis tasks. Users can access these services for data analysis and data mining of any OGC-compliant online data source. Moreover, the GeOnAS can integrate new Web services dynamically to provide users with an open and fully extensible environment. Users who have their own geospatial processing service and would like to use it to perform data analysis can integrate that service into the GeOnAS to build a unique system. If the service is registered into the catalog service, other users will benefit from it. Hence, the more users are involved, the more powerful the GeOnAS becomes.

Overall, the GeOnAS provides value by its efficiency and flexibility in using geospatial Web services to integrate distributed spatial data and processes over the Web within the context of SOA.

6.5 Spatial Web Portal and Geobrowser*

6.5.1 Spatial Web Portal (SWP)

The rapid development of the WWW provides an effective and efficient method to share information worldwide. Currently, massive Geoinformation is available over the WWW, such as data sets, applications, and OGC Web services, as discussed in Section 7.4. However, most of the Geoinformation resources are distributed and heterogeneous, so it is hard for the users to find information of interest to them. Even though the search engines such as Google and Yahoo provide a powerful way to locate online resources, the limitations are obvious when applied to spatial searches due to the unique characteristics of Geoinformation. How can distributed Geoinformation resources be integrated into a single application to serve the Geoinformation community in an effective and intuitive manner? The SWP (Yang et al. 2007) has been proposed to address this challenge.

A Web portal often acts as a gatekeeper to the Internet. Users may begin their sessions on the WWW by visiting a portal and obtain information like news, weather, stock quotes, e-mail, and entertainment (Telang et al. 2004). Generally, a Web portal is actually a Web site that integrates all kinds of information from diverse sources and presents them in a series of Web pages that have a consistent look and feel with access control and procedures for multiple applications. Web portals have been widely used over the Internet and are widely accepted by the users due to the highly integrated information, the intuitive user interface, and the ease of sharing information; examples include the Yahoo Web site (http://m.www.yahoo.com/) and the Microsoft MSN Web site (http://www.msn.com/).

As indicated by the name, an SWP is a Web portal that deals with Geoinformation by integrating the Geoinformation resources (data sets, services, applications, and models) and presenting them in a unified manner. As a result, an SWP inherits all of the characteristics of a Web portal but focuses on the geospatial field and, therefore, has its own unique characteristics. According to Lowe (2004), the SWP is a Web portal that tries to answer the challenge of integrating and presenting data from paper records, isolated files, geodatabases, GIS software, nonspatial databases, and nonspatial applications (such as content management systems, Lowe 2004). The architecture of SWP (Figure 6.10) illustrates the components of an SWP and its relationship with other services, communities, and SWPs.

A mashup might be used in an SWP or as a stand-alone application to combine and integrate different resources such as data, services, or applications so as to create a new service or a new application. Mashup technology can be used as an efficient and effective method to build a robust SWP on both the

* This section is contributed by Zhenlong Li and Wenwen Li, supported by NASA projects (NNX07AD99G and SMD-08-0768) and FGDC project (G09AC00103).

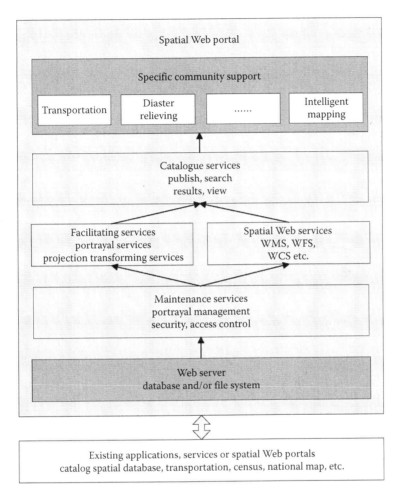

FIGURE 6.10
Spatial Web portal architecture. (Adapted from Yang, C. et al. 2007. *PE&RS* 73(6):691–698.)

server side and the client side. For example, if we need to develop a CSW client that supports the synchronous discovery of multiple CSW services in an SWP: On the server side, mashup techniques can be used to construct a common query string for multiple CSW servers by using an adapter and then fetching the resulting fragments from all of the servers; on the client side, mashup techniques are again used to collect these resulting fragments and combine them into a single result page for displaying.

The SWP includes two characteristics: (1) the capabilities of integration and unified presentation of Geoinformation resources and (2) geospatial interoperability, which is well demonstrated by Geospatial One-Stop (GOS 2009), an SWP that serves as a public gateway for the GIS to access, share, and interoperate the Geoinformation resources. The SWP relies on IT standards, for example, the XML, HTTP, and Web services. Thus, no matter whether the Geoinformation resources are hosted on a Windows OS with SQL Server or a Unix OS with an Oracle database, an open source alternative, or any other combination, it makes no difference for the users. This advantage is an essential factor to facilitate the geospatial interoperability. For example, in an SWP, users can fetch the satellite imagery data from Machine A with Windows Operation System (OS), fetch the DEM from Machine B with Mac OS, and use a visualization application hosted on Machine C with Unix OS to overlay DEM data to the imagery data to render a topographic scene. More important, the visualization application can also interact with other SWPs based on the XML and HTTP standards (Yang et al. 2007). In this scenario, the visualization application can and usually is a Geobrowser, which is detailed in the following section.

6.5.2 Geobrowser

A typical Geobrowser is a client in a Web browser or a stand-alone application that is capable of accessing georeferenced data and providing multidimensional visualization to end users. It aims at improving a user's experience by providing better maps, faster services, more local context, and high resolution data. As an important module of SWPs, the study of Geobrowsers has spawned incredible interest in recent years. It utilizes the cutting-edge information visualization technology, which amplifies cognition by increasing human mental resources, reducing search times, improving recognition of patterns, increasing inference making, and increasing monitoring scope (Card et al. 1999).

6.5.2.1 Architecture

Figure 6.11 demonstrates a basic architecture of the Internet-based Geobrowser. At the remote server side, spatial data such as digital maps,

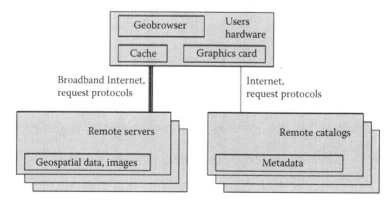

FIGURE 6.11
The architecture of Geobrowser.

satellite imageries, statistics data, DEM, and other data for scientific analysis are stored. Meanwhile, metadata, which is used to describe spatial data, is also stored remotely in catalogs. Once a data request comes, a remote server initiates the catalog service to find the exact match of spatial data to the request. Then, the real data is sent back to the client. Rendering and visualization are executed at the client side. To obtain a better displaying effect, special types of hardware such as high-performance graphic cards and IT technologies such as mutlilevel caches are adopted at the client side.

The history of Geobrowsers can be traced back to 1993, when the first Web-based mapping client, Xerox PARC Map Viewer, was developed by Xerox Palo Alto Research Centre. The PARC map viewer provided basic zoom and pan operations and was able to search for predefined geographic locations (Putz 1994). In 1994, the first online atlas called the National Atlas of Canada was released by Natural Resources Canada, and it provided online mapping for cities, towns, villages, and hamlets in Canada. In 1996, MapQuest (owned by AOL) started to serve the public for address matching and routing services combined with an online map. It was also the first popular free online mapping service. The early success of MapQuest has attracted more attention from the geospatial community. Since then, governments, universities, and industries have endeavored to advance Geobrowser research, for example, when the United States Geological Survey (USGS) coordinated and created the online National Atlas of the United States of America (1997). In 1996, ESRI entered into the Web mapping business and developed a series of software for internet mapping service, such as MapObjects Internet Map Server (1998) and ArcIMS3.0 (2000).

In 1998, the former U.S. Vice President Al Gore proposed a more visionary concept, Digital Earth (DE), for describing a virtual representation of the Earth on the Internet that is interconnected with the world's digital knowledge archives (Gore 1998). The DE vision has propelled the Geobrowser research to a completely new phase: a 3D virtual globe that provides fast speed and near-real user experience to explore and view our planet. An emerging 3D Geobrowser is Google Earth developed by Google, which displays satellite images and digital maps at various resolutions to visually see the Earth's surface. The utilization of a DEM makes available 3D terrain display. Meanwhile, Google Earth helps users manage their own data, including 3D spatial data through KML. The KML also allows people to share 3D models (such as buildings) on Google Earth produced with Google's SketchUp product. Google Earth could also act as a WMS client. Recently, a new feature has been added that monitors traffic speeds at loops located every 200 yards in real time. Other well-known 3D Geobrowsers include Microsoft's Virtual Earth and NASA's WorldWind. Among them, Google Earth and WorldWind are stand-alone models and Virtual Earth is a Web-based service. All of the efforts put into 3D Geobrowser development are aimed at trying to present the world on a single computer. In the next sections, architecture and characteristics of Geobrowser (mainly on 3D) as well as key technologies are introduced. Table 6.1 compares the popular Geobrowsers in aspects of source code openness, visualization dimension, and performance. We see in the table that current 2D Geobrowsers do not achieve as good a performance as the 3D Geobrowsers in rendering vector graphic data due to the Web browser limitations (Hudson-Smith et al. 2007).

6.5.2.2 Supporting Technologies

Since SWPs are always dealing with huge amounts of Geoinformation, in order to provide users a real-time service, advanced techniques should be adopted in terms of data storage, internet transmission, and client-side rendering and visualization.

Image pyramid provides a flexible and convenient multiresolution data storage method to handle multiple scales of data processing. It structures the original image into a sequence of copies in which both the density and resolution are decreased by sampling data regularly. When a request for a certain extent comes in, spatial data with lower resolution (smaller size) will be returned to the client quickly. When the user continuously zooms into the area and wants to view detailed information, the resolution scale of data will be determined at the server side and will be sent back to the client. In this way, users will not need to wait a long time for the original high-resolution data to be loaded.

On-demand tile server is able to leverage base maps as well as cache local tiles in Geobrowsers to provide better performance. The idea is to divide huge spatial data into small tiles in order to speed up in-memory data loading and processing. The tile server serves up a pre-rendered map, computes the needed tile sets, and generates tiles on demand.

Progressive transmission makes immediate feedback of spatial data requests available by providing lower detail data first with further refinements later.

TABLE 6.1

Comparison of Popular Geobrowsers

Geobrowser	Proprietary/ Open Source	Visualization Dimension	Performance
Google Earth	Proprietary	3-D	Good
Google Earth Plug-in	Proprietary	3-D	Good
Map Quest	Proprietary	2-D	Intermediate
Google Maps	Proprietary	2-D	Intermediate
Microsoft Live Search Maps	Proprietary	3-D	Intermediate
Microsoft Virtual Earth	Proprietary	2-D/3-D	Good
ArcGIS Explorer	Proprietary	2-D	Good
OpenLayers	Open Source	3-D	Intermediate
NASA Worldwind	Open Source	3-D	Goods

In the raster data world, the progressive transmission could either refine the image resolution or refine the image detail by sending further coefficients obtained by its core transformation. The refine strategy could be utilizing an image pyramid discussed above or some lossy or lossless compression algorithm.

There are also some IT techniques that are useful to improve the performance of Geobrowser, such as distributed data and server management, intelligent multithreading task scheduling, and client cache (Yang et al. 2005).

6.5.2.3 Current and Future Development

The GOS as described in Section 6.8 is an SWP that serves as a public gateway for the Geoinformation to access, share, and interoperate the Geoinformation resources. Earth Information Exchange (Yang et al., 2008) is a standard-based, Web-services enabled SWP, aiming at serving the public by sharing spatial data, products, services, and knowledge. UniPortal (http://eie.cos.gmu.edu/ WMSUniPortal/2009) is another SWP that focuses on the seamless integration of thousands of OGC Web services.

The SWP and Geobrowser techniques have exerted important roles for both education and commercial use. In the future, global standards for the performance and operational protocols that would allow for ubiquitous and free access of information about the planet should be further developed (Foresman 2004), and the intergovernmental cooperation should be strengthened for the sustainable development of DE research.

6.6 Online Visualization*

6.6.1 Introduction

This new century has seen Geoinformation-related services becoming widely used by the general public in their daily activities. With the advent of Internet, more and more people have got in touch with some sort of digitized Geoinformation, thanks to such Web based map portals as Google Maps, Microsoft Virtual Earth, and other products. The GIS, as the underpinning technology, provides tools and methods for the implementation and operation of these Web based map portals. Among all critical features of a GIS system, visualization might be the most utilized one that helps to facilitate the information transformation process while representing geographic features and their relationships as a set of intelligent maps and views.

---

* This section is contributed by Haihong Yang and Jing Li.

In general, visualization is any technique for creating images, diagrams, or animations to communicate a message. Visualization through visual imagery has been an effective way of communicating both abstract and concrete ideas. With respect to Geoinformation, visualization techniques can turn georeferenced demographic, social, economic, natural, environmental, or political data into visible and intuitive formats so as to demonstrate hidden spatial patterns or moving trends more effectively.

Accessing digital maps has become a trivial task for people since the emergence of the Internet. Focusing specifically on visualization of Geoinformation in Web based platforms, this section discusses more details of this field in terms of technical architectures, representative visualization tools, and common concerns for Geoinformation online visualization.

6.6.2 Technological Architectures of Geoinformation Online Visualization

There are many phases involved in turning geospatial-related facts from pure pictures, signals, or observations into intuitive, highly generalized, and integrated information for visualization use by people. The diagram (Figure 6.12) depicts the basic workflow of Geoinformation transformation from data preparation to visualization.

From the diagram just referred to, we can see that online visualization is just one of many ways to convey the information in a spatial context to users. However, online visualization has proved to be the most popular and effective manner to dispatch Geoinformation. A user, nonetheless, perceives a Web based mapping system for visualization from new perspectives such as

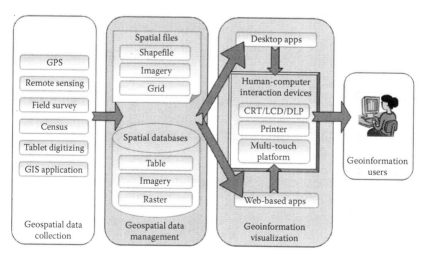

FIGURE 6.12
Work flow of Geoinformation visualization.

performance, responsiveness, interactivity, customizability, and multimode viewing support, besides other conventional ones such as information richness, fidelity, and accuracy. These concerns for user experience in Geoinformation online visualization are greatly related to the architectures chosen in support of a Web based mapping system.

The Internet and Internet-enabled Web applications are built on the client/ server computing framework. The past few years have seen a trend moving from traditional client/server architecture to a Web based model called *smart clients*. Smart clients are Internet-connected tools that allow a user's local applications to interact with server-based applications through the use of Web services. Smart clients are distinguished from standard browser-based applications in that they can work with data even when they are not connected to the Internet. With respect to Geoinformation online visualization, our discussion will be focused on thin-client mapping solutions and smart-client mapping solutions in three architectures, as described below.

6.6.2.1 Architecture I: Standard Browser-Based Mapping for Geoinformation Online Visualization

In this architecture, the maps shown on the Web pages are just image documents that are either hosted or dynamically rendered on the server but delivered to the client's browser per user's mapping requests. The maps shown to the end users are in such common image formats as JPEG, GIF, PNG, and so on. When a map is expected to display on the Web page, either pre-rendered image files or dynamically-generated image files are downloaded to the client's browser. This type of image file serving mechanism could utilize either static file downloads or dynamic file generation, or both methods combined to suit specific mapping needs. More and more state-of-the-art mapping platforms have started to support the composite of a series of adjacent images (a.k.a. tiles), or multiple layers of map documents from different map sources to be fused and shown for the geographic area of interest in a Web page. Options are available for online GIS developers to choose regarding the strategies of fusing in consideration of performance, transparency control, labeling support, and other factors.

Many Internet map servers take a single-frame dynamic generation approach as a solution. For example, the OGC WMS is built upon such an architecture. Other mapping servers including ESRI's ArcIMS and ArcGIS Server also support this kind of on-demand map generation for Web based online visualization. A representative Web based mapping application supported with ArcIMS map services is the refugee resettlement mapping tool (Figure 6.13), which is sponsored by the U.S. Department of Labor, Employment and Training Administration, and the U.S. Department of Health and Human Services, Office of Refugee Resettlement. This tool maps out major refugee serving cities across the United States and the corresponding One-stop Career Centers in those service areas. The georeferenced

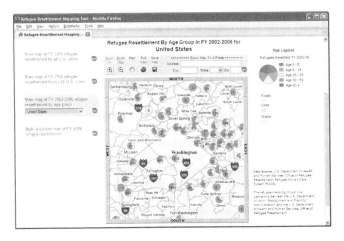

FIGURE 6.13
(See color insert following page 144.) Single-frame dynamic loading mapping example—refugee resettlement mapping tool.

refugee resettlement information is published as an online map service backed by ESRI's ArcIMS Server.

The pros and cons of this approach are

Pros

- Flexible extent and scale manipulation, arbitrary extent, and scale level can be specified.
- Fine-grained control of layer visibility.
- User-specified symbology can be applied by dynamically binding the mapping styles with the underlying spatial data.
- Resource efficient in terms of minimal usage of hard disk spaces on the server side and little demand of computing power on the client side.

Cons

- Performance tends to downgrade dramatically when the number of layers involved for mapping increases, due to the time increased for image generation.
- High demand for server computing power. It needs one round of map generation for each mapping request even if the requested maps are for an identical geographic area.
- Hard to achieve seamless zooming or panning effects on the client side. Tend to pose perceivable lapse when map redraws.
- Need a scheduled routine on the server to clean up the temporary files.

Another approach of standard browser-based mapping for Geoinformation visualization aims at improving the user experience by minimizing the lapse of map refreshing operation perceived by users. With the help of AJAX (Asynchronous JavaScript and XML) technologies, the tile-based dynamic loading approach has become a trend for online mapping tools. Google Maps (Figure 6.14) is one of the representative tile-based mapping systems. For Google Maps, tiles are pre-rendered at a series of zoom levels with each succeeding zoom level dividing the map into 4^N tiles, where N refers to the zoom level. Tiles exist for different map types, because Google Maps supports normal viewing mode, physical viewing mode, satellite viewing mode, or hybrid modes. Due to concerns of the high demand for storage space, tiles do not necessarily cover all regions at all zoom levels. For example, many areas of the Pacific Ocean do not appear at high zoom levels in Google Maps (Google 2009).

The screenshot (Figure 6.15) shows that at least six tiles are needed to fill in the whole map area, but unfortunately no satellite imagery is available for the viewing area at current zoom scale in Google Maps.

Microsoft Virtual Earth (Figure 6.16) is another representative tool using this kind of tile-based mapping approach for Geoinformation online visualization. One striking feature of Virtual Earth is that it includes "bird's eye" aerial imagery taken at 45-degree angle view to show building façades and entrances in addition to imagery taken from satellite sensors, aerial cameras, and 3D city models and terrain. Similar to Google Maps, Virtual Earth allows developers to create applications that layer location-relevant data on top of the Virtual Earth map imagery as pushpins (a.k.a. markers),

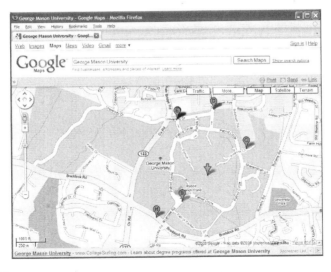

FIGURE 6.14
Tile-based dynamic loading mapping example. (Data from Google Maps.)

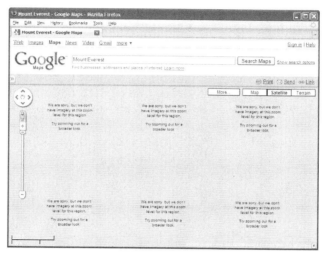

FIGURE 6.15
Tile-based dynamic loading mapping example showing unavailable tiles. (Data from Google Maps.)

simple lines, or simple polygons. Other online mapping tools that support the tile-based mapping approach include Yahoo Maps and MapQuest.

There are tools available that help in the generation of static tiles. For example, ESRI's ArcCatalog can be used to generate tiles for an ArcGIS Server map service (this process is also called *map caching* for this product). Another popular tool to generate tiles for big pictures with high resolution is Zoomify (http://www.zoomify.com), which can create a pyramid of image tiles for

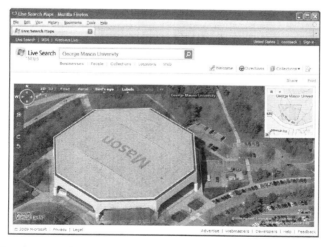

FIGURE 6.16
Tile-based dynamic loading mapping example. (Data from Microsoft Virtual Earth.)

viewing them in a flash-based Web component. The pros and cons of the tile-based mapping approach are as follows:

Pros

- Fast response to a user's map request. Theoretically, the map serving time is just the total of file loading time for all pre-rendered image documents to cover the current viewing area.
- Rich (but static) symbology and mapping styles can be provided to the client without the worry of performance penalty. Also, the details of the map for each zoom level can be maximized without the concern of visualization performance penalties.

Cons

- Can only be viewed in a few pre-defined zoom scales
- Usually, it is a very time-consuming process to generate all tiles for all zoom levels in advance
- Requires a lot of disk space on the server to hold all tile files
- Hard to control fine-grained layer visibility

Please note that the strategies taken for standard browser-based mapping systems are not one way or another as just discussed. Many Internet mapping systems are adopting some sort of hybrid solutions in an attempt to maximize the benefits of both approaches while minimizing their drawbacks. No matter what kind of strategies are taken for the standard browser-based mapping architecture, the biggest shortcoming of this architecture is the lack of comprehensive support of graphics-related operations (e.g., freehand drawing of common graphics such as circles, lines, rectangles, or polygons) and animation, all due to the limitation of standard features available in existing Web browsers. Responding to these limitations, vendors have come up with different solutions to provide users with the capability to draw basic graphics and render animations within a Web browser. This leads us to the second architecture of Geoinformation online visualization.

6.6.2.2 Architecture II: Rich Browser-Based Mapping for Geoinformation Online Visualization

In this architecture, raw spatial data are retrieved from servers and streamed to the client's browser, where maps are rendered within a built-in graphics display engine. These built-in visualization engines for browsers are usually called plug-ins. The plug-ins in browsers relieve the burden of servers by taking more computing power from the client side while using the server only for minimal data support. The most common browser plug-in is Java runtime environment (JRE), which enables Java applets to be executed inside a browser.

Other browser plug-ins related to visualization include Scalable Vector Graphics (SVG) Viewer, Macromedia Flash Player, and the latest Microsoft Silverlight. All these browser plug-ins can be characterized as Rich Internet application (RIA) solutions for Geoinformation online visualization.

A Java applet is an applet delivered to the users in the form of Java bytecode. Java applets are used to provide interactive features (including graphics and animation) to Web applications that cannot be achieved by HTML. Java applets can run in a Web browser with the support of JRE. Some Web-based mapping systems use Java applets as one of their mapping solutions. For example, an old version of ESRI's ArcIMS (version 8x) allows the developers to choose the Java Viewer template as one way to publish maps, which puts map rendering logic and other graphic components inside a Java applet while relying on the servers to stream spatial data to it.

As an open standard under development by the World Wide Web Consortium (W3C) since 1999, SVG is a family of specifications of XML-based file format for describing 2D vector graphics, both static and dynamic (interactive or animated). The SVG recognizes shape types for 2D graphics including rectangles, circles, ellipses, lines, polylines, polygons, symbols, and path elements. These graphical objects can be grouped, styled, and transformed using stylesheets such as XSLT and CSS. The DOM for SVG, which includes the full XML DOM, allows for straightforward and efficient vector graphics animation via scripting. This makes SVG useful for dynamic Web sites with graphical content, especially for Web mapping, which needs graphical representation for Geoinformation (Cheung and Shea 2004). All major modern Web browsers can render SVG markup directly except Windows Internet Explorer (as of October 2008). To view SVG files in Internet Explorer, users have to download and install a browser plug-in, SVG Viewer sponsored by Adobe Systems. Below is a screenshot of a mapping tool that is built upon an open source SVG-based mapping framework, GeoClient (Figure 6.17).

Adobe Flash Player might be the most installed browser plug-in currently, because it is available for most common Web browsers to display flash-based contents at no charge. One representative flash-based mapping platform is worldKit, which is a lightweight flash-based mapping solution for the Web as an open source project. The applications built upon worldKit rely on a shockwave file (worldkit.swf) as the core mapping engine and some XML files (config.xml, rss.xml) to configure the appearance and functionality and feed data by RSS to the maps (WorldKit 2009).

Microsoft Silverlight is a programmable Web browser plug-in that enables features such as animation, vector graphics, and audio-video playback, which characterize rich Internet applications. It is claimed to be compatible with multiple Web browser products used on Microsoft Windows and Mac OS X operating systems. Silverlight integrates multimedia, graphics, animations, and interactivity into a single runtime environment. It is being designed to work in conjunction with XAML (used for marking up the vector graphics and animations) and is scriptable with JavaScript. The GIS industry has

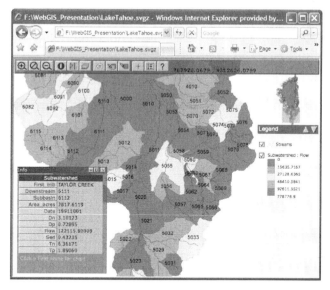

FIGURE 6.17
(See color insert following page 144.) Scalable vector graphics-based mapping tool. (Data from GeoClient.)

welcomed this new rich Internet application solution and started to develop Silverlight-based mapping platforms. For example, ESRI has released ArcGIS API for Microsoft Silverlight, which allows for the integration of ArcGIS Server and Microsoft Virtual Earth services and capabilities in a rich interactive mapping application. With ArcGIS API for Microsoft Silverlight, ArcGIS Server resources (e.g., maps, address locators, GIS tasks, and spatial analysis) can be leveraged in Silverlight-enabled applications (ESRIb 2009).

The pros and cons of this kind of rich browser-based mapping architecture for Geoinformation online visualization are as follows:

Pros

- Spatial data are streamed and most often cached in the memory of a client's machine. Spatial query and spatial analysis operations can be done on the client side without further requests sent to the servers.
- Some common mapping operations might respond relatively faster, such as zoom in or out, pan, and extent or scale change.
- Complicated graphics operations and animations can be carried out inside a browser.
- Can achieve much more interactivity for better system performance and user experience.
- Interaction with other components of a Web page (though limited by local access control policy and other browser settings) can be achieved.

Cons

- Need to install a specialized plug-in package (with the SVG as an exception for non-Internet Explorer Web browsers)
- Depending on how large the mapping area is and how complex the mapping contents are, the data transferred and held on the client's machine might be too big to handle
- Require online map developers to deal with either a specific markup language (e.g., SVG) or a dedicated authoring tool (e.g., Flash Professional) or new programming platforms (e.g., Microsoft Silverlight) to implement the graphics rendering logic for the raw spatial data

To mitigate the limitations incurred when using browsers, most browser plug-in vendors also release their stand-alone versions called *players* or *viewers*. For example, a ShockWave file can be played back either inside a Web browser or within a stand-alone ShockWave player application. The stand-alone viewer mapping solution is the third architecture we are going to discuss with regard to Geoinformation online visualization.

6.6.2.3 Architecture III: Stand-Alone Viewer Mapping for Geoinformation Online Visualization

Stand-alone map viewers run as independent applications in a client's machine while retrieving spatial data from the server via Web services. These stand-alone viewer applications can be categorized as smart clients as discussed earlier, because they can also consume local data files and, therefore, work offline. A number of proprietary or open source map viewers for Geoinformation online visualization are available in the GIS industry, including ArcGIS Explorer, Google Earth, NASA World Wind, and others.

ESRI's ArcGIS Explorer is a lightweight desktop client that lets the users connect to 2D and 3D data published on the Web or to fuse such data with their local data. Users of ArcGIS Explorer can perform a wide variety of queries on their maps and customize their map displays. Though freely downloadable on the Web, it can operate in a stand-alone environment and does not need to connect to a server (ESRIc 2009).

Google Earth is a virtual globe that shows map and geographic information upon the Earth's surface. Google Earth displays satellite images of varying resolution of the Earth's surface, allowing users to visually see things like cities and houses looking perpendicularly down or with bird's eye view. Google Earth allows easy and highly interactive exploration of geographic data for public users to overlay and explore geographic data, from many sources simultaneously, in ways that have been limited to the GIS professionals who own the data and work with specialized software. Below is a

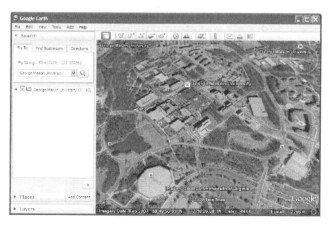

FIGURE 6.18
Stand-alone map viewer example. (Data from Google Earth.)

screenshot of Google Earth (Figure 6.18), which shows the campus of George Mason University from an oblique angle.

The pros and cons of the stand-alone map viewer architecture for Geoinformation online visualization are as follows:

Pros

- Can take full use of the computing power of the machine where such a stand-alone map viewer is installed. This usually means that it can take full advantage of the graphic acceleration mechanisms and other features that the client's machine provides, especially when doing 3D graphics rendering.
- Some viewers provide customization capabilities with programmable APIs.
- Better interactivity and performance.
- Supports advanced graphics operations and animation.

Cons

- Need to install a proprietary or open source viewer application.
- Cannot be easily integrated with other Web based applications (some viewers have been converted into browser plug-ins to minimize such concern, e.g., Google Earth plug-in as mentioned earlier).
- Some viewers have their own proprietary data formats and customized programming languages. This restricts the integration of data for applications and analyses.

Stand-alone map viewers offer users better mapping performance and interactivity. However, they might pose interoperability challenges due to the use of specific data formats for spatial data representation and transfer and the use of dedicated vendor-specific services. Overall, we have concerns about Geoinformation online visualization no matter which architecture is chosen to support an online mapping system, as discussed in the next paragraph.

6.6.3 Challenges of Geoinformation Online Visualization

Online visualization of Geoinformation, with the support of standard Web browsers or stand-alone map viewers, helps users investigate georeferenced facts and spatial relationships from a collaborative perspective without the worry of formal GIS training or a dedicated GIS system management need. However, to make an online mapping system accessible and, therefore, usable, there are several challenges that should be solved when designing and developing it for Geoinformation visualization.

A mapping tool's *performance* regarding visualization is usually judged by its response time for initial map loading and subsequent map refreshing operations. There are many factors that affect the responsiveness of an online mapping system varying with different system architectures such as feature density for display, communication bandwidth between server and clients, and the server processing time for map requests. Generally speaking, static tile-based mapping systems have a quick and predictable performance on comparison with single-frame or dynamic tile-based mapping systems. For online mapping systems that need a browser plug-in or a stand-alone map viewer, the size of raw spatial data streamed into a client's machine must be taken into consideration, because it impacts the client-side rendering performance and resource usage.

In the scientific domain, visualization requirements change with multi-dimensional issues. Besides 2D images or features, 3D or 4D representation is a necessary tool to understand the geospatial sciences. The visualization intensity and complexity might exceed the capacity of the client due to the increasing data volume. Under such circumstances, a robust computing environment is required at the server side.

The *reliability* of an online visualization tool is one of the important criteria in evaluating its quality. An online mapping system should produce mapping results that a user expects. Elements that control the dynamics of visualization involve layer visibility, symbology, and dynamic generation of graphics to highlight spatial query or analysis results. The system should respond to the changes of such elements in a meaningful and expected way. Otherwise, explanations or messages should be prompted to the user as feedback.

Considering the framework of online visualization, the success of an online Geoinformation visualization system depends on the reliability of client application, the reliability of server, and the reliability of client/server communication. For example, client applications should respond quickly to the users' manipulation, send requests to servers, and display visualization results in a timely fashion. Servers are responsible for effectively processing the requests sent by clients while maintaining the robustness of each component involved and the interactions among components. The design of the communication mode between the server and client should count in extreme situations, such as the limitation of network bandwidth, interruption of message delivering, and other factors.

Interactivity means how flexible the users can manipulate a mapping system for a customized display by defining their own viewing areas, zoom levels, and mapping styles. The interfaces of predefined or *ad hoc* queries are also considered essential for the users to interact with an online mapping system; specific features can be identified with user-specified searching criteria. On the other hand, geocoding and service locating functions are also important to location-sensitive business intelligence tools. Regarding 3D visualization, it is essential for an online mapping tool that enables a user to easily rotate the globe and adjust the viewing altitude and tilting angle to see the area under investigation. In general, the interactivity of an online mapping system is determined by its level of flexibility and the number of adjustable mapping elements.

Information richness is conventionally one aspect in the field of cartography in terms of generalization, thematic overlay, and symbology definition. One rule of thumb is that if a static tile-based approach is used to support the online visualization, as many possible pieces of information can be put inside the map, as long as the map is not so crowded that a map user can not discern any distinctive features. For maps that are dynamically generated, care must be taken when selecting layers to display due to the concern of performance and layers blocking each other. Scale-dependent display technique, along with transparency control, can help in meeting the need of information richness for Geoinformation visualization while maintaining acceptable performance.

The *scalability* for Geoinformation online visualization is a many-fold issue. It could mean how large a number of users the system can support to access the mapping tool concurrently. It could also mean how wide the extent of geographic area the mapping tool can cover. For a general map portal like Google Maps, massive hardware and software resources are required to handle over hundreds of thousands of concurrent requests. On the contrary, for some mapping tools that are intended to be used only inside an Intranet, one or two dedicated map servers might suffice. It is an intimidating task for a small firm to support a mapping tool like Google Maps, considering its extent of geographic coverage (with worldwide satellite imagery and street-detailed aerial photos for most large cities of North America). Scalability should be kept in mind when determining an application hosting environment, Web programming framework, and underlying spatial data management infrastructure for Geoinformation online visualization.

Interoperability indicates the communication between different computer systems in terms of seamless access and sharing of data structures across multiple hardware platforms, operating systems, and application software. Geoinformation visualization on the Web calls for open access to geographic data and software functionality using widely adopted, practical standards. Online mapping systems, therefore, are expected to provide appropriate open application programming interfaces and support key data interchange formats and Web services standards to ensure relevant GIS and information technology interoperability between systems over wired and/or wireless networks. Standards such as Geography Markup Language (GML) are developed to promote the interoperability of online visualization system, which allows Web-based mapping tools the abilities of representing and sharing geospatial data in a common and interchangeable format. More online mapping platforms are expected to support such open spatial data representation standards.

6.7 Quality Issues in GII*

The GII includes the Geoinformation data and services, which are the qualitative characteristics of these data and functions compared with the final users' requirements. The first dimension is the GII itself, whereas the second is the quality issues of this GII, which act as an umbrella to cover all of its nonfunctional aspects (Onchaga 2003; 2005; Wu et al. 2005; Wu and Zhang 2007). This is very similar to the commercial area in which customers buy merchandise. A customer first selects a product for his purpose. Then, he considers the appropriate quality of the merchandise to purchase. Usually the merchandise with same functions and higher quality costs more.

In terms of the quality issues in GII, two types of research issues can be identified, that is, spatial data quality and the Quality of Geospatial Information Service (QoGIS). Without the support of high-quality data, high QoGIS is impossible. Even when high-quality data are available, it is still difficult to obtain high QoGIS. Quality is a very subjective topic, however, and the quality of an object changes when evaluated by different people. It is very difficult to give an explicit and operational definition. The research of quality includes its definition, contents, factor model, evaluation procedure, propagation, control and management, and so on. *Quality evaluation* is defined here as a numerical value describing an object. A quality propagation model describes how the quality of parts is synthesized in a series of processes. A quality propagation model helps to control and manage the synthesized quality and answers the following two questions: (1) how can we reach a preset quality with minimum costs? (2) how do we optimize available resources to reach the highest quality goal?

Definitions of quality are situated in temporal and community contexts (Reeves and Bednar 1994). As a compromise between international communities, the ISO 9000 series of standards give a widely recognized definition: quality is the degree to which a set of inherent characteristics fulfils requirements

---

* This section is contributed by Huayi Wu and Hanwu Zhang.

(ISO 2005). This definition implies three aspects of quality. First, quality is about a set of characteristics. A product may have many characteristics, but when considering its quality, only some of these characteristics may be considered. Second, these characteristics must be inherent and not imposed by stakeholders. For example, price is not a kind of characteristic and it should not be considered as an element of quality. Third, how good these characteristics are measured by the degree that they fulfill customers' requirements. Since different people have different requirements, a product or service may be of different qualities when evaluated by different people.

According to the above analysis, quality has two general entities: subject and object. The subject of a quality is the consumer of this quality. The object of a quality is the product or service that fulfils the requirements of its subjects. One object may have more than one subject. For example, a geographic catalog service provided in a Web page may be of high quality to final users, but it may be of low quality if someone requires integrating this service into his own portal; however, this service does not provide standard interfaces. Therefore, when a quality is evaluated, the subject is a prerequisite of the evaluation result.

6.7.1 Quality of Spatial Data

Spatial data is a data type that records spatial distributed objects, states, or phenomena that describe the environment of the Earth surface by quantity, quality, distribution features, relations, and evolving laws. Spatial data is the essential element of a geographic information project, because all the procedures including acquisition, processing, storage, analysis, and dissemination are working on spatial data. Most of the investment in a geographic information project is spent on data.

The quality of spatial data can be evaluated by measuring how well a set of inherent characteristics of spatial data fulfils clients' requirements. These requirements or characteristics form a hierarchical structure that is referred to as element model of spatial data quality, a general model including most quality elements with possible extension and tailoring for any specific applications. Table 6.2 shows an element model of spatial data quality synthesized from FGDC standards, NCGIA core curriculum, and other references.

TABLE 6.2

Element Model of Geospatial Data Quality

	Accuracy	Precision	Completeness	Consistency
Feature			*	
Spatial	*	*	*	*
Temporal	*	*	*	*
Thematic	*	*	*	*
Format				

Considering the general requirements of spatial data, the above model considers accuracy, precision, completeness, and consistency as the most important characteristics of spatial data. Although spatial data usually involves spatial, temporal, and thematic information, the accuracy factor can be further described by the three sub-factors, that is, spatial, temporal, and thematic.

6.7.2 Quality of Geospatial Information Service

The Quality of Geospatial Information Service (QoGIS) is the degree to which a set of inherent characteristics of a service matches a customer's requirements. However, geospatial information service is unique, and the element model of QoGIS is fundamental for evaluation.

Geospatial information service is a many-faceted product and it has several levels of customers. The first level is those who use a geospatial information service as a software tool. It is installed, deployed, maintained, and upgraded. In this aspect, the quality of geospatial service is the same as the software quality. The ISO/IEC 9126 (2001) and ISO/IEC 14598 (1998–2000) specify the software quality using six major elements, that is, functionality, reliability, usability, efficiency, maintainability, and portability. These six elements are further split into 21 sub-elements as shown in Figure 6.19.

The second level is the quality of a geospatial information service as an executing instance. This quality depends on the computing and networking environments when the instance is executed. For this quality, customer's requirements include five elements, that is, security, transaction, stability, performance, and usability as shown in Figure 6.20.

The third level is the quality of a service's result, whose element tree depends on the function of the service. The result can be a map, a transformed data set, a best route, a spatial decision, or other processing results. Taking WMS as an example, the result is a raster map, so its quality elements can be, but are not limited to, resolution, color schema, information delivered, label, legend, and so on.

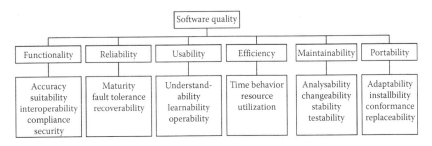

FIGURE 6.19
Element model of software quality.

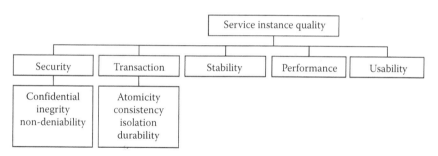

FIGURE 6.20
Element tree of service instance quality.

6.7.3 Measurement, Evaluation, and Other Issues of Geospatial Quality

To evaluate quality is to find a percentage of which the object's elements meet the customer's requirements in two aspects: (1) measure an object's elements, and (2) compare the measured elements with customers' requirements. Although there is usually more than one element measured, an overall evaluation would be helpful for comparing the two aspects through a weighted structure to consider all elements. The overall quality evaluation is a global indicator of how good the quality is.

The ultimate purpose of Geoinformation and services is to support decision making. To assure the quality of the final decision, obtaining the knowledge about how elements' qualities are propagated within the workflow of Geoinformation processes is essential. Quality usually decreases with any processing, and geospatial quality is a complicated and application-oriented issue, so the propagation mechanism is not a pure mathematical problem. Probability theory, numerical approach, and fuzzy theory have been used in modeling the propagation of mapping errors. Theoretically, quality of Geoinformation services composition is determined by those services that form the composition. However, many issues still remain open for modeling the quality of service composition.

6.8 NSDI/GOS/FEA GeoProfile*

From the SDI Cookbook, the SDI "provides a basis for spatial data discovery, evaluation, and application for users and providers within all levels of government, the commercial sector, the non-profit sector, academia and by citizens in general" (GSDIA 2009). An SDI provides a framework consisting of

* This section is contributed by Doug Nebert, Qunying Huang, Jing Li, Min Sun, supported by FGDC project (G09AC00103).

spatial data, metadata, tools, and users that facilitate scientific and public uses (Masser 2005). The core elements of an SDI include policies and institutional arrangements (governance, data privacy and security, data sharing, and cost recovery); people (training, professional development, cooperation, and outreach); data (digital base map, thematic, statistical, and place names); and technology (hardware, software, networks, databases, and technical implementation plans) (Williamson et al. 2003).

6.8.1 Components of a Spatial Data Infrastructure

The U.S. FGDC has identified the key components required by a community to enable an Spatial Data Infrasturcture (SDI). These components include

- Metadata—structured descriptions of data and services that support search (discovery), evaluation for fitness for use, and detail sufficient to support end-user applications of the data or services.
- Data—both core base map type data, known as "Framework," and thematic data (other "GeoData") may be registered with an SDI.
- Services—provide end users with the ability to discover, access, and process the data, as described in metadata through a catalog service.
- Standards—relevant to all the components, these provide agreements that improve interoperability and exchange of data for multiple purposes.
- Partnerships—the governance and agreements between producers and with users for the collective creation and maintenance of the data sets and their access mechanisms.

Figure 6.21 shows this general SDI architecture.

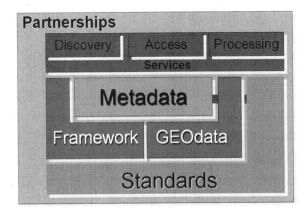

FIGURE 6.21
Overview of the elements and status of SDI.

6.8.2 Metadata

Typically, an early task in establishing an SDI is to inventory what data exists within the community and to capture basic information about its characteristics, quality, formats, and accessibility and to make this information searchable through an online catalog. This basic structured information is known as metadata. An early standard for metadata was published in June 1994 by the FGDC. An international standard (ISO 19115/19139) now exists and is being adopted or profiled by most countries. Metadata can apply to data, services, and other resource types.

Metadata provides documentation of existing internal Geoinformation resources within an organization (inventory). It permits structured search and comparison of held Geoinformation resources by others (catalog). It provides end users with adequate information to take the resource and apply it in an appropriate context (documentation).

The ISO 19115 and TS19139 provide an international standard for metadata and its encoding, respectively. Metadata from ISO 19115 are expressed in XML according to ISO 19139 to promote a common terminology and format for the metadata. The XML metadata may include reference to an XML Style Sheet (XSL) to present the XML data in a browser. Metadata may also be presented in a HTML or X-HTML format that can be parsed and presented.

Metadata describes data and service resources for order, access, or local use. Metadata is used to describe all types of data, with an emphasis on "truth in labeling." This means that in most SDI implementations, data of varying detail and accuracy may be documented and made available, as described in the metadata.

6.8.3 Framework

Framework data are common-use data layers that include the elements of a base map, as they are used to provide context and orientation to additional geographic data. These framework data layers tend to be of a consistent representation, detail, resolution, and information content over the entire geographic extent of the community. Viewed through map services or downloaded as data, they can be used for visualization or for analysis with other data in desktop or Web GIS.

In the United States, 11 abstract data content standards are being promulgated through the ANSI process as American National Standards. These framework layers include elevation, orthoimagery, hydrographic data, governmental unit boundaries, cadastral, geodetic control, and transportation. In Europe, the INSPIRE initiative is defining three sets of basic and thematic data that will be developed and served by European member countries for multiple uses. It defines a core set of framework data that will provide a consistent coverage of Europe (INSPIRE 2009). In both the U.S. and European contexts, each standard data theme (layer) also includes

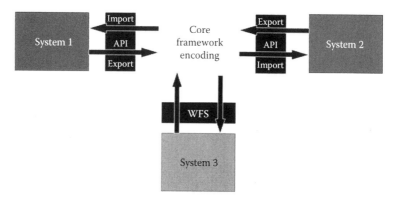

FIGURE 6.22
Interoperability with common framework data encoding.

guidance on its encoding in XML using the OGC Geography Markup Language that can be served over the Web using OGC Web Feature Services (Figure 6.22).

6.8.4 Services

Services in SDI can be categorized as supporting discovery, data access, and geospatial processing. These services interact with the metadata and on the data repositories themselves, and they are often presented to the end user through a portal (Maguire et al. 2005).

6.8.4.1 Discovery

An SDI requires catalog services to help users discover and interact with data, using metadata to support simple and advanced searches. Commercial search engines are still not able to support spatial and temporal search on spatial metadata, thus specialized catalogs of spatial metadata are created to fill this need. Often, the search of the SDI community catalog is the main capability presented in an SDI Web portal.

Spatial Web portals, or Geo-Portals, centralize access to SDI resources. These may include locating data and services, support online map viewing, download of data, link to related Web sites, and providing helper applications for others to access. In the United States, the GOS Portal also allows self-organizing communities to post and manage selected content and share data collection plans and requirements to support partnerships and collaborations. The GOS Portal (Figure 6.23) features a search portlet that lets users search for, evaluate, find, and obtain spatial data, a set of community pages organized by topic, and a featured topic area where current data, events, or topics can be emphasized.

Search — enables
users to find
geospatial data

Communities —
highlight
authoritative data
sources

Featured
resources —
highlight maps,
applications and
Web sites of
current interest

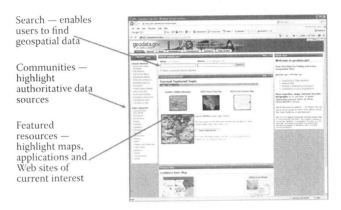

FIGURE 6.23
GOS portal.

Users may contribute metadata to a community catalog in several ways. They may enter metadata into a form on the central catalog and have it managed there, upload metadata as XML to the catalog, or register their existing metadata collection or service to be harvested into or searched by the catalog (Figure 6.24).

6.8.4.2 Access

A second category of services provides standardised access to Geoinformation. This may be made via static files on ftp or via Web services. These services deliver raw spatial data, not maps, through standard files or encoding formats. Data access services may, in turn, be used by processing services to create maps or perform analyses.

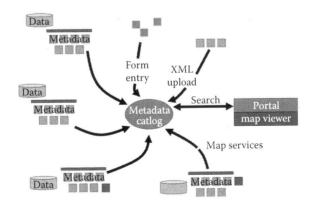

FIGURE 6.24
Metadata publication options.

6.8.4.3 Process

A third class of services provide additional processing on Geoinformation. These include capabilities that extend and enhance the delivery of data through processes applied to raw data. Examples include Web mapping, symbolization, coordinate transformation, topologic overlay, or routing services.

Figure 6.25 uses an informal modeling notation to depict the interaction of common data resources and functions present in an advanced SDI (Nebert 2004). The relationships between objects is described by the words, and the directionality is implied by the arrows, for example, "Metadata are made searchable through Catalog Services."

6.8.5 Standardization

Standardization makes SDI work; standards should affect every SDI activity to promote maximum use and access of shared information. Standards include specifications, formal standards, and documented practices (Figure 6.26). The SDI Cookbook identifies a Recommended Minimum Software Standards Suite for SDI, current in 2009, as shown in Figure 6.26 (GSDIA 2009).

Although not comprehensive, this set of standards will encourage the discovery, access, and use of spatial data and maps by focusing on standards that are developed, adopted, and implemented by the GIS industry and the users of commercial and open-source software who support these capabilities. Adjacent regional or national SDIs that adopt a common standards suite will be able to support cross-border data discovery, analysis, and visualization.

6.8.5.1 Enterprise Architecture

Enterprise architecture (EA) defines a framework in which one describes the current and future state of mission-oriented work processes and helps justify the investments (personnel, data, and applications) of an "enterprise." It is a precursor to electronic government (e-gov) and business process re-engineering. Geospatial capabilities often support many activities within an enterprise, known as *lines of business*, so the inclusion of these capabilities in the design of systems is important to improve the uptake of standard approaches and reusable solutions for many application areas.

The goals of EA can be summarized in five points:

1. To provide a structured approach to business process assessment and re-engineering
2. To support classification of business processes, data concepts, standards, and services (components and interfaces)
3. To provide a reference framework to store models that explains the services and data behind them

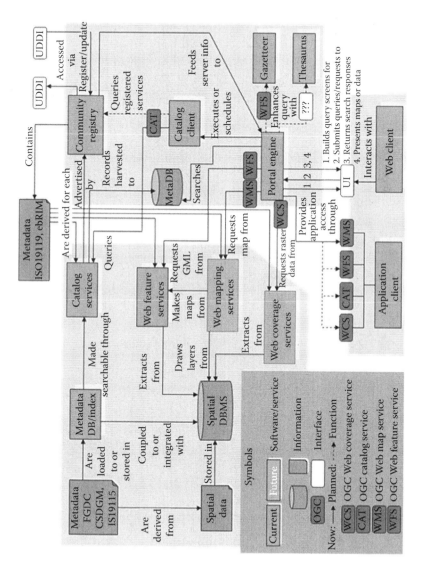

FIGURE 6.25
(See color insert following page 144.) Data and service interaction in an SDI.

2009 Recommended Minimum Software Standards Suite for Spatial Data Infrastructure (REMSSS-SDI 2009)	
SDI Core Standards	OGC Abbreviation for Standard*
OGC Web Map Service, Version 1.3	WMS
OGC Web Feature Service, Version 1.1 or OGC Web Feature Service (Transactional)	WFS or WFS(T)
OGC Filter Encoding ,Version 1.1 (used in conjunction with WFS)	Filter
OGC Geography Markup Language Version, 3.2.1	GML
OGC KML, Version 2.2	KML
OGC Web Coverage Service, 1.1.2	WCS
plus	
OGC Catalogue Service, Version 2.0.2 HTTP protocol binding, CS-W and	CAT CSW
OpenGIS Catalogue Services Specification 2.0.2 - ISO Metadata Application Profile (1.0.0) (Note: supports ISO Metadata Standard 19115:2003 and ISO DTS 19139:2006)	CAT2 ISO AP
or	
OGC Catalogue Service, Version 2.0 Z39.50 protocol binding and	CAT Z3950
OpenGIS Catalogue Services Specification 2.0.1 - FGDC Application Profile supporting FGDC Content Standard for Digital Geospatial Metadata (CSDGM, 1998)	FGDC AP
SDI-REMSSS 2009 Supplemental Standards	
OGC Styled Layer Descriptor, Version 1.0	SLD
OGC Web Map Context, Version 1.1	WMC
OpenGIS Sensor Model Language, Version 1.0.0	SensorML
OpenGIS Sensor Observation Service, Version 1.0.0	SOS

FIGURE 6.26
Candidate "SDI 1.0" standards.

4. To improve efficiency within an organization

5. To identify potential for services and data reuse for multiple purposes

Figure 6.27 shows the flow and detail of the framework of the U.S. Federal Enterprise Architecture (FEA). It contains those components (business, data,

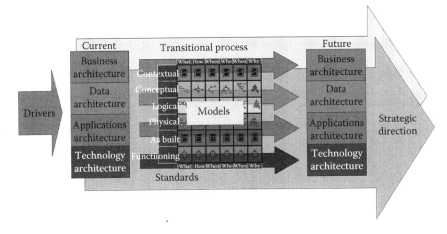

FIGURE 6.27
(See color insert following page 144.) Conceptual process diagram for the U.S. FEA.

applications, and technology) that must be considered together in designing enterprise solutions (The Chief Information Officers Council 1999).

The FEA is built using a set of interrelated reference models, designed to facilitate cross-agency analysis and the identification of duplicative investments, gaps, and opportunities for collaboration within and across agencies (Executive Office of the President 2007). In other words, the reference models comprise a common taxonomy and ontology for describing and managing important IT resources across the federal government in response to mission or business needs. The reference model consists of five parts:

1. *Business reference model (BRM):* Defines mission-critical lines of business, business processes, and functions

2. *Performance reference model (PRM):* Developed to define measures of business performance

3. *Technical reference model (TRM):* Identifies and describes the technology (components, interfaces) used to achieve the BRM

4. *Service reference model (SRM):* Defines the types and instances of services required to support processes

5. *Data reference model (DRM):* Defines the data or information concepts, structures, definitions, and values or enumerations required by the BRM in the context of the TRM

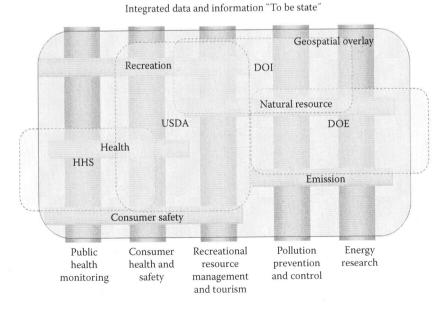

FIGURE 6.28
Cross-cutting nature of geospatial capabilities in government.

6.8.5.2 *Geospatial Profile for Enterprise Architecture*

Geographic information is used in a majority of business settings in and outside of government. However, geographic information and services are often not addressed consistently within and between organizations. Interoperability among providers and consumers of geographic data and services requires a common understanding of semantics and functional capabilities. Thus, development of common multijurisdictional approaches to the use of geographic information and services should be considered when developing an EA.

The Office of Management and Budget has defined several cross-government "Lines of Business" (LoBs), one of which includes Geoinformation. The Geospatial LoB aims at optimizing federal geospatial-related investments to reduce the cost of government and, at the same time, improving service quality to the public (Geospatial Line of Business 2009). Figure 6.28 illustrates the various vertical (internal) lines of business that are common to multiple agencies and that have the potential for integration using location (geospatial overlay).

As illustrated in Figure 6.29, Geoinformation resources and activities comprise a cross-cutting of a horizontal segment, shown against the vertical lines of business, that supports core mission areas and business services (Architecture and Infrastructure Committee 2009).

FEA Geospatial Profile. The FEA Geospatial Profile provides guidance to government agencies to highlight the role of geospatial capabilities within agency lines of business. It is intended for use by federal business planners and chief architects involved in budget planning and submission. Inclusion of standardized geospatial capabilities is recommended in an agency's enterprise architecture. The Geospatial Profile has the following specific objectives:

1. Promotes the consideration of location as a component of agency business activities and processes
2. Supports the description and sharing of Geoinformation resources across organizations
3. Builds spatial considerations into all aspects of agency enterprise architecture development
4. Reinforces existing guidelines, standards, and policies established by OMB, FGDC, and ISO for development, management, and use of Geoinformation resources
5. Measures performance in the use and management of Geoinformation resources

Version 2.0 of the FEA Geospatial Profile follows the Federal Segment Architecture Methodology (FSAM) and expands existing taxonomies and

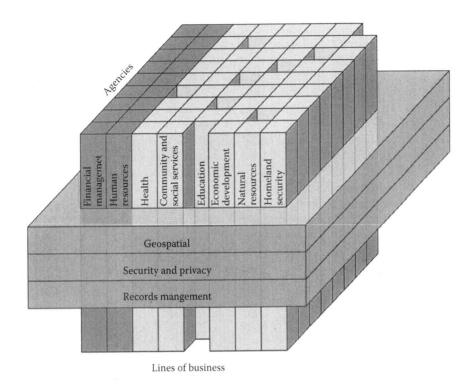

FIGURE 6.29
Geospatial capabilities as a cross-cutting enterprise segment.

approaches to recognize the activities and products of the geospatial community.

Reference Model in Geospatial Profile. Five reference models are used in the US FEA to provide context for the development of agency architectures. These include

1. The PRM, which is used to evaluate how performance or success can be measured, in this case through the incorporation of location in agency data.
2. The BRM, which provides the mission requirements, lines of business, business areas, and business processes based on a common taxonomy and approach. Geospatial capabilities are inserted here with respect to the use of spatial analysis on location data within the enterprise (i.e., Where? How far? What is near?).
3. The DRM, which provides guidance on the management of data content, structure, and description. The geospatial domain has well-known data formats and metadata that support DRM, as documented by the Profile. Spatial data in the DRM includes: administrative and

operational data (business data maintained and used by federal agencies), other spatial data (spatial data with national or regional coverage that do not fit into the context of the NSDI framework themes set forth in OMB Circular A–16), and nationally significant spatial data (core spatial visualization and analysis component of many geospatially enabled business functions) (Geospatial Line of Business 2009). The DRM provides an architectural pattern for sharing and exchanging data with the use of framework data content standards (ANSI/INCITS), the use of standard formats and services (OGC GML, WFS), and the use of spatial data formats. The data description refers to the FGDC Content Standard for Digital Spatial Metadata (CSDGM) in active use. Spatial metadata should be organized by a common schema that is recommended use of ISO 19115, known as the North American Profile, and its XML encoding (ISO TS 19139) for the format and transfer of spatial metadata.

4. The SRM, which provides a taxonomy of service domains and service types that has been expanded to include geospatial processes.

5. The TRM, which defines the standards that are relevant to the execution of the activity. In this case, the list of geospatial standards is made available to the agency architect as a checklist to promote interoperability and reuse.

6.8.6 Partnerships

Partnerships extend our capabilities. Proper governance of the community is essential through the declaration of a variety of roles and responsibilities. To be most effective, national governments and nongovernmental organizations should partner with other levels of government and sectors to promote multilateral coordination (Jacoby et al. 2002). The government or a foundation may be able to fund agencies with "seed" funding to further existing efforts toward common goals. The SDI participation should also include the contribution of citizens through voluntary geospatial information (Goodchild 2005). Partnerships extend local capabilities in technology, skills, logistics, and data and contribute to a more accessible *commons* of shared resources.

6.8.7 Summary and Conclusion

As a framework for sharing spatial data and metadata by users and their software, SDI has been successfully implemented in a number of domains. For example, as a pilot project of the Infrastructure for Spatial Information in Europe (INSPIRE), SDIGER examines the hydrological settings of major river basins in Europe. The GOS has provided public access to a variety of Geoinformation that supports spatial analysis. The general components of such frameworks should include metadata, framework data standards, common Web services, standardization on core capabilities, and effective parternships (Rajabifard and Williamson 2001). This new paradigm enables the users to discover, access, and use spatial data over a computer network.

The success of an SDI often requires the formal coordination of the activities and the various participants. An SDI is more than the sum of its parts, as it requires attention, planning, and coordination of the roles and responsibilities in support of a broad range of end users. When establishing an SDI, one should also consider the principles that link the components as a whole, such as central versus coordinated data management, development of national policies, participation and membership, issues of open or controlled access, and effective collaboration between different organizations. Success of an SDI is fundamentally enhanced by the availability of data, its ease of use, flexibility to support many uses, and funded contributions and maintenance (Kok and Loenen 2004).

This chapter provides an overview of the Geoinformation infrastructure with illustrations and examples. The infrastructure research is building toward the CI research in the geospatial fields (Yang et al. 2010). A well-implemented infrastructure would require (a) the full development of different aspects discussed, (b) the mature of supporting computing platforms, and (c) well-designed middleware. Such an infrastructure should be able to provide both generic information and Geoinformation specific functionalities toward a spatial cloud computing arena.

In the process of building the GII, software will continue to play a significant role, and end users or most users will still be exposed to software rather than just services. Therefore, the evolution of moving from software to services will continue with the two co-existing for multiple years before we can finally achieve what has been envisioned in GCI and spatial cloud computing or the plug-and-play stage of GII.

References

Alameh, N.S. 2001. Scalable and extensible infrastructures for distributing interoperable geographic information services on the internet. PhD dissertation. Massachusetts Institute of Technology, Boston, MA, 132pp.

Architecture and Infrastructure Committee, Federal Chief Information Officers Council and Federal Geographic Data Committee. 2009. *FEA Geospatial Profile Version 2.0*, http://www.fgdc.gov/library/FEA%20Geospatial%20Profile%20 v%202%200%20draft%205%206%20011609.doc/download (accessed January 16, 2009).

Arctur, D., D. Hair, G. Timson, E.P. Martin, and R. Fegeas. 1998. Issues and prospects for the next generation of the spatial data transfer standard (SDTS). *International Journal of Geographical Information Science*, 12(4):403–425.

Atkins, D. et al. 2003. Revolutionizing Science and Engineering Through CI: Report of the National Science Foundation Blue-Ribbon Advisory Panel on CI: National Science Foundation.

Ballinger, K. 2001. Web Services Interoperability and SOAP. http://msdn.microsoft.com/library/default.asp?url=/library/en-us/dnsoap/html/soapinteropbk-gnd.asp (accessed December 22, 2009).

Bambacus, M., C. Yang, J. Evans, M. Cole, N. Alameh, and S. Marley. 2007. ESG: An interoperable portal for prototyping applications. *URISA Journal*, 19(2):15–21.

Batty, M. 1999. New technology and GIS. In *Geographical Information Systems: Principles and Technical Issues: Volume 1*, eds P. Longley, M.F. Goodchild, D.J. Maguire, D.W. Rhind, pp. 309–316. Wiley and Sons, New York.

Beaujardiere, J. 2004. OGC Web Map Service Interface, version 1.3.0, OGC 03–109r1. http://portal.opengeospatial.org/files/index.php?artifact_id=4756&passcode=b9mnkb6rr7uc1hs1t1ue (accessed September 6, 2009).

Bishr, Y. 1998. Overcoming the semantic and other barriers to GIS interoperability. *International Journal of Geographical Information Science*, 12(4):299–314.

Booth, D., H. Haas. F. McCabe, E. Newcomer, M. Champion, C. Ferris, and D. Orchard. 2004. Web Service Architecture. http://www.w3.org/TR/ws-arch/ (accessed December 22, 2009).

Brodie, M.L. 1992. The promise of distributed computing and the challenge of legacy Information systems. In *Proceedings of the IFIP WG2-6 Database Semantics Conference on Interoperable Database Systems (DS-5)*, Lorne, Victoria, Australia, pp. 1–31, North-Holland, Amsterdam.

Brodie, M. 2007. Semantic technologies: Realizing the services vision. *IEEE Intelligent Systems*, 12(5):12–17.

Card, S., J. MacKinlay, and B. Shneiderman. 1999. *Readings in Information Visualization*, 716pp. Morgan Kaufman Publishers, San Francisco, CA.

Cheung, I.K.L. and G.Y.K. Shea. 2004. The development of algorithms for on-demand map editing for internet and mobile users with GML and SVG. In *Proceedings of the 20th ISPRS Congress Istanbul*, Turkey, Vol. 35, pp. 158–162.

Choicki, J. 1999. Constaint-based interoperability of spatiotemporal databases. *Geoinformatica*, 3(3):211–243.

Coppock, J.T. and D.W. Rhind. 1991. The history of GIS. In *Geographical Information Systems: Principles and Applications*, Vol. 1, eds D.J. Maguire, M.F. Goodchild, and D.W. Rhind, pp. 21–43. John Wiley and Sons, New York.

Cruz, I., A. Rajendran, W. Sunna, and N. Wiegand. 2002. Handling semantic heterogeneities using declarative agreements. In *Proceedings of the 10th ACM International Symposium in Geographic Information Systems*, pp. 168–174. ACM Press, New York.

Dadi, U. and L. Di. 2007. Data independence and geospatial web services. In *Geoinformatics 2007 Conference*, San Diego, CA.

Data Mining and Image Processing Toolkits. http://datamining.itsc.uah.edu/adam (accessed December 22, 2009).

Deem, M. 2002. Microsoft SOAP Toolkit Type Mappers. http://msdn.microsoft.com/library/default.asp?url=/library/en-us/dnservice/html/service06042002.asp (accessed December 22, 2009).

Devogele, T., C. Parent, and S. Spaccapietra. 1998. On spatial database integration. *International Journal of Geographical Information Science*, 12(4):335–352.

Di, L., P. Zhao, W. Yang, G. Yu, and P. Yue. 2005. Intelligent geospatial web services. In *2005 IEEE International Geoscience and Remote Sensing Symposium*, pp. 1229–1232, Seoul, Korea.

Di, L., P. Zhao, W. Hang, Y. Wei, and X. Li. 2007. GeoBrain Web Service-based Online Analysis System (GeOnAS). In *NASA Earth Science Technology Conference 2007*, College Park, MD.

Duke, A., J. Davies, M. Richardson. 2005. Enabling a scalable service-oriented architecture with semantic web services. *BT Technology Journal*, 23(33):191–201.

Earth Information Exchange. 2009. http://eie.esipfed.org (accessed December 22, 2009).

Egenhofer, M. and M.F. Goodchild. 1997. Interoperating Geographic Information Systems. http://www.ncgia.ucsb.edu/conf/interop97/i20prop/i20prop.html (accessed December 22, 2009).

Egenhofer, M. 2002. Toward the semantic geospatial web. In *Proceedings of Tenth ACM International Symposium on Advances in Geographic Information Systems*, pp. 1–4. New York, USA.

Erl, T. 2005. *Service-oriented Architecture: Concepts, Technology, and Design*, 760pp. Prentice-Hall, Englewood Cliffs, NJ.

ESRIa. http://www.gis.com/whatisgis/index.html (accessed March 16, 2009).

ESRIb. Web Mapping APIs. http://www.esri.com/getting_started/developers/silverlight.html (accessed March 16, 2009).

ESRIc. ArcGIS Explorer. http://www.esri.com/software/arcgis/explorer/index.html (accessed March 16, 2009).

Executive Office of the President. 2007. *FEA Consolidated Reference Model Document Version 2.3*. http://www.whitehouse.gov/omb/assets/fea_docs/FEA_CRM_v23_Final_Oct_2007_Revised.pdf (accessed October 20, 2007).

Fielding, R. 2000. Architectural styles and the design of network-based software architectures. PhD disseration, University of California, Irvine, CA, Irvine, USA, 180pp.

Fonseca, F.T., M.J. Egenhofer, P. Agouris, and G. Câmara. 2002. Using ontologies for integrated geographic information systems. *Transactions in GIS*, 6(3):231–257.

GeoBrain Processing Web Services. http://geobrain.laits.gmu.edu:81/grassweb/manuals (accessed December 22, 2009).

GeoBrain Online Analysis System. http://geobrain.laits.gmu.edu:81/OnAS/ (accessed December 22, 2009).

Geographic Resource Analysis Support System. http://grass.itc.it/ (accessed December 22, 2009).

Geospatial Line of Business. http://geolob.wik.is/ (accessed December 22, 2009).

Geospatial Ontology. http://geobrain.laits.gmu.edu/ontology (accessed December 22, 2009).

Global Spatial Data Infrastructure Association (GSDIA). http://www.gsdi.org (accessed December 22, 2009).

Goodchild, M.F., M.J. Egenhofer, and R. Fegeas. 1997. Interoperating GISs. http://www.ncgia.ucsb.edu/conf/interop97/report.html (accessed December 22, 2009).

Goodchild, M.F. 2007. Citizens as voluntary sensors: Spatial data infrastructure in the World of Web 2.0. *International Journal of Spatial Data Infrastructures Research*, 2:24–32.

Google. Google Maps API. http://code.google.com/apis/maps/documentation/overlays.html (accessed December 22, 2009).

Gore, A. 1998. The Digital Earth: Understanding our planet in the 21st century. Given at the California Science Center, Los Angeles, California. http://portal.opengeospatial.org/files/?artifact_id= 6210 (accessed December 22, 2009).

GOS. http://geodata.gov (accessed December 22, 2009).

ISO. 2000. ISO/IEC 14598: Information Technology—Evaluation of Software Products.

ISO. 2001. ISO/IEC Standard 9126: *Software Engineering—Product Quality*, Part 1.

ISO. 2005. International Standard, *Quality Management Systems—Fundamentals and Vocabulary*.

ISO/TC211. 2005. ISO 19119:2005 *Geographic Information—Services.*

Jacoby, S., J. Smith, L. Ting, and I. Williamson. 2002. Developing a common spatial data infrastructure between state and local government—An Australian case study. *International Journal of Geographical Information Science*, 6(4):305–322.

Jones, C.B., A.I. Abdelmoty, and G. Fu. 2003. Maintaining ontologies for geographical information retrieval on the web, In *Ontologies, Databases, and Applications of Semantics for Large Scale Information Systems (ODBASE03)*, Lecture Notes in Computer Science, Vol. 2888, pp. 934–951.

Laurini, R. 1998. Spatial multi-database topological continuity and indexing: A step towards seamless GIS data interoperability. *International Journal of Geographical Information Science*, 12(4):373–402.

Lowe, J. 2004. Geospatial Web Portals. http://www.geospatial-solutions.com/geospatialsolutions/article/articleDetail.jsp?id=101549&pageID=1&sk=&date= (accessed December 22, 2009).

Lo, C.P. and A.K.W. Yeung. 2002. *Concepts and Techniques of Geographic Information Systems*. Prentice-Hall, Englewood Cliffs, NJ, 532pp.

Kok, B. and B. Van Loenen. 2004. How to assess the success of National Spatial Data Infrastructures? *Computers, Environment and Urban Systems*, 29:699–717.

Ma, K.L., A. Stompel, J. Bielak, O. Ghattas, and E.J. Kim. 2003. Visualizing very large-scale earthquake simulations. In *Proceedings of the 2003 ACM/IEEE Conference on Supercomputing*, Washington, DC, 48pp.

Maguire, D.J. and P.A. Longley. 2005. The emergence of geoportals and their role in spatial data infrastructures. *Computers, Environment and Urban Systems*, 29(1):3–14.

Martell, R. 2005. OGC™ Catalogue Services—ebRIM (ISO/TS 15000-3) profile of CSW. Open Geospatial Consortium, Inc., 62pp.

Masser, I. 2005. *GIS Worlds: Creating Spatial Data Infrastructures*, 388pp. ESRI Press, Redlands, CA.

McIlraith, S., T. Son, and H. Zeng. 2001. Semantic Web Services. *IEEE Intelligent Systems*, 16:46–53.

NASA Earth Science Gateway. http://esg.gsfc.nasa.gov (accessed December 22, 2009).

Nebert, D., A. Whiteside, and P. Vretanos. 2007. OpenGIS Catalogue Service Specification OGC 07–006r1. http://www.opengeospatial.org/standards/cat (accessed December 22, 2009).

Nedovic-Budic, Z. and K. Pinto. 2000. Information sharing in an interorganizational GIS environment. *Environment and Planning B: Planning and Design*, 27:455–474.

Newcomer, E. and G. Lomow. 2005. *Understanding SOA with Web Services*, 444pp. Addison-Wesley, Reading, MA.

NIST (National Institute of Standard and Technology). 1994. Federal Information Processing Standard Publication Spatial Data Transfer Standard. US Department of Commence, Washington, DC.

OASIS. 2002. OASIS/ebXML Registry Information Model v2.0. http://www.oasis-open.org/committees/regrep/documents/2.0/specs/ebrim.pdf (accessed December 22, 2009).

OASIS. 2004. Introduction to UDDI: Important Features and Functional Concepts. http://www.uddi.org/pubs/uddi-tech-wp.pdf (accessed December 22, 2009).

Onchaga, R. 2005. On quality of service and geo-service compositions. In *Proceedings of 8th Conference on Geographic Information Science*, pp. 519–528, Estoril, Portugal.

Open Geospatial Consortium (OGC). 1998. *A Request for Technology in Support of a Web Mapping Technology Testbed*, 1pp. The Open GIS Consortium Inc, US. http://www.opengeospatial.org/pressroom/pressreleases/233 (accessed June 11, 2010).

OWL-S. 2004. Semantic Markup for Web Services—W3C Member Submission 22. http://www.w3.org/Submission/2004/SUBM-OWL-S-20041122/ (accessed December 22, 2009).

OWL. 2004. Web Ontology Language Reference. http://www.w3.org/TR/2004/REC-owl-ref-20040210/ (accessed December 22, 2009).

Semantic Web for Earth and Environmental Terminology (SWEET). http://sweet.jpl.nasa.gov/sweet/ (accessed December 22, 2009).

Sonnet, J., 2005. Web Map Context Documents OGC 05–005. http://www.opengeo-spatial.org/standards/wmc (accessed December 22, 2009).

Peng, Z.R. and M.H. Tsou. 2003. *Internet GIS: Distributed Geographic Information Services for the Internet and Wireless Network*, 679pp. John Wiley & Sons, New York.

Peng, Z.R. and C. Zhang. 2004. The roles of geography markup language, scalable vector graphics, and web feature service specifications in the development of internet geographic information systems. *Journal of Geographical Systems*, 6:95–116.

Putz, S. 1994. Interactive information services using World-Wide Web hypertext. *Computer Networks and ISDN Systems*, 27(2):273–280.

Rajabifard, A. and I.P. Williamson. 2001. Spatial data infrastructures: Concept, SDI hierarchy and future directions. In *Proceedings of Geomatics'80*, Tehran, Iran.

Rajasekaran, P., J. Miller, K.Verma, and A. Sheth. 2004. Enhancing web service description and discovery to facilitate composition. *In Proceedings of the 1st International Workshop on Semantic Web Services and Web Process Composition (SWSWPC 2004)*, in conjunction with the *2004 IEEE International Conference on Web Services (ICWS'2004)*. San Diego, CA.

Reeves, C.A. and D.A. Bednar. 1994. Defining quality: Alternatives and implications. *The Academy of Management Review*, 19(3):419–445.

Schut, P. 2007. OpenGIS Web Processing Service OGC 05–007r7. http://www.open-geospatial.org/standards/wps (accessed December 22, 2009).

Shi, X. 2007. Semantic web services: An unfulfilled promise. *IEEE IT Professional*, 9(4):42–45.

Sivashanmugam, K., A. Sheth, J. Miller, K. Verma, R. Aggarwal, and P. Rajasekaran. 2003. Metadata and semantics for web services and processes. In *Datenbanken und Informationssysteme: Festschrift zum 60*, eds W. Benn, P. Dadam, S. Kirn, and R. Unland, pp. 245–271. Geburtstag von Gunter Schlageter.

Smith, B. and D. Mark. 1998. Geographic categories: An ontological investigation. *International Journal of Geographic Information Science*, 15(7):591–612.

Snell, J. 2002. Web Services Interoperability. http://www.xml.com/pub/a/2002/01/30/soap.html (accessed December 22, 2009).

Snell, J. and T. Glover. 2003. Portability and interoperability. http://www-106.ibm.com/developerworks/webservices/library/ws-port/ (accessed December 22, 2009).

Sondheim, M., K. Gardels, and K. Buehler. 1999. GIS interoperability. In *Geographical Information Systems: Principles, Techniques, Applications and Management, Second Edition.* eds P.A. Longley, M.F. Goodchild, D.J. Maguire, and D.W. Rhind, pp. 347–358. John Wiley & Sons, New York.

SWRL. 2003. A Semantic Web Rule Language Combining OWL and RuleML. http://www.daml.org/2003/11/swrl/ (accessed December 22, 2009).

Telang, R., T. Mukhopadhyay. 2004. Drivers of Web portal use. *Electronic Commerce Research and Applications*, 4:49–65.

The Chief Information Officers Council. 2002. Geospatial One Stop Best Practices White Paper. http://www.cio.gov/Library/documents_details.cfm?id=Geospatial%20One%20Stop%20Best%20Practices%20White%20Paper,%20May%202,%202002&structure=Information%20Technology&category=Best%20Practices (accessed December 22, 2009).

The Chief Information Officers Council. 1999. *Federal Enterprise Architecture Framework Version 1.1.* http://www.cio.gov/documents/fedarch1.pdf (accessed December 22, 2009).

The Open Group, 2006. Definition of SOA. http://www.opengroup.org/projects/soa/doc.tpl?CALLER=doc.tpl&gdid=10632 (accessed December 22, 2009).

Tost, A. 2003 Web services interoperability. http://www.ibm.com/developerworks/webservices/demos/wsid/ (accessed December 22, 2009).

Tu, T., H. Yu, L. Ramirez-Guzman, J. Bielak, O. Ghattas, K.L. Ma, and D.R. O'Hallaron. 2006. From mesh generation to scientific visualization: An end-to-end approach to parallel supercomputing. In *Proceedings of the 2006 ACM/IEEE Conference on Supercomputing*, Tampa, FL.

Uitermark, H. 1996. The integration of geographic databases. In *Proceedings of Second Joint European Conference*, pp. 92–95. IOS Press, Amsterdam.

Vogels, W. 2003. Web services are not distributed objects. *IEEE Internet Computing*, 7(6):59–66.

Vretanos, P.A. 2005. *OGC Web Feature Service Implementation Specification, version 1.1.0.* OGC 04–094. http://portal.opengeospatial.org/files/index.php?artifact_id=8339&passcode=b9mnkb6rr7uc1hs1t1ue (accessed December 22, 2009).

W3C. 2004a. Web Services Glossary. http://www.w3.org/TR/2004/NOTE-ws-gloss-20040211/ (accessed December 22, 2009).

W3C. 2004b. Web Services Architecture. http://www.w3.org/TR/ws-arch/ (accessed December 22, 2009).

Webopedia. http://www.webopedia.com/DidYouKnow/Hardware_Software/2006/thin_client_applications.asp (accessed December 22, 2009).

Whiteside, A. and J.D. Evans. 2008. *OGC Web Coverage Service (WCS) Implementation Standard, version 1.1.2 07–067r5.* http://portal.opengeospatial.org/files/index.php?artifact_id=27297&passcode = b9mnkb6rr7uc1hs1t1ue (accessed December 22, 2009).

Williamson, I.P., A. Rajabifard, and M. Feeney. 2003. *Developing Spatial Data Infrastructures: From Concept to Reality*, 358pp. Taylor & Francis, UK.

WorldKit. http://worldkit.org (accessed December 22, 2009).

Wu, H., H. Zhang, X. Liu, and X. Sun. 2005. Adaptive Architecture of Geospatial Information Service over the Internet with QoGIS Embedded. In *ISPRS Workshop on Service and Application of Spatial Data Infrastructure*, XXXVI(4/W6), Hangzhou, China.

Wu, H. and H. Zhang. 2007. QoGIS: Concept and Research Framework. *Geomatics and Information Science of Wunan University*, 5:385–388.

Yang, C., Y. Cao, M. Kafatos, J. Evans, and M. Bambacus. 2006. Interoperable Web Portal Supporting Spatial Data Infrastructure. *Journal of Geographic Information Sciences*, 12(1):38–43.

Yang, C., J. Evans, M. Cole, N. Alameh, S. Marley, and M. Bambacus. 2007. The emerging concepts and applications of the spatial web portal. *PE&RS*, 73(6):691–698.

Yang, C., W. Li, J. Xie, and B. Zhou. 2008. Distributed geospatial information processing: Sharing earth science information to support Digital Earth. *International Journal of Digital Earth*, 1(3):259–278.

Yang C., D. Wong, R. Yang, M. Kafatos, and Q. Li. 2005. Performance improving techniques in WebGIS. *International Journal of Geographical Information Science*, 19(3):319–342.

Zhang, C., W. Li, M. Day, and Z.R. Peng. 2003. GML-based interoperable geographical database. *Cartography*, 32(2):1–16.

Zhang, C. and W. Li. 2005. The roles of web feature and web map services in real-time geospatial data sharing for time-critical applications. *Cartography and Geographic Information Science*, 32(4):269–283.

Zhang, C., W. Li, and T. Zhao. 2007. Geospatial data sharing based on geospatial semantic web technologies. *Journal of Spatial Science*, 52(2):35–49.

Zhao, P., G. Yu, and L. Di. 2006. Geospatial web services. In *Emerging Spatial Information Systems and Applications*. ed. B. Hilton, pp. 1–35. Idea Group Publishing, Hershey.

Zyperski, C. 2002. *Component Software: Beyond Object-Oriented Programming*, 589pp. Addison-Wesley Professional, Reading, MA.

7

Geoinformation Knowledge Representation and Applications

Rob Raskin, Naijun Zhou, and Wenwen Li

CONTENTS

This chapter addresses how spatial knowledge can be effectively captured and put to use. The primary motivation is to enable the automated discovery, reasoning, and integration of heterogeneous sources of Geoinformation.

Spatial knowledge is defined first (Section 7.1). Knowledge about geographic space is effectively captured for both machine and human use via an ontology (Section 7.2). As the prototypical knowledge representation mechanism, an ontology can effectively encode common sense knowledge and make it accessible to a machine. Ontology examples (Section 7.3) are presented next to provide specific examples of the capture of domain-specific knowledge. The subsequent three sections describe applications of spatial knowledge. Knowledge-based discovery (Section 7.4) is the ability to find needed data, both at the data layer level (with the assistance of an ontology) and within a layer (using data mining methods). Knowledge-based reasoning (Section 7.5) is the ability to infer results that are not explicitly stated. If we declare that object A is north of B and B is north of C, we can conclude that A is north of C, provided that the "north of" relation is declared a transitive property. Finally, knowledge-based fusion (Section 7.6) is the ability to merge layers or perform some type of layer algebra in a meaningful way. The results are relevant to support future intelligent agent-based tasks.

7.1 Spatial Knowledge

The field of geography studies the human and physical dimensions of geographic space, and spatial knowledge is the knowledge about geographic space. In particular, spatial knowledge contributes to Naïve Geography, which is the body of knowledge that people have about the surrounding geographic world. Research of spatial knowledge can be traced back to the 1950s. In the 1990s, we entered into a data-rich era with more diverse and detailed spatial data due to (1) less expensive equipment and methods of data collection and processing; (2) advanced technologies of data dissemination and delivery (e.g., Internet, Web services, and volunteered geography) (Goodchild 1997, 2007); and (3) increasing use of georeferenced data from allied fields such as public health, location-based services, homeland security, emergency response, and management. Increasing data availability and diverse applications of spatial data provide unprecedented opportunities and challenges for the research of spatial knowledge.

Spatial knowledge is "the product of spatial thinking and reasoning about the world's natural and human phenomena." It includes the following aspects (Golledge 2002):

- Spatial primitives: arrangement, organization, distribution, pattern, shape, hierarchy, distance, direction, orientation, regionalization, categorization, reference frame, and spatial association
- Advanced concepts derived from these primitives
- Formal linking of the primitives into theories and generalizations

The sources of spatial knowledge are direct environmental experiences, representations, and languages (Montello and Freundschuh 1995). Spatial knowledge can be acquired from a person's direct interaction with the environment. Such direct experiences are generated and stored in the human mind and are referred to as cognitive maps in psychology. Spatial knowledge can also be derived from static or dynamic representations of reality. An observation source (researcher, surveyor, and sensor) generalizes and represents direct observed knowledge as maps, diagrams, and animation, either in digital or analog format (paper map, GIS data). Oral and written languages, such as natural, mathematical, and gestural languages, are common media to transfer knowledge among humans. The direct experiences and representations are the major forms of spatial knowledge in Geoinformation Science.

Figure 7.1 illustrates three sources of spatial knowledge of a mountain. Through field observations, a person forms first-hand spatial knowledge in his or her mind. To record and share his or her knowledge, he or she may choose to represent the knowledge as a map or a database, or through oral or written languages. To produce a map of the mountain, he or she may determine a vector data model to represent the mountain boundary (location), describe the features of the mountain (name, height, slope, orientation, temperature, vegetation, etc.), and explain how he or she collected and represented the knowledge. The map becomes second-hand spatial knowledge for map users. Map users are expected to recover the knowledge representation and discover the spatial knowledge encoded in the map.

Second-hand knowledge encoded in the form of maps and databases often becomes a primary source of spatial knowledge (Barkowsky 2002), as made

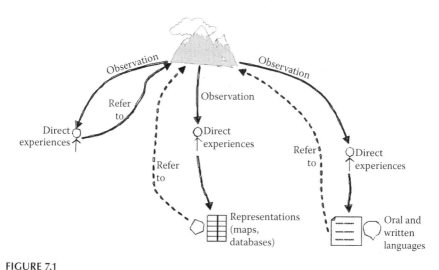

FIGURE 7.1

Three sources of spatial knowledge about a mountain: (i) direct experiences; (ii) representations as maps and databases; and (iii) oral and written languages. Both (ii) and (iii) are sources of second-hand spatial knowledge.

available in GIScience. In fact, geospatial data are typically more faithful than our direct experiences with geographic space; for example, it is common to say, "when I get home, I want to look at the route on a map, to see where I went" (Egenhofer and Mark 1995). Although direct experiences are first-hand and usually offer the most complete and precise knowledge, the full reality may be more complex and may lie well beyond a human's mental interpretation. Consequently, second-hand spatial information assisted by computational methods can provide more accurate and complete knowledge.

Thus, GIScience investigates two types of knowledge: direct knowledge of geographic space and indirect knowledge through geospatial data and other second-hand descriptions. Knowledge of geospatial data includes how to determine an appropriate scale and generalization, how to classify qualities into categories, how to choose suitable data, and how data are transformed; a proper interpretation may require the use of metadata and context for the search, transform, interoperation, and application of geospatial data (Gahegan et al. 2000).

7.1.1 How Knowledge Differs from Data and Information

It is possible to distinguish between geospatial *data, information*, and *knowledge*. Some ambiguity exists between these classifications, but some generalizations are clear (Figure 7.2). Data correspond to raw output values from observations, such as the brightness values obtained from remote sensing instrumentation. Geospatial data may be numeric or non-numeric, and the geospatial parameter may be independent variable, dependent variable, or both; in any configuration, minimal processing is applied to these values. The earliest era of computing produced *data systems*, that is, computers with abilities to store and retrieve data values.

Information adds value to data through some type of summarization, categorization, or association. A GIS operates at the level of information, because data have been georeferenced to some common coordinate system and made accessible to others using this common geospatial perspective. That is, a GIS user can query from the perspective of a desired location and learn about its

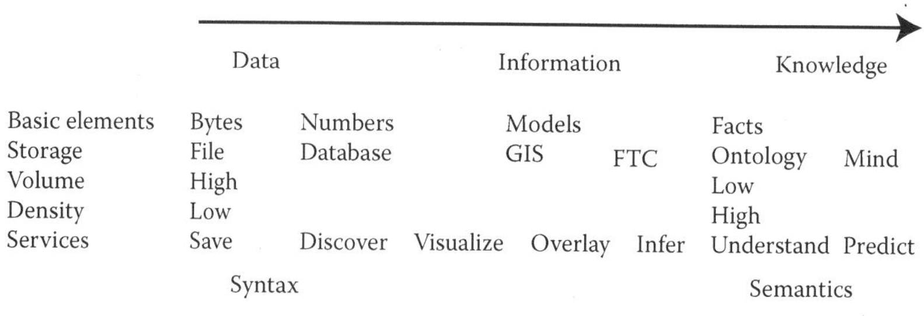

FIGURE 7.2
Data to Knowledge according to several dimensions.

multiple features and attributes. Time plays an analogous unifying theme, but such an approach has not been employed in commercial GIS to date. A GIS belongs to the class of *information systems,* as space enables multiple layers to be associated.

What a GIS does not offer is any interpretation of what these layers mean. Numeric values from all application domains are treated interchangeably, with no way to distinguish them other than through documentation that is human readable but not machine readable. Knowledge provides the added context through which to interpret information. Some add-on GIS data models (e.g., ArcHydro) provide some of these capabilities for targeted domains (but not globally in the ArcGIS environment). A GIS that is fully knowledge enabled would have layers semantically registered, analogous to the spatial registration procedure currently deployed for georeferencing the layer.

7.2 Ontologies

Formal representation of knowledge is achieved through the use of an *ontology*. An ontology enables the discovery, representation, management, reasoning, and sharing of data and knowledge. The concept of ontology involves the fields of philosophy, artificial intelligence, computer and information science, linguistics, and application domains such as geography. Due to its interdisciplinary nature, a wide range of interpretations of the term *ontology* exist, but there are two broad definitions (Guarino and Giaretta 1995):

Ontology as a philosophical discipline. Ontology as a classical study explicitly inaugurated by Aristotle investigates the nature of all things in reality. It answers questions of whether a being exists, how a being exists (what substance reality is made of), and what features and relations these things have. Ontological research aims at establishing a catalog of everything in the world and how it works, with only one ontology in the world beyond any science, society, and culture. Ontology serves as a philosophical or mathematical discipline, providing meta-theories for understanding the world. It does not study the phenomena of a specific domain and is "not a description of how we conceptualize the world but rather a description of the world itself" (Johnston 2009; Smith 1998). Our personal knowledge and conceptualizations of reality have no implications with this classical notion of reality (Agarwal 1995; Smith 1998).

Ontology in information science and artificial intelligence (AI). In the modern usage, an ontology is "a formal, explicit, specification of a shared conceptualization ... the objects, concepts, and other entities that are assumed to exist in some area of interest and the relationships that hold among them" (Gruber 1993). In this definition, *formal* means the conceptualization and representation will be standardized and machine readable; *conceptualization* refers to abstraction of the real world; and *shared* implies that the ontology is able to facilitate the sharing of conceptualization and knowledge. Multiple ontologies exist based on the various conceptualizations provided by different groups or domains. These ontologies offer theory and methods for domain-specific applications and provide "a statement of the necessary and sufficient conditions for something to be a particular kind of entity within a given domain" (Smith 1998). An ontology provides a "logical theory accounting for the intended meaning of a formal vocabulary" to address particular applications (Guarino 1998). This notion is based on our conceptualizations and provides approximations to the truth; in this sense, there are many ontologies corresponding to our various (incomplete and imperfect) conceptualizations of the world.

7.2.1 Ontologies as Triple Statements

Unless otherwise specified, we will use the term *ontology* in its modern sense—as a formal shared understanding of concepts and their relationships to one another. An ontology is typically expressed in an XML language so as to be both human readable and machine readable (as a compromise to both). The W3C has standardized Resource Description Framework (RDF) (Brickley and Guvha 2004) and OWL (Dean and Schreiber 2004) for this purpose. Use of these language standards enables anyone to extend or specialize an ontology developed by others. Large, centralized ontologies are difficult to maintain; instead, the paradigm of many smaller ontologies has emerged as the model to accommodate scalable growth.

The RDF and OWL describe classes, properties, relationships, restrictions, and constructs to build complex classes as knowledge of spatial objects and geographic space. The RDF defines the basic primitive of the *class* (a noun). A class can be declared to be a subclass of other classes, in which case it inherits all attributes of the parent class(es). The RDF property is the mechanism (a verb) to relate classes to one another in more general ways. An ontology consists of a collection of facts expressed as *triple* assertions, in a *subject-verb-object* form, such as

Flood	subClassOf	WeatherPhenomena
TIFF	subClassOf	FileFormat
Soil Type	subClassOf	PhysicalProperty
Ocean	subClassOf	WaterBody
Atmosphere	hasPrimarySubstance	Air
Thermometer	measures	Temperature

In the above list, the *subClassOf* property is built in to the language, whereas *hasSubstance* and *measures* are user-defined properties in the ontology. Using the property relation, a subclass can be distinguished relative to its (more general) parent class. Thus, *Atmosphere* is a *PlanetaryLayer*, but it is distinguished from other layers in that it contains *Air* as its primary substance.

An *individual* (for instance) is analogous to a subclass that cannot be further subclassed; so, *PacificOcean* is an individual (instance) of *Ocean*, because there can be only one PacificOcean. Further facts can be asserted. The statement *PacificOcean borders Canada* uses the property *borders* to relate *PacificOcean* and *Canada*. The OWL provides additional constructs to increase semantic expressivity; for example, properties can be specified to be symmetric, transitive, and/or functional (*borders* is symmetric, but not transitive or functional). Equivalence of two classes (or two properties or two individuals) can be declared to account for vocabulary differences across communities, wherein two terms refer to the same concept. Numerical values or value ranges are defined by the W3C xsd specifications (Carroll and Pan 2006), although users can extend the predefined numeric types (e.g., to represent a numerical interval).

An ontology can include "statements about statements." These assertions may address degree of belief, quality, uncertainty, provenance, security, authoritativeness, versioning, timestamp, and so on. All ontologies are assumed to follow the *open world* assumption, wherein facts not stated are not necessarily assumed to be false. In contrast, the more rigid *closed world* assumption (adopted in the database world) implies that all facts about the world have been included, and any fact not explicitly stated is necessarily false.

Ontologies are more semantically expressive than taxonomies (such as land use types classification systems) and thesauri (such as WordNet). A taxonomy classifies and stores entities hierarchically using only *is-a* relations to relate the entities. A thesaurus is a controlled vocabulary organized in a structure where basic relationships among terms (e.g., equal, subset, superset, and associative) are recorded.

7.2.2 Spatial Ontologies

A spatial ontology specifies the existence of geographic space and objects residing in this space. When objects are represented as spatial data, a spatial ontology also investigates the linguistic and spatial characteristics of the data itself (Casati and Varzi 1997). Since the late 1990s, the ontology and spatial ontology concepts received special attention in GIScience (Agarwal 2005; Frank 1997). The fundamental questions to be addressed by a spatial ontology are (1) conceptual issues concerning what would be required to establish an exhaustive ontology of the geography domain, (2) representational and logical issues relating to the choice of appropriate methods for formalizing ontologies, and (3) implementation issues regarding how an ontology influences the design of spatial information systems (Mark et al. 2005). Specifically, a spatial ontology can serve the following purposes (Frank 1997; Winter 2001):

- Provide a formal base of the model of the reality; enhance our capability of modeling locations, processes, and properties; and offer choices beyond the current simplified models of vector and raster, time stamp and snapshot, and mathematical and statistical spatial process.

- Encode human-oriented knowledge and build a formal and ontology-based GIS instead of a system built by a computer.
- Support spatial knowledge discovery and effective and efficient use of spatial data and computational methods for decision making.
- Enable interoperability and facilitate and enhance the capability of GIS as an integrator of data and functions from multiple sources and multiple conceptualizations.

Some outstanding research issues in spatial ontologies include the integration of multiple ontologies (ontology engineering); representation languages for spatial concepts, methods, and tools of reasoning; standardization of vocabularies; and metadata standards.

7.3 Spatial Ontology Examples

This section highlights some spatial ontologies developed to date in GIScience. Although formally represented in OWL or other ontology languages, ontology content can be effectively communicated through visual diagrams. These pictures necessarily present only the highlights of the ontology, to avoid complex spaghetti diagrams.

7.3.1 Spatial Feature Ontologies

Spatial feature ontologies capture semantic descriptions of the features of the spatial world and may include concepts such as space, place (complete enclosure, territory, home, neighborhood, and region), topological features (surface, interior, edge, and side/end), properties of spatial features (width, breadth, nearness, density, and color), and location (Mark et al. 1996). The USGS *The National Map* project developed the Topographic Feature Ontology, which includes a taxonomy of features (elevation, ecology, division, etc.), feature attributes, feature relations, parts and wholes, and processes (Varanka 2009). A hydrology ontology has been developed at UK Ordnance Survey; for example, a channel is defined as

A channel is a kind of topographic object.

A channel has a linear form.

A channel enables flow of water.

A channel has exactly one bed as the part of the channel.

A channel has at least two banks as the part of the channel.

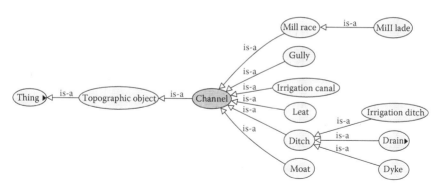

FIGURE 7.3
The ontology of a channel as a subconcept of Topographic Object and a super-concept of gully, moat, ditch, mill race, irrigation canal, and leat. (Adapted from UK Ordnance Survey.)

The channel ontology as a subset of a hydrology ontology is represented graphically in Figure 7.3 and in OWL in Table 7.1.

7.3.2 Geometry Ontology

A geometry ontology aims at standardizing the description and storage of locations as spatial data. The OGC Simple Features (OGC 2006) specification is a well-accepted geometry ontology. This specification includes a set of geometric primitives, complex features composed from the primitives, and their relations in a hierarchy (Figure 7.4).

7.3.3 Spatial Relation Ontologies

The spatial relations of spatial objects are topological relations (adjacent, inside, disjoint, contain, and equal) and other measures (e.g., near, close). The interpretation of a spatial relation may be context dependent, and a spatial relation ontology can clarify ambiguities and assist in understanding. The meaning of *near* in regional and global data may be defined with inconsistent distance definitions, ranging from several miles or less to thousands of miles.

A spatial relation ontology can include the treatment of synonyms (in, inside, and within), provide spatial context of relational scale measures (regional, national, and global), facilitate the mapping of spatial terms onto corresponding geometries (e.g., region *a* is in region *b* implies that the interior of *a* belongs to the joint of the boundary and interior of *b*), and enable the reasoning of spatial relations (e.g., if *a* is in *b* and *b* is in *c*, then *a* is also in *c*) (Egenhofer 2002). For example, the *TransitiveProperty* of OWL (*owl:TransitiveProperty*) enables such inference : If College Park is spatially in Prince George's County and Prince George's County is spatially in the State of Maryland, then College Park is spatially located in Maryland. Kuhn (2002) developed a formal ontology for spatial relations of boat, boathouse, and water including container, surface,

TABLE 7.1

Class Channel Represented in OWL as a Topographical Object with Linear Extent
That Enables the Flow of Water and Has Bed and Bank as Parts

```
<owl:Class rdf:about = "#Channel">
  <owl:equivalentClass>
    <owl:Class>
      <owl:intersectionOf rdf:parseType = "Collection">
        <rdf:Description rdf:about = "#TopographicObject"/>
        <owl:Restriction>
          <owl:onProperty rdf:resource = "#hasLinearForm"/>
          <owl:someValuesFrom rdf:resource = "#LinearForm"/>
        </owl:Restriction>

        <owl:Restriction>
          <owl:onProperty
rdf:resource = "http://www.ordnancesurvey.co.uk/ontology/Topography/
v0.1/Topography.owl#enables"/>
          <owl:someValuesFrom>
            <owl:Class >
              <owl:intersectionOf rdf:parseType = "Collection">
                <rdf:Description rdf:about = "#Flow"/>
                <owl:Restriction>
                  <owl:onProperty
rdf:resource = "http://www.ordnancesurvey.co.uk/ontology/Rabbit/v1.0/
Rabbit.owl#of"/>
                  <owl:someValuesFrom rdf:resource = "#Water"/>
                  </owl:Restriction>
              </owl:intersectionOf>
            </owl:Class>
          </owl:someValuesFrom>
        </owl:Restriction>
      </owl:intersectionOf>
    </owl:Class>
  </owl:equivalentClass>

<rdfs:subClassOf>
  <owl:Restriction>
    <owl:onProperty
rdf:resource = "http://www.ordnancesurvey.co.uk/ontology/
MereologicalRelations/v0.2/MereologicalRelations.owl#hasPart"/>
    <owl2:onClass rdf:resource = "#Bed"/>
  </owl:Restriction>
</rdfs:subClassOf>

<rdfs:subClassOf>
  <owl:Restriction>
    <owl:onProperty
rdf:resource = "http://www.ordnancesurvey.co.uk/ontology/
MereologicalRelations/v0.2/MereologicalRelations.owl#hasPart"/>
    <owl2:onClass rdf:resource = "#Bank"/>
  </owl:Restriction>
</rdfs:subClassOf>
</owl:Class>
```

Source: Adapted from AUK Ordnance Survey.

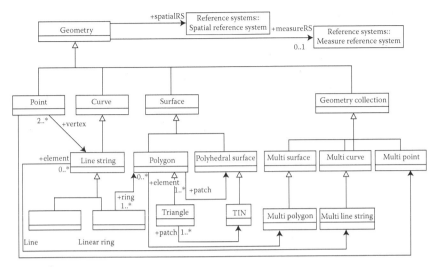

FIGURE 7.4
OGC Simple Features. (Adapted from OGC. 2006. OpenGIS Simple Feature. http://www.opengeospatial.org/standards/sfa.)

contact, and path. Boathouses are houses used to store boats (container) and are located at the edge of a water body (contact) (WordNet).

7.3.4 Spatial Attribute Ontologies

The properties of spatial objects are stored as data attributes. Attributes are often determined and used by individual agency, domain, and data provider. In order to explain and share the data, an agreement of the attributes (ontology) is necessary within a domain, a community, and an agency. The ontology of land use and land cover categories is one of the most active research topics, primarily because land use and land cover is a very common type of spatial data. The USGS Anderson Land Use/Cover classification system (Anderson et al. 1976) is one of the widely accepted standards and can be used as an ontology of land categories. Another important ontology of land categories is the ISO 19115 Topic Category classifications (Table 7.2).

7.3.5 Semantic Web for Earth and Environmental Terminology

An example of a large OWL spatial ontology set is the Semantic Web for Earth and Environmental Terminology (SWEET) (Raskin and Pan 2005). The SWEET 2.0 includes seven spatial ontologies that represent spatial coordinates, directions, distribution, extent, objects, scale, and general spatial concepts. The SWEET includes several additional geography modules to represent landforms, climate zones, soil types, biomes, administrative regions, and so on. The spatial components are part of a much larger set of 4000 concepts in 140 modular ontologies that serve as an upper-level

TABLE 7.2

ISO Topic Categories Related to Land Use and Land Cover

Category	Description	Example Land Use in the Category
Farming	Rearing of animals and/or cultivation of plants	Agriculture, irrigation
Biota	Flora and/or fauna in natural environment	Wetlands, vegetation
Boundaries	Legal land descriptions	Administrative boundaries, voting districts
Economy	Economic activities, conditions, and employment	Forestry, fisheries
Environment	Environmental resources, protection, and conservation	Waste storage, nature reserves
Intelligence/Military	Military bases, structures, and activities	Training grounds, military transportation
Inland Waters	Inland water features, drainage systems, and characteristics	Rivers and glaciers, wetlands
Society	Characteristics of society and culture	Housing, parks
Structure	Man-made construction	Museums, factories
Transportation	Means and aids for conveying persons and/or goods	Roads, airports
Utilities/Communication	Energy, water and waste systems, and communications infrastructure and services	Hydroelectricity, telecommunication

ontology for Earth system science, including its data and applications. With such an integrated collection, it is possible to begin to make headway toward the goal of machine understanding of georeferenced concepts.

A Plate Tectonics ontology based on a portion of SWEET is shown in Figure 7.5.

7.3.6 Other Ontologies

Some examples of large, high-level ontologies include Cyc (Matuszek et al. 2006), Basic Formal Ontology (BFO) (Arp and Smith 2008), and Descriptive Ontology for Linguistic and Cognitive Engineering (DOLCE) (Gangemi et al. 2003). Cyc was the first ontology of its kind and has been partially released into the public domain. The BFO is intended to be a complete ontological theory of reality. It includes two components: Snapshot (SNAP) entities represent the state of reality at a moment (i.e., snapshot), such as regions, qualities, functions, and so on. (Figure 7.6), and SPAN entities for temporal aspect of reality can be indexed by a time interval (Grenon and Smith 2004) (Figure 7.7). There are five subcategories of spatial SNAP entities: features, artifacts, agents, places, and qualities. The main kinds of spatiotemporal changes in SPAN entities are

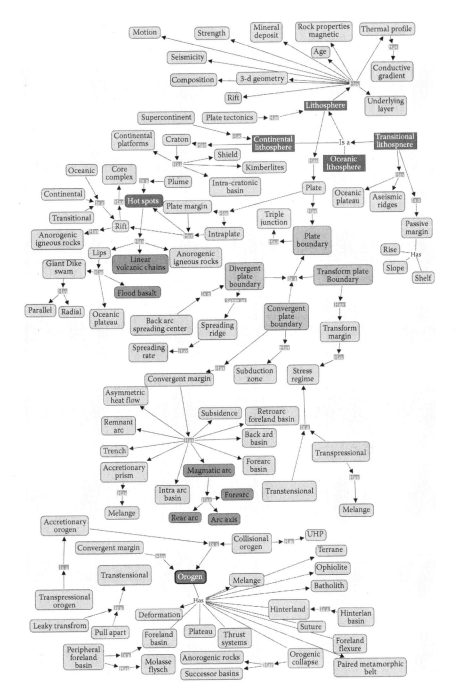

FIGURE 7.5
Plate Tectonics ontology. Note that "is a" is equivalent to the OWL subclass property. Some concepts shown are repeated to improve readability.

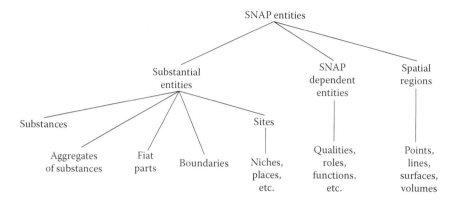

FIGURE 7.6
The main formal categories of SNAP entities. (Adapted from Grenon, P. and B. Smith. 2004. SNAP and SPAN: Prolegomenon to geodynamic ontology. In *Spatial Cognition and Computation*.)

Substantial Changes, wherein spatial objects are created and destroyed, and *Locational Changes* for qualitative, structural, and morphological changes.

In addition, spatial ontology repositories (such as the Marine Metadata Interoperability Project and the Open Ontology Repository) are being developed as a forum of open ontologies that can be adopted widely as *ad-hoc* usage.

7.4 Spatial Knowledge Discovery

Noesis (Ramachandran et al. 2006) is an atmospheric science semantic search and discovery application that exploits the associations between science

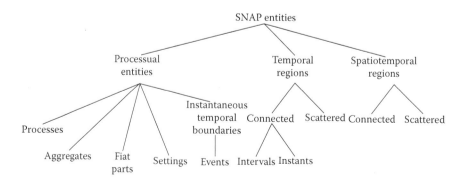

FIGURE 7.7
The taxonomy of SPAN entities. (Adapted from Grenon, P. and B. Smith. 2004. SNAP and SPAN: Prolegomenon to geodynamic ontology. In *Spatial Cognition and Computation*.)

terminologies and a search query. The Noesis uses the SWEET ontology (Raskin and Pan 2005) as its knowledge base and Jena and Pellet as its back-end reasoning engines. Given a query term, the reasoning engine invokes reasoning methods to infer related terms, such as those broader, narrower, or related. In this way, the user query can be expanded in a meaningful manner, and the efficiency of the search engine is dramatically improved (Li et al. 2008).

Knowledge of geographic space can be discovered from spatial data using qualitative and quantitative methods such as spatial analysis, spatial data mining, reasoning, and geovisualization. As the core of geographic study, knowledge discovery is one of the fast-growing areas of GIScience (Yuan et al. 2005). Knowledge discovery in databases (KDD) provides an indirect source of spatial knowledge. The KDD has been described as "the non-trivial process of identifying valid, novel, potentially useful and ultimately under-standable patterns in data" (Fayyad et al. 1996) that are distilled into knowl-edge through interpretation (Miller and Han 2009). The tasks of KDD include data warehousing, data selection, cleaning, preprocessing, trans-formation and reduction, data mining, evaluation and interpretation, and consolidation and use of the extracted knowledge. The KDD process is iter-ative, and the tasks may need to be refined during the iterations. Data min-ing is the application of computational, statistical, or visual methods to identify and discover the patterns, models, or relations hidden in the data. To some extent, the terms *data mining* and KDD are interchangeable; however, data mining is the central technology, whereas KDD is the whole process of knowledge discovery. The KDD usually deals with massive amounts of data and must be inductive instead of deterministic and deduc-tive. Some results are unknown and novel at the start of the KDD process, and valid and useful conclusions can be made only after the discovery process (Yuan et al. 2005).

Research of spatial knowledge discovery and spatial data mining is grow-ing rapidly; however, these areas still lie in the exploration stage, and suc-cessful applications are not common in geography and GIScience. Table 7.3 provides a summary of spatial data mining tasks and representative tech-niques (Mennis and Guo 2009; Miller and Han 2009; Yuan et al. 2005). With regard to Gollege's (2002) three aspects of spatial knowledge, only geometric primitives have been investigated with a certain degree of success. The dis-covery of advanced concepts, theories, and generalization still needs sub-stantial efforts. The major barriers of the research and application of spatial knowledge discovery include the lack of complete understanding of the human-environmental interaction to guide KDD processes and explain KDD results, and the lack of spatially aware algorithms and tools to discover novel and meaningful knowledge.

For a comprehensive introduction to spatial knowledge discovery and spatial data mining, see Fayyad et al. (1996), Miller and Han (2009), and Yuan et al. (2005).

TABLE 7.3

Spatial Data-Mining Tasks and Techniques

Tasks	Description	Techniques
Classification	Group objects into classes based on their nonspatial variables, locations, and spatial relations among objects	Decision trees Neural networks Maximum likelihood estimation Support vector machines
Spatial regression Spatial prediction	Predict a variable at a specific location based on the variables at nearby locations	Regression
Spatial association rule mining	Discover qualitative and quantitative spatial relations between objects	Association rules Bayesian network
Spatial clustering	Discover whether locations are clustered or a variable of locations is clustered	Cluster analysis
Outliner analysis	Detect objects that are inconsistent with other objects in terms of variables or location	Outlier detection
Spatial summary	Provide a summary of objects in terms of their locations and/or variables	Spatial statistics
Spatiotemporal trend analysis	Detect a regular change of one or more nonspatial variables in space	Trend analysis
Geovisualization	Explore, synthesize, and analyze data through visual and interactive tools	Geovisualization

7.5 Knowledge-Based Reasoning

It is a common experience to use spatial and temporal reasoning methods to infer information of the environment and to explore the consequence of changing our locations and settings (Egenhofer and Mark 1995). Reasoning of spatial knowledge requires a common set of syntactic and semantic conventions. Syntactic conventions include the rules for combining symbols to form valid and formal expressions. Semantic conventions specify how the expressions are interpreted and executed for reasoning (Bench-Capon 1990).

Reasoning is the cognitive process of looking for reasons for beliefs, conclusions, actions, or feelings (Kirwin 1995). Based on the facts defined in a knowledge base and logical reasoning, new facts can be inferred. The study of reasoning stems from philosophy, which studies how reasoning enables conclusions to be drawn and why some reasoning approaches are more efficient or appropriate than others. In psychology, researchers tend to study how people recognize, learn, and perform reasoning. After the invention of the digital computer in the mid-twentieth century, *reasoning like a human* has become a long-term dream and goal for computer science and AI research. Based on insights from both philosophy and psychology, AI researchers can generate abstractions and build a computational model to simulate the human learning process on a computer and imitate human intelligence. One of the most dramatic accomplishments was the performance of the famous chess computer Deep Blue, which beat the world chess champion Garry Kasparov in 1997. After two convincing victories in 2005 and 2006, it appears that AI chess programs can now defeat even the strongest chess players (Hoekenga 2007).

Knowledge-based reasoning consists of two main research branches: *logical* reasoning and *numerical* reasoning. Logical reasoning conducts inductive or deductive reasoning in explicitly defined logics using a set of inference rules. Numerical reasoning relies on statistical analysis and neurological and evolutional theory to simulate how the human brain reasons based on existing knowledge.

7.5.1 Logical Reasoning

Logical reasoning has three sub-branches: deductive, inductive, and abductive reasoning. Deductive reasoning determines conclusions based on inference rules and preconditions. Inductive reasoning determines the rule from the knowledge base itself, based on given preconditions and conclusions. Abductive reasoning determines the preconditions that support the conclusion based on given logic rules. Logical reasoning has been extensively studied (Amir and McIlraith 2005; Eiter and Gottlob 1995; Gilbert 1965; Hendricks 2005; Johnson-Laird and Byrne 1991; Josephson 2001; Mayer et al. 1995; Menzies 1996), but it has received renewed interest, as several reasoning tools have been developed to work directly on OWL files, including Racer (Harrslev and Moeller 2003), Jena (Carroll and Dickinson 2004), Fact++ (Tsarkov and Horrocks 2006), and Pellet (Sirin et al. 2007). These reasoning tools rely on forward chaining or backward chaining strategies to realize deduction, induction, or abduction.

Forward chaining is a form of logical reasoning that progresses from what is known toward a desired solution. Using available data, a forward chaining system infers new knowledge by applying the inference rules until a goal is reached. The execution cycle is to (1) select a rule, the preconditions of which match the current state of the system; (2) change the system state to match the conclusions of the selected rule by executing the rule; and (3) repeat the above steps until there is no rule to apply. Forward chaining can be categorized as a data-driven approach, and the algorithm can be summarized as

Forward Chaining Algorithm

```
1. Given a Knowledge Base(KB) and a goal statement (α)
2. for each sentence s in KB do
3. (p₁ ∧ p₂. . . ∧ pₙ ⇒ q) ← Standardize-Apart (s)
4. for each substitution θ such that (p₁ ∧ p₂. . .∧ pₙ)θ =
   (p₁ ∧ p₂. . .∧ pₙ)θ, q'← subst(θ,q)
```

```
5. if q' is new information to the KB, add q' to New;
6. if φ ← Unify(q',α) is not fail, then a solution is
   found/validated, return φ
7. Merge KB ← KB ∪ New
8. Loop step 2–7 if the set of new inferred knowledge(New)
   is not empty, else return false
```

For example, suppose the goal is to ascertain the color of a pine tree. Given that pine is a woody plant with branches, assume that the rule space contains the following rules:

1. $Plant(X) \wedge Woody(X) \wedge Has(, branches) \Rightarrow Tree(X)$

2. $Plant(X) \wedge Flowering(X) \Rightarrow Flower(X)$

3. $Tree(X) \Rightarrow Green(X)$

Rule 1 would be searched, because we can find satisfactory substitutions (X = pine) for the conditions based on the given data. Then, new information *Tree(pine)* is added to the knowledge base. Next, Rule 3 is selected, because its antecedent *Tree(X)* matches the knowledge we inferred. Now, a new consequent *Green(pine)* is added. No more knowledge can be inferred from this information, but through unification, we can accomplish our goal of determining the color of pine, which is green.

In contrast with forward chaining, backward chaining starts with the goal to be achieved and repeatedly breaks it into subgoals that are easier to solve with the available data and the inference rule space. An inference engine with a backward chaining algorithm continues to search the inference rules to find one with a consequence matching the current goal. If the precondition of that rule cannot be confirmed to be true using the existing knowledge base, the precondition must be added to a list of unsolved goals (subgoals), which should be validated using other rules and data. The backward chaining algorithm is

Backward Chaining Algorithm

```
1. BC(KB, goals, θ)
2. if empty(goals), then return {θ}
3. q' ← Subst(θ, first (goals))
4. for each sentense s in KB, where
      Standardize Apart (s) = (p₁ ∧ p₂.. ∧ pₙ ⇒ q) and
      θ' ← Unify(q, q') succeeds
5. ans ← BC(KB, rest(goals), θ∪θ')
6. return ans
```

Backward chaining is an iterative process. In the above algorithm, *ans* is a set of satisfied substitutions from which all goals and subgoals can be achieved. For the same pine color example, the goal in backward chaining is to find evidence (knowledge) to support *Green(pine)*. First, Rule 3 is searched and selected from the rule space, because its conclusion matches the goal to determine the pine's color. As the condition [*Tree(pine)*] of the goal *Green(pine)* is not supported directly by the available data, it is inserted in the goal list as a subgoal. The rule space is traversed again and then Rule 1 is selected. The preconditions of Rule 1 match the available data *Plant(pine)*, *Woody(pine)*, and *Has(pine, branches)*; therefore, a chain of reasoning demonstrating that the color of pine is green has been established using Rules 1 and 3 and the given knowledge.

7.5.2 Numerical Reasoning

Although logical reasoning is widely used in expert systems and robotic research, it is based on an explicitly defined knowledge base and is unable to handle uncertainty. Numerical reasoning has been proved to perform efficiently in the presence of uncertainty and to serve as an effective tool to learn from data and explore implicit knowledge and regularities.

The genetic algorithm (GA) (Holland 1975; Goldberg 1989) is an important branch of numerical reasoning, which uses models from biological evolution to guide computer simulations. It is an automatic reasoning process used to find exact or approximate solutions to global optimization problems. A typical GA algorithm works as follows:

Genetic Algorithm

```
1. Generate a random initial population
2. Evaluate the fitness of individuals in the population,
   if the ideal individual is found, finish and exit.
3. Initial an empty population P,
4. SELECT individuals i and j from old population
5. CROSSOVER between i and j
6. MUTATE i and j seperately and add new i and j to P
7. if P is not full, go loop 4-6, else replace old
   generation with P.
8. go to 2
```

Initially, a knowledge base is represented by the *population*, and each fact in the knowledge base is an *individual*. A fitness function is used to measure how well the individuals achieve the goal. In other words, has the learning process entailed the exact knowledge we want? This learning process is carried out by iterative selection, crossover, and mutation, in analogy to biological adaptation and evolution. Among these operators, selection is used to determine which individuals will be chosen for later breeding. Available selection algorithms include fitness proportionate selection (FPS) and tournament selection (TS).

The FPS selects an individual based on the distribution of possibility, which equals the normalized fitness value of individuals. Thus, the higher the fitness of an individual, the greater the chance it will be selected. In contrast, the fitness value in TS does not dictate the selection mechanism. In TS, N individuals are picked randomly, and the most fit individual is selected.

A value of N that is too small will make the system wander aimlessly and find a solution very slowly. A value of N too large will reach a solution quickly, but it might be suboptimal. In practice, $N = 7$ is a proper size to balance out the above effects. Another process for genetic-based learning is crossover, which allows sexual reproduction (gene modification among more than one individual). Typical algorithms for crossover include one-point crossover, two-point crossover, and *cut and splice*. In a one-point crossover, two individuals are broken at the crossover point, and the pieces are swapped to generate new individuals. In a two-point crossover, the individuals are cut into three pieces, and the middle piece is swapped. In a third crossover method, cut and splice, two individuals have different crossover points. Compared with the previous two crossover algorithms, cut and splice will lead to different lengths in the newly generated individuals. Another method of reproduction is mutation, which changes an arbitrary bit in a gene sequence (individual) to enhance the diversity and reduce the similarity in individuals. Without mutations, the system tends to get stuck at a local optima and does not continue learning.

Reinforcement learning is a type of numerical reasoning that learns new strategies to guide how a computer agent takes actions in an environment. The environment is essentially a Markov model, containing a set of states (S) connected by transitions or actions (A) and a set of scalar rewards (R). Each time an agent performs an action, a state transition occurs and a reward (reinforcement) is received (which might depend only on the current or previous actions). The learning process fills up a reward table of the Markov model, in which the number of rows equals the number of total states, and the number of columns equals the number of actions. Each cell of the table stores the reward value for each possible action in that state. If the agent takes an action i in state S_j, it will get a reward R_{ij} by taking the chosen action. The reinforcement learning process can be described mathematically as follows: At each state t, the intelligent agent will choose the best action, represented by the highest reward (based upon the current move plus any expected future rewards). The detailed algorithm is as follows:

Reinforcement Learning Algorithm

```
1. Intialize the reward table (n × m) which includes
   every action (m) at each state (n).
2. Loop for each move of the intelligent agent:
3.    S ← initial state of the agent
4.    For each time the agent needs to make a decision
      on where to move always try to best action
          Q(s,a) ← (1−α)Q(s,a) + α(r+γ maxₐ'(s',a'))
          s' ← s
5. Return reward table
```

The above learning process uses an active learner who allows modification of the policy on the fly as the intelligent agent learns. The computer agent can become more intelligent by gaining new knowledge through continuous learning.

7.5.3 Ontology-Based Reasoning

Most knowledge representations are declarative languages such as frame languages and first-order logic. With the advent of the Internet, Web services, and semantic Web, the ontology has become the primary mechanism of knowledge representation and management to enable automated reasoning.

Ontology-based knowledge engineering has been explored extensively in computer science (Gomez-Perez et al. 2007). Ontology languages are primarily based on description logic and are weak in reasoning. Rule-based languages provide syntactic and semantic regulations to represent rules and support rule-based reasoning (Breitman et al. 2007; Kashyap et al. 2008). These languages, such as declarative languages RuleML and SWRL, enhance ontology languages with knowledge reasoning functions. Ontology-based data interoperation also can be achieved with rule-based languages that compare and integrate ontologies based on facts, knowledge, and rules.

The RuleML is a markup language in XML syntax for publishing and sharing rule bases. It builds a hierarchy of rule sublanguages upon XML, RDF, XSLT, and OWL. Datalog logic is the foundation for RuleML. Datalog defines *facts* corresponding to explicit rows of relational tables and *rules* corresponding to tables defined implicitly by views. The RuleML Lite has been developed as a RuleML subset compatible with RDF and OWL-DL, and it covers binary Datalog facts, rules, and queries. The SWRL is proposed to combine OWL and RuleML. It is an extension of OWL with if-then statements. Open source tools such as Protégé SWRLTab support the editing and execution of SWRL rules.

7.5.4 Spatial Reasoning Applications

Existing transportation planning modeling tools have critical limitations with respect to assessing the benefits of ITS deployment. To solve this problem, intelligent reasoning has been employed to improve transportation navigation and ensure faster emergency response. For example, Sadek et al. (2003) used a case-based reasoning algorithm to increase the benefit of diverting traffic away from incident locations. Bouamrane and Beldjilali (2004) proposed to utilize spatial reasoning to discover the regulations of an urban transportation network; Bouzid (2003) demonstrated an online transportation scheduling system that can delegate a new transportation task to a truck that is moving toward a destination. The integration of reasoning in the above systems realizes a suitable decision-making process.

The ability of robots to recognize natural objects in a visual scene is important, especially when a robot must navigate autonomously through space and perceive the environment as automatically as possible (Hois et al. 2008). This capability is key to military RECON (Reconnaissance) intelligence and is the first step of any military action, whether in defense or attack mode (Kennedy et al. 2007). Currently, there are many knowledge bases containing various encodings of spatial knowledge, and these knowledge bases can be comprehensively used to perform spatial reasoning tasks (Schultz et al. 1999). Based on the knowledge base and reasoning algorithms discussed earlier, a robot could recognize unknown terrain and unknown objects when conducting a special task (Montemerlo et al. 2003).

Much more future research is required to improve both the efficiency and effectiveness of the reasoning algorithm for Geoinformation applications. Meanwhile, combining logical and numerical reasoning and utilizing the best of both methodologies may prove to be a fruitful new direction for knowledge-based reasoning.

7.6 Knowledge-Based Fusion

Data fusion at the simplest level is the ability to create new meaningful data from existing layers. Fusion typically applies both discovery and reasoning methods to combine the results in some meaningful way, often in real time. For example, if MassFlux is defined as the product of a mass layer and a wind layer, a new MassFlux layer can be generated using map multiplication on a pair of layers that are of type mass and wind, respectively. Any needed unit conversions could be carried out with the assistance of a units ontology.

With an ontology as mediator of diverse spatial data, ontology-based spatial information systems have been developed. Fonseca et al. (2002) designed an ontology of spatial concepts with *hierarchy* and *role* of each concept; for instance, the ontology of *lake* is hierarchically under the ontology of *water body*, and the role of *lake* may be *transportation*. Ontology-based concept integration is achieved through a comparison of the *roles* and *hierarchies* of two concepts in the same or different ontologies, respectively. Wiegand and Zhou (2005) developed an ontology-based Web query system to address semantic heterogeneity in land use categories (Figure 7.8) by enhancing an XML database query system, Niagara, with an ontology subsystem to query distributed XML land use data sets. This system allows users to pose a query on an ontology of diverse land use categories; the semantic relations between the ontology and individual land use categories are recorded in an ontology mapping agreement. An XML-QL query is processed by the ontology subsystem that consults the ontology agreements, based on which the ontology-based query is rewritten into a set of subqueries for individual spatial data sets.

Noy (2004) provides a comprehensive review of the use of an ontology to achieve semantic integration across heterogeneous databases. It is possible to

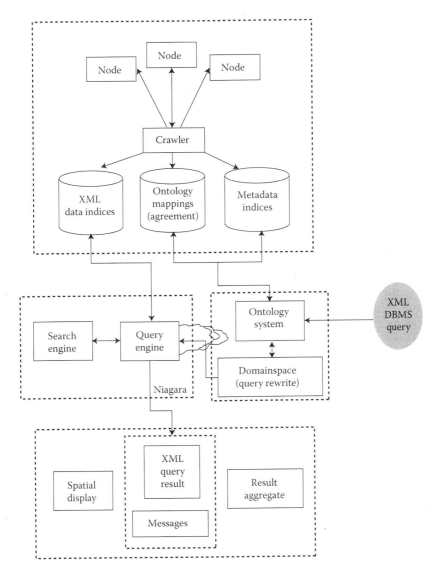

FIGURE 7.8
An ontology-based land use data query system. (Adapted from Wiegand, N. and N. Zhou. 2005. An ontology-based geospatial web query system. In *Next Generation Geospatial Information*, eds P. Agouris and A. Croitoru, pp. 157–168. Taylor and Francis, Balkema.)

treat a taxonomy (e.g., land use categories) as an ontology, where data interoperability is implemented with automated technologies such as

- Concept analysis by comparing the keywords and/or subconcepts (Kokla and Kavouras 2001)
- Document classification by measuring concept similarity (Zhou 2003)

- Natural language processing (Zhou and Wei 2008)
- Structure aware methods (Cruz and Sunna 2008)
- Distinguishing features of concepts (Rodriguez et al. 2004), and so on

7.6.1 Future Potential of Knowledge-Based Fusion

Two hypothetical examples illustrate what may be possible in the future with suitable knowledge-based fusion methods.

Example 7.1: Intelligent Agent

The AI machines are expected to be sufficiently intelligent to implement more integrated e-services for humans. For example, in 2001, Dr. Tim Berners Lee presented a great insight for a future intelligent agent (Lee et al. 2001):

When a phone rang, all the local devices that had a volume control at Pete's house were *turned down automatically*. Pete answered the phone and it was from his sister Lucy about *setting up an appointment* for their mom to attend a series of physical therapy sessions. Then both of Pete's and Lucy's intelligent agents started to *negotiate* the available time based on their schedule, *looked up* the list of providers who are in-plan for Mom's insurance within a 20-mile radius of her home, and *ranked* them by the quality of service from previous patients. After it *presented* them with a plan, Pete found the plan to be not suitable because the recommended hospital was too far away from Mom's place, and he would be driving back in the middle of rush hour. He set his intelligent agent to *redo the search* based on these preferences, and to *assist* Lucy's agent automatically by *sharing the data* the agent already retrieved. The new plan produced a much closer hospital and earlier time. The details of the new plan were *shared* with Lucy's agent and the process was accomplished automatically.

This example is a vivid description of how a future intelligent agent could assist our daily lives. Reasoning is at the core of the intelligent agent system. Several reasoning processes are emphasized in the above example, including "turn down devices automatically because the phone rang" and "negotiating between doctor's time and Lucy/Peter's schedules." To realize an optimal solution, each intelligent agent must have complete information; for example, the phone must know which devices are to be turned down, based upon understanding which devices are local, with a volume control, and presently turned on. This state information constitutes the knowledge from which the intelligent agent will train the system and draw conclusions using the process of knowledge-based reasoning.

Example 7.2: Global Warming Detection

Consider the following query:

> Find data which demonstrates global warming at high latitudes during summertime and plot warming rate.

Solving this query requires breaking down the problem into substeps, each with interpretable actions:

"Global warming" = Find trend of increasing temperature over large spatial scales

"High latitude" = Limit to latitudes >60 degrees

"Summertime" = Limit to June-Aug (NH) and Jan-Mar (SH)

"Find data" = Locate data sets using catalogs, then access and read it

"Plot warming rate" = Display temperature vs time

Although this level of understanding seems a daunting task, it is not impossible, given sufficient semantics in the knowledge base and appropriate discovery and reasoning algorithms.

References

Agarwal, P. 2005. Ontological considerations in GIScience. *International Journal of Geographical Information Science*, 19:501–536.

Amir, E. and S. McIlraith. 2005. Partition-based logical reasoning for first-order and propositional theories. *Artificial Intelligence Conference on Temporal Logic*, 162(1–2):49–88.

Anderson, J., E. Hardy, J. Roach, and R. Witmer. 1976. A land use and land cover classification system for use with remote sensor data. *Geological Survey Professional Paper 964*. http://landcover.usgs.gov/pdf/anderson.pdf.

Arp, R. and B. Smith. 2008. Function, role, and disposition in basic formal ontology. *Proceedings of Bio-Ontologies Workshop (ISMB 2008)*, pp. 45–48. Toronto.

Barkowsky, T. 2002. Mental representation and processing of geographic knowledge: A computational approach. Springer, Berlin.

Bench-Capon, T. 1990. *Knowledge Representation: An Approach to Artificial Intelligence*. Academic Press, New York.

Berners-Lee, T., J. Hendler, and O. Lassila. 2001. The Semantic Web. *Scientific American*. http://www.sciam.com/article.cfm?id=the-semantic-web.

Bouamrane, K. and B. Beldjilali. 2004. A case-based reasoning for regulation of an urban transportation network. *CSJM*, 12(3):424–429.

Bouzid, M. 2003. On-line transportation scheduling using spatio-temporal reasoning. In: *Proceedings 10th International Symposium on Temporal Representation and Reasoning*. Cairns, Australia.

Breitman, K., M. Casanova, and W. Truszkowski. (2007). *Semantic Web: Concepts, Technologies and Applications*. Springer, London.

Brickley, D. and R.V. Guvha. 2004. *RDF Vocabulary Description Language 1.0: RDF Schema, W3C Recommendation*, http://www.w3.org/TR/rdf-schema.

Carroll J.J. and J.Z. Pan. 2006. XML schema datatypes in RDF and OWL, W3C Working Group Notes, http://www.w3.org/TR/swbp-xsch-datatypes.

Casati, R. and A. Varzi. 1997. Spatial entities. In *Spatial and Temporal Reasoning*, ed. O. Stock, pp. 73–96. Kluwer Academic Publishers, Dordrecht.

Cruz, I. and W. Sunna. 2008. Structural alignment methods with applications to geospatial ontologies. *Transactions in GIS*, 12(6):683–711.

Dean, M. and G. Schreiber. 2004. OWL Web Ontology Language Representation, W3CRecommendation, http://www.w3.org/TR/owl-ref.

Egenhofer, M. and D. Mark. 1995. Naïve geography. In *COSIT'95 Lecture Notes in Computer Science*, Vol. 988, eds. A. Frank and W. Kuhn, pp. 1–15. Springer-Verlag, Berlin.

Egenhofer, M. 2002. Toward the semantic geospatial Web. *10th ACM International Symposium on Advances in Geographic Information Systems*, pp. 1–4. ACM Press, Virginia.

Fayyad, U., G. Piatetasky-Shapiro, and P. Smyth. 1996. From data mining to knowledge discovery in database. *AI Magazine*, Fall, 37–54.

Fonseca, F., M. Egenhofer, C. Davis, and G. Camara. 2002, Semantic granularity in ontology-driven geographic information systems. *Annals of Mathematics and Artificial Intelligence*, 36(1–2):121–151.

Frank, A. 1997. Spatial ontology: A geographical point of view. In *Spatial and Temporal Reasoning*, ed. O. Stock, pp. 135–153. Kluwer Academic Publishers, Dordrecht.

Gahegan, M., B. Brodaric, and M. Takatsuka. 2000. Designing a framework to support the semantic interoperability of geospatial information. *First International Conference on Geographic Information Science*, Savannah, GA, October 28–31.

Gangemi, A., N. Guarino, C. Masolo, and A. Oltramari. 2003. Sweetening WordNet with DOLCE. *AI Magazine*, 24(3):13–24.

Gilbert, H. 1965. The inference to the best explanation. *The Philosophical Review*, 74(1):88–95.

Goldberg, D.E. 1989. *Genetic Algorithms in Search, Optimization, and Machine Learning*. Addison-Wesley, Reading, MA.

Golledge, R. 2002. The nature of geographic knowledge. *Annals of the Association of American Geographers*, 92:1–14.

Gomez-Perez, A., M. Fernandez-Lopez, and O. Corcho. 2007. *Ontology Engineering*. Springer-Verlag, Berlin, Germany.

Goodchild, M. 1997. Towards a geography of geographic information in a digital world. *Computer, Environment, and Urban Systems*, 21(6):377–391.

Goodchild, M. 2007. Citizens as sensors: The world of volunteered geography. *Journal of Geography*, 69(4):211–221.

Grenon, P. and B. Smith. 2004. SNAP and SPAN: Prolegomenon to geodynamic ontology. *Spatial Cognition and Computation*, 4(1):69–103.

Gruber, T. 1993. Toward principles for the design of ontologies used for knowledge sharing. In *International Workshop on Formal Ontology*, eds. N. Guarino and R. Poli. Padova, Italy.

Guarino, N. and P. Giaretta. 1995. Ontologies and knowledge bases: Towards a terminological clarification. In *Towards Very Large Knowledge Bases: Knowledge Building and Knowledge Sharing 1995*, ed. N. Mars, pp. 25–32. IOS Press, Amsterdam.

Guarino, N. 1998. Formal ontology and information systems. In *Formal Ontology in Information Systems*, ed. N. Guarino, pp. 3–15. IOS Press, Trento, Italy.

Haarslev, V. and R. Moeller. 2003. Racer: A core inference engine for the Semantic Web. In *Second International Workshop on Evaluation of Ontology-Based Tools* (EON-2003), Sanibel Island, FL.

Hendricks, V.F. 2005. *Thought 2 Talk: A Crash Course in Reflection and Expression*. Automatic Press/VIP, New York, ISBN 87–991013–7–8.

Hoekenga, B.C. 2007. Mind over machine: What Deep Blue taught us about chess, artificial intelligence, and the human spirit. Master thesis, 49pp. MIT DSpace: http://hdl.handle.net/1721.1/42144.

Hois, J., M. Wünstel, J.A. Bateman, and T. Röfer. 2008. Dialog-based 3D-image recognition using a domain ontology. *Spatial Cognition V Reasoning, Action, Interaction,* 107–126.

Holland, J. 1975. *Adaptation in Natural and Artificial Systems.* University of Michigan Press, Ann Arbor.

Johnson-Laird, P. and R. M. J. Byrne. 1991. *Deduction.* Lawrence Erlbaum Associates, Hillsdale, NJ.

Johnston, R. J. 2009. *The Dictionary of Human Geography* (5th ed.). Blackwell, London.

Josephson, J.R. and S.G. Josephson (eds). 1995. *Abductive Inference: Computation, Philosophy, Technology,* Cambridge University Press, Cambridge, UK.

Kashyap, V., C. Bussler, and M. Moran (eds). 2008. *The Semantic Web: Semantics for Data and Services on the Web,* 122–123. Springer, Berlin.

Kennedy, W.G., M.D. Bugajska, M. Marge, W. Adams, B.R. Fransen, D. Perzanowski, A.C. Schultz, and J.G. Trafton. 2007. Spatial representation and reasoning for human–robot collaboration. In *Proceedings of the 22nd National Conference on Artificial intelligence,* pp. 1554–1559. Vancouver, British Columbia, Canada.

Kirwin, C. 1995. 'Reasoning'. In *The Oxford Companion to Philosophy,* ed. Ted Honderich, p. 748. Oxford University Press, Oxford.

Klyne, G. and J.J. Carroll. 2004. Resource description framework (RDF): Concepts and abstract syntax. http://www.w3.org/TR/2004/REC-rdf-concepts-20040210/, February 2004.

Kokla, M. and M. Kavouras. 2001. Fusion of top-level and geographical domain ontologies based on context formation and complementarity. *International Journal of Geographical Information Science,* 15:679–687.

Kuhn, W. 2002. Modeling the semantics of geographic categories through conceptual integration. In *Geographic Information Science: Second International Conference,* eds. M. Egenhofer and D. Mark, pp. 108–118. Boulder, CO.

Li, W., C. Yang, and R. Raskin. 2008. A semantic enhanced search for spatial web portals. In *Technical Report of Semantic Scientific Knowledge Integration of AAAI Spring Symposium,* pp. 47–50. AAAI Press, Merlo Park, CA.

Mark, D., M. Egenhofer, and K. Hornsby. 1996. Formal models of commonsense geographic worlds. *Report on the Specialist Meeting of Research Initiative 21 Specialist Meeting.* October 30–November 3, 1996. San Marcos, Texas.

Mark, D., B. Smith, M. Egenhofer, and S. Hirtle. 2005. Ontological foundations for geographic information science. In *A Research Agenda for Geographic Information Science,* eds. R. McMaster and L. Usery, pp. 335–350. CRC Press, Boca Raton, FL.

Matuszek, C., J. Cabral, M. Witbrock, and J. DeOliveira. 2006. An introduction to the syntax and content of Cyc. In *Proceedings of the 2006 AAAI Spring Symposium on Formalizing and Compiling Background Knowledge and Its Applications to Knowledge Representation and Question Answering.* Stanford, CA, March 2006.

Mennis, J. and D. Guo. 2009. Spatial data mining and geographic knowledge discover—An introduction. *Computers, Environment and Urban Systems,* 33:403–408.

Miller, H. and J. Han. 2009. Geographic data mining and knowledge discovery: An overview. In *Geographic Data Mining and Knowledge Discovery,* eds. H. Miller, and J. Han, pp. 1–26. CRC Press, Boca Raton, FL.

Montello, D. and S. Freudschuh, 1995. Sources of spatial knowledge and their implications for GIS: An introduction. *Geographical System,* 2:169–176.

Montemerlo, M., N. Roy, and S. Thrun. 2003. Perspectives on standardization in mobil robot programming: the Carnegie Mellon Navigation (CARMEN) Toolkit. In *Proceedings of the IEEE/RJS International Conference on Intelligent Robots and Systems* (IROS 2003), pp. 2436–2441. Las Vegas.

Noy, N. 2004. Semantic integration: A survey of ontology-based approaches. *SIGMOD Record*, 33:65–70.

OGC. 2006. OpenGIS Simple Feature. http://www.opengeospatial.org/standards/sfa.

Ramachandran, R., S. Movva, X. Li, P. Churukuri, and S. Graves. 2006. Noesis: Ontology-based scoped search engine and resource aggregator for atmospheric science. American Geophysical Union Fall 2006 Meeting.

Raskin, R. and M. Pan. 2005. Knowledge representation in the semantic web for Earth and environmental terminology (SWEET). *Computer & Geoscience*, 31(9):1119–1125.

Rodriguez, M., M. Egenhofer, and R. Rugg. 2004. Comparing geospatial entity classes: An asymmetric and context-dependent similarity measure. *International Journal of Geographical Information Science*, 18:229–256.

Sadek, A., S. Morse, J. Ivan, and W. El–Dessouki. 2003. Case-based reasoning for assessing intelligent transportation systems benefits. *Computer-Aided Civil and Infrastructure Engineering*, 18(3):173–183.

Schultz, A.C., W. Adams, and B. Yamauchi. 1999. Integrating exploration, localization, navigation, and planning with a common representation. *Autonomous Robots*, 6:293–308.

Sirin, E., B. Parsia, B.G. Cuenca, A. Kalyanpur, and Y. Katz. 2007. Pellet: A practical OWL-DL reasoner. *Journal Web Semantics*, 5(2):51–53.

Smith, B. 1998. An introduction to ontology. In *The Ontology of Fields: Report of a Specialist Meeting of NCGIA*, eds. D. Peuquet, B. Smith and B. Brogaard. June 11–13, Bar Harbor, Maine.

Tsarkov, D. and I. Horrocks. 2006. FaCT++ description logic reasoner: System description. In: *Proceedings of the International Joint Conference on Automated Reasoning* (IJCAR 2006). Lecture Notes in Artificial Intelligence, Vol. 4130, pp. 292–297. Springer, Berlin, Heidelberg, New York.

Varanka, D. 2009. Landscape features, technology codes, and semantics in U.S. national topographic mapping databases. *The International Conference on Advanced Geographic Information Systems & Web Services* (GEOWS), Cancun, Mexico, February 1–7, 2009.

Wiegand, N. and N. Zhou. 2005. An ontology-based geospatial web query system. In *Next Generation Geospatial Information*, eds. P. Agouris and A. Croitoru, pp. 157–168. Taylor and Francis, Balkema.

Winter, S. 2001. Ontology: buzzword or paradigm shift in GI science? *International Journal of Geographic Information Science*, 15(7):587–590.

Yuan, M., B. Butternfield, M. Gahegan, and H. Miller. 2005. Geospatial data mining and knowledge discovery. In *A Research Agenda for Geographic Information Science*, eds. R. McMaster and E. Usery, pp. 365–388. CRC Press, Boca Raton, FL.

Zhou, N. 2003. A study on automatic ontology mapping of categorical information. In *Proceedings of National Conference on Digital Government Research*, pp. 401–404. May 18–21, Boston, MA.

Zhou, N. and H. Wei. 2008. Semantic-based data integration and its application to geospatial portals. *Fifth International Conference on Geographic Information Science Extended Abstract*. September 23–26, Park City, Utah.

8

Environmental and Related Applications

Joan L. Aron, Sheryl Luzzadder-Beach, Min Sun, Fahui Wang,
Laura E. Jackson, Greg Susanke, Edward Washburn, John White,
David Wong, Chaowei Yang, and Stephen Young

CONTENTS

8.1 Introduction

Most introductory textbooks on GIS include a set of standard topics essential to the early development of GIS. Since GIScience and technology have experienced tremendous advancements during the past several decades, those textbooks are insufficient or even obsolete to discuss thoroughly topics pertinent to the current computing environments in which most GISs operate. In addition, these texts fall short of addressing both the theoretical and real world challenges that GIS developers and users have to deal with these days. One of the intentions of this written volume is to fill this void to provide more advanced and in-depth discussions on relevant topics and issues. One of these topics is to evaluate how GIS are being used in various academic disciplines and sectors in the society. The specific objective is to demonstrate how GIS are used in various types of applications to deal with real world problems.

This chapter and the next review GIS applications found in both public sectors and academic disciplines. Clearly, GIS are widely adopted in different aspects of the society, from environment management and monitoring to economic planning, from a stand-alone system on a desktop computer dealing with a very specific and routine application to a general application being embedded within a popular Web application to serve a large pool of public users. To expect this book to cover such a diverse and broad range of applications is unrealistic. Instead of providing a broad base coverage, we review several environmental applications of the broadly defined GI Science and technology. We report the recent efforts of the U.S. EPA in using GIS and related technologies not just to monitor air quality but also to inform the public of the most updated conditions available to the government. Public health has traditionally been the research domain of medical and clinical experts, whereas recent development acknowledges the importance of various environment characteristics in affecting health outcomes, such as the roles of vegetation and land cover in the spread and recent outbreaks of vector-borne diseases (e.g., Dister et al., 1997 for Lyme disease; Allen and Wong, 2006 for West Niles Virus) and how the layout of urban landscape may encourage or discourage the development of a healthy lifestyle (e.g., Frank et al., 2006). In addition, the public health concerns have been expanded to include the health of the environment. In this chapter, a section will portray how remote sensing or Earth Observing (EO) technology can assist the monitoring of heat waves, wildlife-associated diseases, and bacterial outbreaks in the ocean. GIS has also been widely adopted in interdisciplinary sciences, such as environmental science, Earth system science, and global change studies. A section in this chapter reviews how global and environmental change studies have been using GIS to enhance our understanding of various spheres of the changing world. Many GIS applications have a bias toward the physical or environmental aspects of the Earth systems, whereas the use of GIS in social science is highly significant. Instead of providing a broad overview of GIS applications in various social science disciplines, the last section in this chapter focuses on the two interdisciplinary areas of crime analysis and health care service provision.

The intention of this chapter is not to provide a comprehensive review of all types of GIS applications. Instead, our emphases are on the use of advanced GI Science technology in the applications (such as air quality monitoring) and the use of GIS in nontraditional academic disciplines (such as global change and crime studies). We hope that these focused reviews can provide readers a relatively updated picture of the roles of GIS played in various aspects of the society so that the readers may develop a better appreciation of the broad impacts of GIS.

8.2 Air Quality and AIRNow*

Air quality is one of the most important aspects of environmental quality. Since humans rely more on fossil fuels now than in the past the concern of pollution is ever increasing. However, other natural events and phenomena also affect air quality. In the United States, the EPA spearheads the effort in monitoring air quality and formulating related policies. In this section, we review some of the recent efforts by U.S. EPA in utilizing advanced geoinformation technologies to assist the monitoring of air quality condition and the dissemination of information.

8.2.1 GEOSS and the Roles of EPA

The Global Earth Observation System of Systems (GEOSS) takes the pulse of the planet by linking data gathered from sensors on board satellites and aircrafts, monitors on the ground, and computational models to support decision making. Besides comprehending the information derived from the massive data collected by GEOSS, decisions have to be made in the context of think globally and act locally to protect human health and the environment. The more we understand the Earth, the better stewards we may become.

To participate in the international effort of GEOSS, the U.S. Group on Earth Observations (GEO) developed a strategic plan for the United States Integrated

---

* Min Sun and David Wong compiled this section based on public domain material provided by Edward Washburn, Stephen Young, Greg Susanke, and John White of the U.S. EPA. The views expressed in this section are those of individual authors and do not necessarily reflect the views and policies of the U.S. EPA.

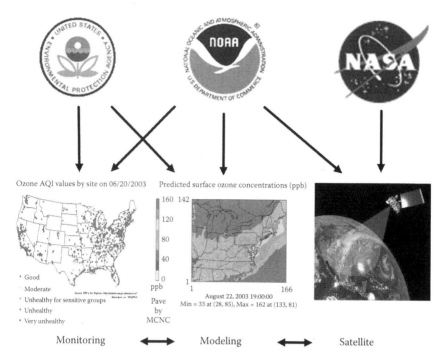

FIGURE 8.1
EPA partnership with other agencies.

Earth Observation System (IEOS) with the near-term objectives on six issues: integrate data management, improve observations for disaster management, global land observing system, sea level observing system, national integrated drought Information system, and air quality assessment forecasting system. With a great deal of experience in measuring, monitoring, and modeling environment, the U.S. EPA is heavily involved and plays a leadership role to support the U.S. GEO and IEOS.

The EPA's "niche" in GEOSS is both a major user and provider of data and models (http://www.epa.gov/geoss/role.html). The EPA provides information linkages from observations and integrates multiple societal benefit areas with users. It needs to connect with state and local environmental agencies to support decisions on assessment, regulatory action, guidance, uncertainty, accountability. The EPA has already established partnerships and collaborations with other federal agencies such as NOAA and NASA on satellite missions, monitoring, and modeling (Figure 8.1).

8.2.2 Air Quality

The U.S. GEO and EPA set up the air quality assessment forecasting system as one of the near-term objectives, because air pollution affects everyone nowadays. Air pollution harms our general health through lung diseases (i.e., asthma) and heart diseases (i.e., congestive heart failure). It might result in various undesirable health outcomes, such as premature deaths; visits to hospitals, emergency rooms, and doctors' offices; absence from school and work; and extended symptom days. High levels of air pollution may also damage forests and lakes. Similar to weather, pollution levels vary from day to day, even from hour to hour. Global climate is changing. Poor air quality lowers our quality of life and creates damages to humans in the future. Currently, most people live in areas that do not meet clean-air standards. Thus, air quality assessment and forecasting system was among the emphases of the U.S. GEO in GEOSS. Since the U.S. EPA has been monitoring air quality for a long time, the Agency plays a critical role in the project of air quality assessment and forecast. The EPA identified five goals in the 2006–2011 EPA strategic plan (U.S. EPA, 2006). They are (1) clean air and global climate change, (2) clean and safe water, (3) land preservation and restoration, (4) healthy communities and ecosystems, and (5) compliance and environmental stewardship. Air quality is among one of them.

As part of the mission and contributions to GEOSS, the U.S. EPA collects a wide variety of data and utilizes these data and those from other agencies (i.e., NOAA, Space of Europe). Using these data, EPA identifies measures that need to be reported on their status and trends and, wherever possible, their impacts on human health and the environment. The past, present, and future air conditions give individuals and organizations a better awareness of environmental quality, which in turn helps make better decisions. Good decisions will lead to positive feedbacks to the human–Earth system.

The EPA has also developed indicators that describe particular aspects of the natural, economic, and social environment, and these indicators can connect the environmental conditions with the quality of life. These indicators will be used in some GEOSS projects, such as AIRNow. The Agency will use data and apply models to compute indicators that reflect current and future environmental conditions and will provide foci to a variety of EO system projects. For example, air quality index (AQI) is an indicator used by EPA to characterize the quality of air at a given location. The AQI indicates not just air quality but also the levels of specific pollutants (i.e., ground-level ozone, particulates, sulfur dioxide, carbon monoxide, and nitrogen dioxide). A high value of AQI means a high level of air pollution. For easy to understand, values of AQI are divided into different classes, and each class is assigned a description and color. Classification is also different according to countries or areas. In the United States, the EPA divides AQI into six ranges. Green is good (AQI: 0–50), yellow is moderate (AQI: 51–100), orange is unhealthy for sensitive groups (AQI: 101–150), red is unhealthy (AQI: 151–200), purple is very unhealthy (AQI: 201–300), and maroon is hazardous (AQI: 301–500). Figure 8.2 shows an ozone air quality index by monitoring data over the entire country. They are measures of air quality with routinely available information.

Ozone AQI values by site on 06/01/2009

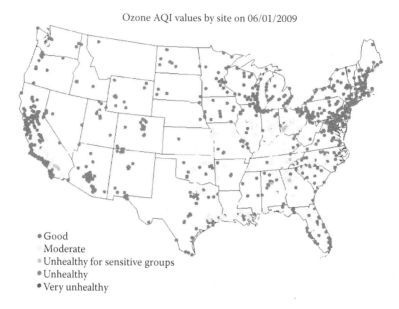

- Good
  Moderate
- Unhealthy for sensitive groups
- Unhealthy
- Very unhealthy

FIGURE 8.2
(See color insert following page 144.) Ozone AQI values by sites (Adapted from EPA's air explorer, http://www.epa.gov/airexplorer/.)

The 2004 Alaskan fire event can be used as an example to demonstrate the power of integrating data from various sources, including models and tools, to report air quality information. The 2004 Alaskan fire had a very extensive impact area, reaching as far as England. Satellite data were gathered continuously during the fire. Computer models were used to fill the gaps in areas in which EPA did not monitor (Figure 8.3). Then, satellite and meteorological data were integrated to determine whether and where particles would impact air quality at the ground level. Results show that the elevated fine particle levels detected on 21 July 2004 in North Carolina were due to the Alaskan fire event. This example took advantage of the strength of each type of data involved.

Air quality information is often provided by models, such as wild fire models and dust storm models. Such models generate smoke or dust plumes, which should be integrated into GIS. Being able to incorporate such 3D or 4D (volumetric and time) models will facilitate not just data management, visualization, and analysis but also decision making.

8.2.3 AIRNow

Under GEOSS is the subsystem of National Air Quality Notification and Forecasting System (AIRNow), which uses AQI to represent air quality information in real time by location. The EPA developed AQI for reporting daily air quality to the hour to address the concerns of air quality and health effects. The AQI was formulated for each of the five major air pollutants regulated by

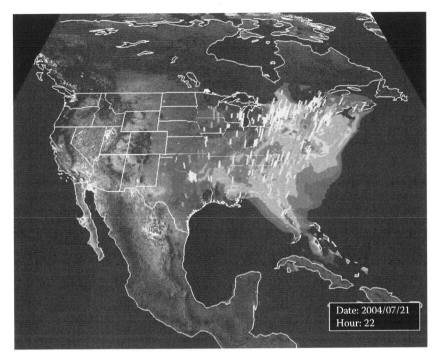

FIGURE 8.3
(See color insert following page 144.) NOAA/EPA Community Multiscale Air Quality (CMAQ) model estimates fine particle values for the Eastern United States.

the Clean Air Act: ground-level ozone (O_3), particle pollution, which is also known as particulate matter (PM), carbon monoxide (CO), sulfur dioxide (CO_2), and nitrogen dioxide (NO_2) (http://airnow.gov/index.cfm?action=topics. about_AIRNow). The EPA, which partners with NOAA and other federal, state, and local agencies; scientific and health research organizations; and media and public outreach groups (Figure 8.4), began to develop the AIRNow Web site in 1997 (Figure 8.5). The site collects, processes, and maps national air quality information so that such information can be easily accessible to the public. Using hourly real-time air quality data compiled and processed by federal referenced or equivalent monitoring techniques and techniques approved by state and local monitoring agencies, AIRNow offers real-time AQI conditions and daily forecasts of AQI for about 300 cities across the United States. The system also provides daily national outlooks, news stories, and the comparison of air quality conditions across different geographical areas (Figure 8.6).

Although AIRNow is a subsystem under GEOSS, it is also a community. The AIRNow community has annual conferences. The community promotes regional cooperation, provides daily interaction with stakeholders, offers support during special events, and continues developing tools to serve the public.

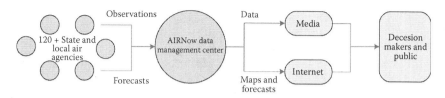

FIGURE 8.4
Participated partners in AIRNow.

The EPA plans to expand the current AIRNow to AIRNow-International, which is intended to promote the sharing of air quality data at the global scale. The new version of AIRNow (AIRNow2.0) will enhance the current U.S. system and support an international system. The package is built upon open components and standards to ensure convenient data exchange. It can easily be adopted by various agencies, public organizations, or among GEOSS members. It supports multiple languages and provides tools and leverages expertise among agencies. The emphases of the system are international collaboration and knowledge exchange (Figure 8.7).

Several remaining steps are needed to finish updating the current version: (1) obtain inputs, need specifications, and requirements from international organizations such as the Shanghai Environment Protection Bureau (EPB) and GEO summit in South Africa; (2) design specifications; (3) build and test the system; and (4) launch. The AIRNow-International has data services capability to facilitate data dissemination and has functions that are more flexible than AIRNow.

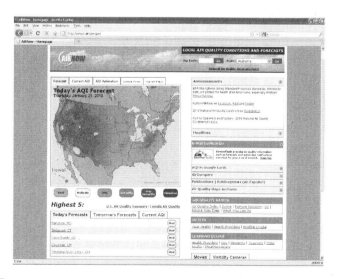

FIGURE 8.5
AIRNow Web site interface. (Adapted from AIRNow, http://www.airnow.gov/, last accessed on December 9, 2009.)

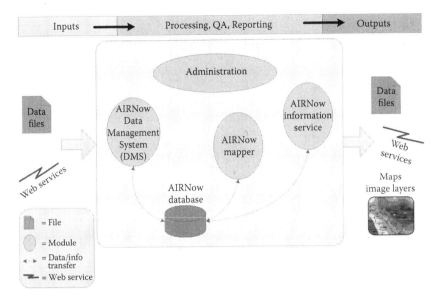

FIGURE 8.6
Conceptual and functional design and data flow.

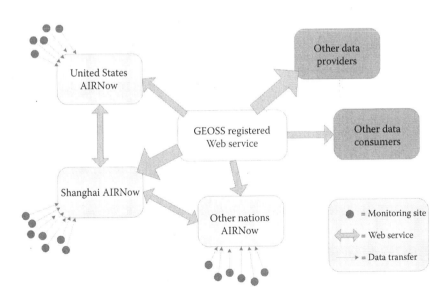

FIGURE 8.7
International knowledge exchange focus of AIRNow-International. (Adapted from White, J.E. 2009. AIRNow-International: The future of the United States real-time air quality reporting and forecasting program and GEOSS participation. 33rd ISRSE—Session: Air Quality and GEOSS—TS34–1, May 6, 2009. http://wiki.esipfed.org/images/2/21/TS34–1_White.ppt. Web site last accessed January 21, 2009.)

The system is much more portable, flexible, and sustainable because open component standards are used. Technically, AIRNow-International consists of the Microsoft SQL Server, GIS-based mapping software (i.e., ESRI products), data management, and service servers (White, 2008). Output formats will conform to standards to promote sharing and interoperability. Examples of format include CSV, KML (Google Earth), NetCDF, GRIB2, Shapefile, RSS and CAP feeds, and Web service (WMS, WCS, WFS) (White, 2009).

Air quality assessment forecasting systems like AIRNow offer concrete societal benefits. Weather forecasting capability is now greatly enhanced. The forecasts of large-scale weather patterns are now much closer to actual observations than those forecasts made in the past. Air quality patterns resulting from weather changes can be presented and predicted at a higher level of accuracy. Decision makers can formulate policies and make decisions related to environmental health with a higher level of confidence.

8.3 Environmental Health*

8.3.1 Introduction

Changes in the environment affect public health. The GEOSS is an international network of satellite- and ground-based sensors for EO that began in 2002. Its purpose is to improve EO to support decisions in human health and other societal concerns (Schmidt, 2005; Group on Earth Observations [GEO], 2009). The GEOSS provides a platform for a better understanding of the environmental factors influencing health, and it may lead to new approaches for environmental and health decision making. The EO can be used to address human health concerns in many ways: projecting occurrence of disease or disease outbreaks; rapid detection and tracking of events; construction of risk maps; targeting interventions; and enhancing knowledge of human health-environment interactions (National Research Council [NRC], 2007a).

This chapter section highlights applications of EO to public health in three thematic areas: heat (heat waves and wildfires); wildlife-associated diseases; and oceans (harmful algal blooms [HABs] and *Vibrio* species of bacteria). The examples are drawn from the United States and several other countries. These topics do not cover all aspects of EO and public health, but they serve to illustrate the application of EO to a variety of public health issues related to infectious diseases and other hazards in both terrestrial and oceanic systems. Examples of decision support include monitoring and forecasting for both short-term emergency response and long-term planning for prevention.

* This section is contributed by Joan L. Aron and Laura E. Jackson of the U.S. EPA, Office of Research and Development. The views expressed in this section are those of individual authors and do not necessarily reflect the views and policies of the U.S. EPA.

The need to consider highly vulnerable populations as well as the effects of global climate change is integral to all of the illustrations.

8.3.2 Heat

8.3.2.1 Heat Waves and Urban Heat Islands

Heat waves are a major threat to public health. In some countries, such as the United States and Australia, heat waves caused more deaths during the twentieth century than any other natural hazard had caused (Kovats and Koppe, 2005). Concern about the effects of heat waves is expected to grow in the twenty-first century. Global climate change is likely to cause an increase in their frequency and intensity (Kovats and Koppe, 2005). As the global population becomes more urbanized, the urban heat island phenomenon will become even more important. Cities experience warmer temperatures than the surrounding countryside due to extensive asphalt, other heat-absorbing surfaces, and intensive energy usage that contributes to ambient heat. Urban heat islands absorb solar energy during the day and radiate heat at night, thus raising the nighttime minimum temperatures; these have been linked to higher levels of mortality during heat events (Luber and McGeehin, 2008). Urban heat islands are also associated with the production of ground-level ozone, which contributes to respiratory illness (Lo and Quattrochi, 2003). Both children and the elderly are particularly vulnerable to the effects of extreme heat and ozone. Individuals in these life stages are less able than the general population to regulate their internal body temperatures to adapt to heat; they are also more likely to have underlying respiratory conditions that heat and ozone can exacerbate.

Remote-sensing observations are useful in the study of heat waves. Figure 8.8 shows the European daytime land surface temperature differences between July 2003 and July 2001 based on data from the MODIS on NASA's Terra satellite. In the summer of 2003, Europe experienced an historic heat wave that killed more than 50,000 people (Brucker, 2005). The elderly were particularly affected. Risk factors that contributed to the heat-related mortality were loss of autonomy and social isolation; many of the affected elderly did not have anyone to assist them. Living directly below the roof of a building, where interior temperatures are the highest, was also a risk factor (Brucker, 2005). Epidemiological studies have shown that poor and minority populations in urban neighborhoods suffer disproportionately; one area of public-health research is to map heat islands with socioeconomic data at the neighborhood level (Luber and McGeehin, 2008).

Data from EO should be linked to a response system. Improved emergency services for heat waves and heat health warning systems can reduce the death toll (Kovats and Koppe, 2005). An effective heat health warning system requires four components: (1) reliable meteorological forecasts; (2) understanding of the relationship between heat exposure and health outcomes; (3) effective response measures within the lead time of the warning (typically one to

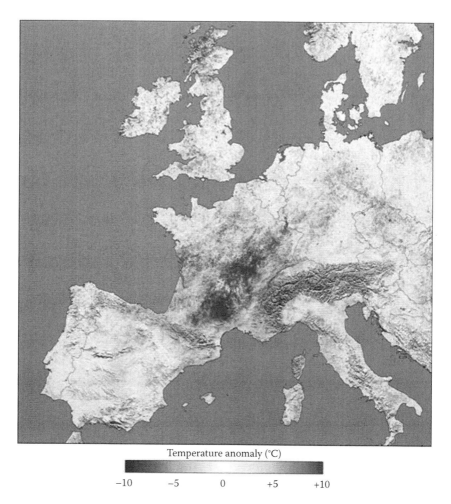

FIGURE 8.8
(See color insert following page 144.) European heat wave: surface temperature anomaly in July 2003. NASA Earth Observatory, August 16, 2003. (Adapted from http://earthobservatory. nasa.gov/IOTD/view.php?id=3714.)

3 days); and (4) institutional capacity and political will to implement the responses (Kovats and Ebi, 2006). Therefore, the adequacy of these systems involves a diverse set of factors ranging from the meteorological threshold used to issue warnings to the estimated costs and benefits per warning.

A public communication strategy, an essential element of a heat health warning system, must take into account characteristics of public perception and response. To evaluate existing systems, telephone surveys were administered within 7 days of the end of heat events in four North American cities during the years 2004 and 2005: Dayton, Ohio, Philadelphia, Pennsylvania Phoenix, Arizona and Toronto, Ontario (Canada) (Sheridan,

2007). The principal finding was that almost all of the respondents were aware of the recent heat event referenced by the interviewer, but only about half stated that they had taken protective measures. Most of these respondents mentioned avoiding the outdoors and keeping hydrated. Some respondents indicated that they limited their use of air conditioning due to concerns about energy costs. People commonly used their own assessment of the heat rather than the official announcements and, in general, did not believe that the weather during heat events could really harm them. Increasing the effectiveness of a public heat health warning system requires overcoming these barriers and moving toward the adoption of protective measures.

Other possible interventions go beyond emergency response by addressing the physical urban infrastructure. Since the built environment magnifies the effect of heat waves, changing urban design to reduce heat radiation and include more trees and green roofs can reduce temperatures both indoors and outdoors (Kovats and Koppe, 2005). Figure 8.9 shows an association between vegetation cover and land surface temperatures in New York City. A study of Guangzhou city in southern China recommended that managers plant more tall trees, taking into account the differential cooling effects of forests, shrubs, and lawns (Weng and Yang, 2004). A study of Xiamen city in southeastern China concluded that a long-term strategy to reduce the heat-island effect should include planting shade trees and using light-colored, highly reflective roof and paving materials (Xu and Chen, 2004).

8.3.2.2 Wildfires

The risk of wildfire depends on multiple factors, especially heat, precipitation, and vegetative cover. This discussion highlights the importance of heat and the use of remote-sensing technology in wildfire tracking and health risk assessment. Figure 8.10a shows that land surface temperatures were elevated in Australia, particularly in the southeast, during a major heat wave in January 2009. One outcome, as depicted in Figure 8.10b, was an outbreak of forest and grassland fires in southeast Australia late in January 2009.

Human health effects of wildfires are not only direct harm from burning but also the consequences of exposure to smoke. Wildfire smoke contains numerous pollutants, such as particulate matter, polycyclic aromatic hydrocarbons, and carbon monoxide. Smoke inhalation can be very damaging to people with asthma, allergies, or other preexisting respiratory conditions. A 2003 study in California found that children with asthma were two to three times more likely to experience symptoms from smoke than nonasthmatic children (Kunzli et al., 2006).

Remote-sensing tools can be used in health studies to assess exposure to smoke. A study of the 2003 California wildfires incorporated smoke information from MODIS satellite imagery with ground-based measurements of particulate matter (PM 2.5) to assess human exposure to wildfire-related PM 2.5 by postal code (ZIP code), which was in turn linked to increased

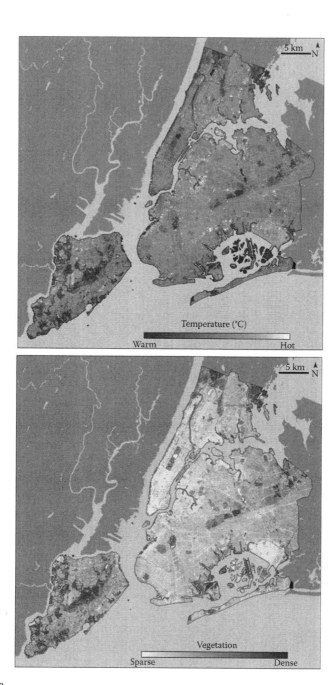

FIGURE 8.9
New York City temperature and vegetation in the summer of 2002. The image is based on data taken on August 14, 2002 from NASA's Landsat Enhanced Thematic Mapper Plus. NASA Earth Observatory, August 2, 2006. (Adapted from http://earthobservatory.nasa.gov/IOTD/view. php?id=6800.)

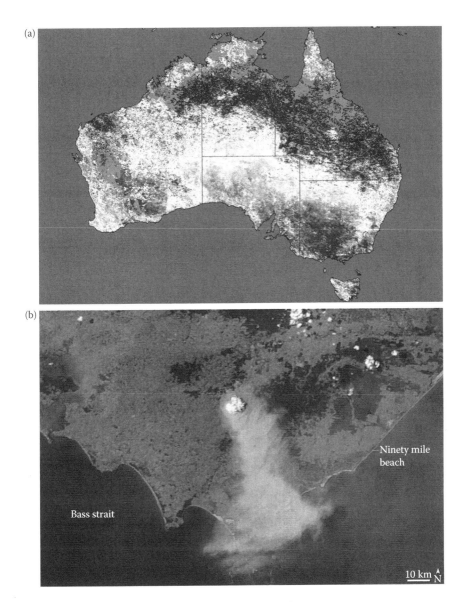

FIGURE 8.10
(See color insert following page 144.) (a) Exceptional Australian heat wave. The image is derived from MODIS on NASA's Terra satellite. NASA Earth Observatory, February 5, 2009. (Adapted from http://earthobservatory.nasa.gov/IOTD/view.php?id=36,900); (b) Fires in southeast Australia. The image is derived from MODIS on NASA's Aqua satellite. NASA Earth Observatory, January 31, 2009. (Adapted from http://earthobservatory.nasa.gov/IOTD/view.php?id=36861.)

respiratory hospital admissions (Delfino et al., 2009; Wu et al., 2006). Studies have also examined the use of MODIS to monitor general particulate matter in urban settings in several countries (e.g., Gupta et al., 2006).

Mitigating health effects from wildfires includes emergency response (dousing flames and evacuating people) and taking personal protective measures (wearing masks, spending less time outdoors, and using air conditioners). For the 2003 California wildfires, health effects related to wildfire smoke exposure were fewer for those who took preventive action than for those who did not (Kunzli et al., 2006). Land-use management strategies can avert outbreaks of fire that endanger people. It may be possible to reduce fire risk through vegetation management and to avoid placing residences, schools, and other buildings in close proximity to fire-prone habitats. Remote sensing can contribute to mitigation strategies by aiding in the prediction of wildfire risks (National Interagency Fire Center, 2009; National Oceanic and Atmospheric Administration, 2009).

8.3.3 Wildlife-Associated Diseases

Environmental change can affect the occurrence of transmissible diseases in humans, especially those that involve wild species. Diseases in this category include familiar scourges from history such as bubonic plague and yellow fever; persistent endemic illnesses like malaria; and many emerging diseases of which avian influenza, West Nile virus, Hantavirus, and Lyme disease are just a few examples. Due to the key role of wildlife, including arthropods (insects, ticks, and mites), in transmitting bacteria and viruses to humans, changes in the extent or condition of wildlife habitat can foreshadow changes in human risk of infectious disease. Since habitat change is conducive to remote sensing, EO data can facilitate the identification of high-risk areas and vulnerable human populations.

Two main drivers of habitat change are conversion for human use and shifting geographic suitability due to climate change. The EO data can be used in spatial models to assess potential future health risks from climate and land-use trajectories and alternative scenarios. Using a small subset of infectious diseases, the following discussion illustrates how land use and climate change can work both independently and together to increase the human risk of wildlife-associated diseases. The EO data that facilitate assessment of these risks are incorporated into each example.

8.3.3.1 Land Use and Land-Cover Change

Exposure to wildlife pathogens commonly occurs with human intrusion into previously isolated natural habitats. Temporary human excursions into intact wildlands may result in pathogen transmission, whereas greater disease risk is associated with long-term human intrusion for residential, agricultural, or other societal purposes. Anthropogenic land-cover change results in habitat loss and fragmentation, which increase human-wildlife interactions at the habitat's edge. Further, these habitat changes alter the wildlife community itself, affecting species composition in ways that can increase pathogen abundance and transmission.

Habitat loss and fragmentation typically degrade environmental quality. Sensitive species may not persist under degraded habitat conditions due to changes in microclimates, the introduction of non-indigenous predators or competitors, or other ecological consequences of habitat disturbance. This loss of sensitive species diminishes native biological diversity. For several infectious diseases including Lyme disease and West Nile virus, biodiversity has been hypothesized to limit pathogen maintenance within the wildlife community (Ostfeld and Keesing, 2000). This theory stems from laboratory analyses indicating that disturbance-tolerant species are more effective reservoirs for the pathogenic agents of these diseases than are disturbance-sensitive species (e.g., Mather at al., 1989; Levin et al., 1995). Studies at local and national scales have correlated biodiversity with both human and host infection rates to support this hypothesis (Allan et al., 2003; Ezenwa et al., 2006; Swaddle and Calos, 2008; Allan et al., 2009).

In the case of tick-borne Lyme disease, the bacterial pathogen (*Borrelia burgdorferi*) is maintained in the wild with varying competency by certain small mammals, birds, and other forest-dwelling vertebrates. Deer play a critical role by hosting adult ticks and transporting egg-laying females around the landscape, but they are not competent *B. burgdorferi* reservoirs. Ticks of the genus *Ixodes* can acquire the pathogen during a blood meal from an infected small vertebrate host and transmit it to any future hosts, including humans. The most competent reservoirs of the pathogen are also highly tolerant of habitat disturbance. In fragmented forests, the native vertebrate community may be diminished in such a manner that these competent reservoir species are the principal remaining members. Applying the biodiversity hypothesis to temperate regions endemic for Lyme disease, the prediction is that forest habitats with diminished native biodiversity would have a higher rate of *B. burgdorferi* infection per host animal than those with more complete representation of native species. Conversely, high native biodiversity would dilute the community infection rate due to the natural abundance of wildlife that is not competent to harbor *B. burgdorferi* infection. This phenomenon has been dubbed "the dilution effect" (Schmidt and Osfeld, 2001), and it is undergoing validation across several multistate study regions in the United States.

Remotely sensed land-cover pattern metrics are often used as indicators of forest condition (e.g., Schumaker, 1996; O'Connell et al., 2000; Riitters et al., 2002; Watts et al., 2007). The appropriate resolution, extent, and classification of the imagery strongly depend on the species and processes of interest. In Lyme disease research, studies have associated disease variables with land-cover metrics derived from 1:2400 aerial photography (e.g., Nicholson and Mather, 1996; Frank et al., 1998), 30-meter resolution Landsat Thematic Mapper imagery (e.g., Glass et al., 1995; Das et al., 2002), and one-kilometer resolution AVHRR imagery (e.g., Kitron and Kazmierczak, 1997; Brownstein et al., 2005). Forests and other land-cover types have been lumped into general categories (e.g., all forest types in one class: Jackson et al., 2006a) and are also very finely categorized (e.g., 17 distinct forest types: Estrada-Pena, 2003). Regardless of forest classification, the land-cover models with the strongest links to human Lyme disease or *in situ* indicators of disease risk (e.g., tick density and infection) include some measure of forest edge (e.g., Dister et al., 1993; Daniel et al., 1999; Jackson et al., 2006a).

The known correlations between forest fragmentation and biodiversity, between forest edge and human presence in the landscape, and growing evidence for the *dilution effect* supply the biological mechanisms behind the association of forest-edge metrics with Lyme disease risk. Figure 8.11 shows the extensive forest edge in two exurban landscapes in Maryland with high rates of Lyme disease among residents.

Determining the relevant areal boundaries of spatial analysis units is important when using EO data to explain disease risk or occurrence. Depending on the circumstances, units such as counties or watersheds may be inappropriate to characterize wildlife habitat or human behavior. Imagery clipped to political boundaries can mask important information such as habitat extent or economic centers within a day's travel. An example of relevant

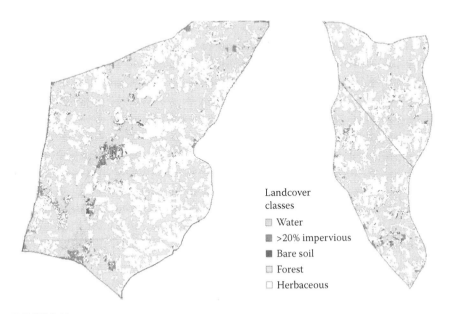

Landcover
classes

▨ Water
▦ >20% impervious
▪ Bare soil
▫ Forest
☐ Herbaceous

FIGURE 8.11
Two exurban Maryland landscapes illustrating the land-cover pattern associated with high Lyme disease rates among residents during 1996–2001. (From 2001 National Land Cover Dataset [Homer, L. et al. 2004. *Photogrammetric Engineering and Remote Sensing*, 70:829–840.] reclassified; Jackson, L.E., Hilborn, E.D. and Thomas, J.C. 2006a. *International Journal of Epidemiology*, 35:315–322. With permission.) Scale: 1 cm ~ 1 km.

boundary delineation can be found in Lyme disease studies, wherein major roads are used to define spatial analysis units (Jackson et al., 2006a,b). Humans are unlikely to cross major roads during casual outings around their homes (where most exposure is presumed to occur); and traffic discourages (although not entirely prevents) deer from crossing major roadways. Therefore, major roads help to circumscribe key actors involved in Lyme disease incidence and the habitats that indicate their movements. The use of well-considered spatial boundaries to study geographic processes improves desirable statistical properties of the data, including within-unit homogeneity and between-unit heterogeneity. This practice minimizes not only noise in the data but also spatial autocorrelation that can artificially inflate the significance of observed associations.

Lyme disease provides a useful illustration of how public health risks may be estimated through EO. In the case of this arthropod-borne disease, remotely sensed metrics of forest fragmentation serve as proxies for wildlife community dynamics thought to operate in disturbed systems and for increased human exposure to natural zones of infection. Land-cover pattern indicators may further inform health care providers, land-use planners, and the public through their application in models of disease risk under alternative scenarios of landscape change. For this reason, spatial analysis units should also relate to the spatial domains of pertinent decisions. Effective disease prevention may include altering landscape development trajectories to minimize the propagation of and exposure to infectious pathogens in degraded ecosystems.

8.3.3.2 Climate Change and Variability

Many infectious diseases are associated with warm climates. Warm environments meet minimum temperature requirements for pathogen survival and reproduction and result in shorter incubation periods within hosts. Wild host species are often constrained by temperature and precipitation. Their survival and reproductive rates vary with climate-sensitive factors such as the availability of food and the onset and duration of the breeding season. The ecological dynamics become more complex when arthropod vectors are involved in a disease cycle; warmer and wetter conditions tend to favor their survival as well. Arthropods have minimum temperature thresholds for overwintering as well as moisture requirements to prevent desiccation (ticks) and provide breeding habitat (mosquitoes). Paradoxically, drought may exacerbate some mosquito-borne diseases by creating remnant pools in formerly flowing water channels. Adding human behavior to multi-species disease systems results in additional climate-mediated risk factors. For example, to seek relief from overheated dwellings, people may spend more time outside in the evenings when mosquitoes are biting. Similar to all biological processes, the maintenance and transmission of wildlife-associated diseases are limited by upper climatic bounds for pathogen, vector, and host survival.

The EO data are widely used to detect spatial gradients in temperature and precipitation and to track how measured values are changing over time. The international scientific community's attention to climate change and its potential for catastrophic effects has led to regional and global models projecting alternative trajectories for multiple environmental phenomena, including those affecting wildlife-associated diseases (e.g., McCarthy et al., 2001; Parry et al., 2007). Incorporation of EO data and models within geographic information systems facilitates identifying vulnerable populations, emergency response, and preventative planning to reduce human exposure in areas of current and future risk (Aron and Patz, 2001).

An overall global warming trend is already shifting wildlife habitat range toward the poles and higher altitudes (Parmesan and Yohe, 2003). In North America and Europe, endemic zones of Lyme disease have been expanding northward with warming temperatures and the accompanying range migration of *Ixodes* ticks (Estrada-Pena, 2002; Epstein, 2005). Increasing minimum temperatures and precipitation levels are projected to exacerbate the expansion, both spatially and seasonally, of many mosquito-borne diseases such as malaria and dengue fever [Patz et al., 1996, 1998, but see Patz et al. (2000) and Phillips (2008) for caveats]. Worldwide, dozens of mosquito species are involved in disease transmission, with widely varying habitat requirements. Warmer temperatures will render existing habitats unsuitable for some species, such as *Culex tarsalis*, the primary mosquito vector in California of St. Louis encephalitis and western equine encephalomyelitis (e.g., Reeves et al., 1994). As a result, climate change is projected to eradicate some diseases from historically endemic areas, even as they migrate into new territory.

Climate variability includes episodic events that take place at time scales from less than a day to more than a decade; these events may become more extreme within the context of long-term global climate change. Episodic warming, droughts, and extreme rainfall can create conditions for local spikes in host and vector populations. In Africa, the availability of satellite-based precipitation trends data has led to an early-warning system for malaria (Figure 8.12). This capability for routine risk surveillance is based on identifying areas of rainfall anomalies that can trigger local proliferation of mosquito populations. The imagery resolution of 0.1 degree latitude × 0.1 degree longitude allows for fairly local estimates of annual risk. See Connor et al. (2006) for further discussion on the use of remotely sensed data to estimate and mitigate malaria risk in Africa.

Episodic extreme precipitation events are also linked to several rodent-associated human diseases. Most rodent species are capable of very high reproductive rates, so population sizes can explode under favorable environmental conditions. Unseasonably heavy local winter-spring precipitation in New Mexico has led to increased incidence of flea-borne plague in humans. The increased moisture results in an abundance of plant and insect food sources for several rodent host species (Parmenter et al., 1999; Epstein, 2005).

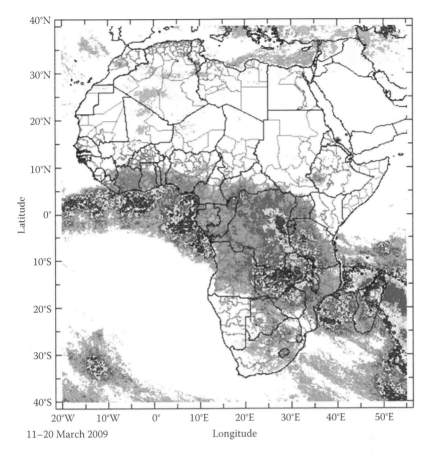

11–20 March 2009   Longitude

FIGURE 8.12
(See color insert following page 144.) 10-day precipitation estimates for malaria-endemic regions of Africa. Continuous tracking enables the identification of rainfall anomalies that signal elevated disease risk due to anticipated increase in mosquito populations. (Adapted from http://iridl.ldeo.columbia.edu/maproom/.Health/.Regional/.Africa/.Malaria/.MEWS/.)

Global-scale episodic conditions preceded the emergence of Hantavirus pulmonary syndrome (HPS) in the Four Corners area of the southwestern United States. Deer mice (*Peromyscus maniculatus*) are the local hosts in this disease system; the pathogen is transmitted indirectly through contact with mouse urine and feces, including aerosolized particles, as well as directly through saliva. In the Four Corners, warming in the Pacific Ocean associated with the cyclic El Nino Southern Oscillation phenomenon is thought to have triggered prolonged early-spring rains. The observed mouse population boom was likely facilitated as well by a prior 6-year drought that eliminated many mouse predators (Epstein, 2005). Tracking El Nino and other long-term climatic cycles may help to anticipate spikes and dips in host populations and habitat availability. Remotely sensed data on vegetation growth has been used to explain local increases in host abundance within the known habitat range [Boone et al., 2000, but see Glass et al. (2000) for caveats].

The complexity of wildlife-associated disease systems coupled with the uncertainty inherent in climate forecasts make precise long-term predictions of disease risks highly unlikely (Hales et al., 1997; Patz et al., 2000). Nevertheless, models depicting general patterns of warming, precipitation, and biotic responses under alternative future climate scenarios are useful to delineate the extent of risk at coarse scales (Figure 8.13). Certainly, EO data are invaluable to understand existing spatial and temporal variability in wildlife-associated diseases and to develop projections within the bounds of expert opinion.

8.3.4 Oceans

Oceans provide many direct benefits, such as food and chemical compounds with biomedical uses, but they also contain toxic agents and microbes that threaten human health (NRC, 2007b). Examples include HABs and *Vibrio* bacteria, which cause cholera and other human illnesses. The EO of the marine environment can detect factors that affect these harmful organisms.

8.3.4.1 Harmful Algal Blooms

Marine algae form the foundation of food webs in oceans. However, some species produce toxins that are harmful to people, causing paralysis, amnesia, nausea, diarrhea, and/or respiratory distress as well as irritation of the skin and eye. Algae can also bloom in such large amounts that the ecological processing of dead algae depletes the water of oxygen, threatening the health of fish and other biological organisms. The HABs are becoming more frequent and affecting more geographic locations around the world (NRC, 2007b).* The U.S. Congress recognized the importance of this issue in passing the Harmful Algal Bloom and Hypoxia Research and Control Act in 1988 and expanding it in 2004.

One heavily affected area is the Gulf of Mexico, where most HABs are composed of *Karenia brevis*. Remote-sensing observations can document their development and extent. Figure 8.14 shows a bloom developing in the Gulf of Mexico off the west coast of Florida on October 30, 2004 and November 21, 2004. The SeaWiFS satellite data indicated the amount of algae in the water through the concentration of chlorophyll, with the highest concentrations shown in darker shades. Observations from MODIS on NASA's Terra satellite supported the identification of an algal bloom by detecting signals of algal fluorescence, which is distinct from the wavelengths of sunlight reflected off the ocean's surface.

---

* The term *red tide* is commonly used to describe HABs, although the link between reddish discoloration and harm is not absolute. Some reddish algal blooms are harmless, whereas HABs do not necessarily cause reddish or even any discoloration (Woods Hole Oceanographic Institute [WHOI], 2008).

GFDL89

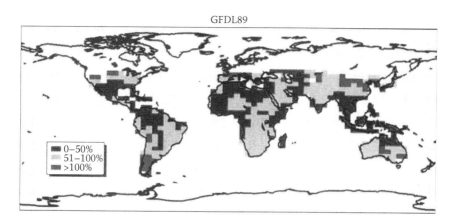

ECHAM1-A

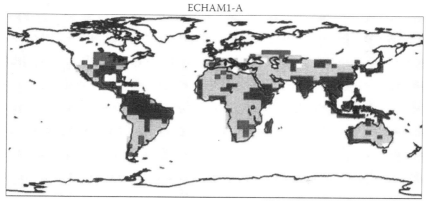

UKTR

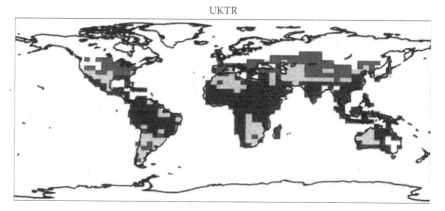

FIGURE 8.13

Three scenarios of dengue average annual epidemic potential compared with baseline climate conditions. (Reproduced from Patz, J.A. et al. 1998. *Environmental Health Perspectives,* 106:147–153. With permission.)

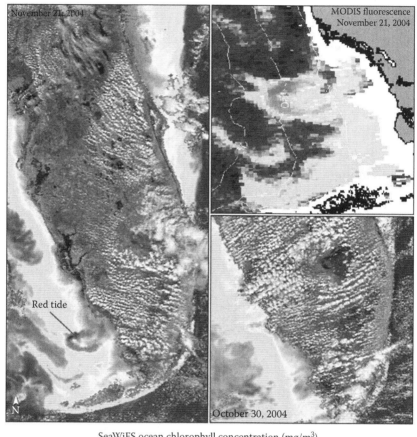

FIGURE 8.14

Red tide near Florida. NASA Earth Observatory, 21 November 2004. (Adapted from http://earthobservatory.nasa.gov/NaturalHazards/view.php?id=14307.)

The Harmful Algal Blooms Observing System (HABSOS) is a joint United States–Mexico effort to gather and disseminate information on the location, extent, and potential for development or movement of HABs in the Gulf of Mexico (National Oceanic and Atmospheric Administration [NOAA], 2008a). The HABSOS utilizes many data layers, including biological sampling of *K. brevis* in the water, measurement of winds by the SeaWinds instrument on the QuikSCAT satellite, monitoring of surface currents by buoys, and MODIS imagery for chlorophyll (ocean color) and SST. More precise and accurate predictions of HABs can help officials to be prepared to make decisions to prevent harmful exposures, such as closing beaches and fishing areas (NRC, 2007b).

The Albemarle-Pamlico Estuary System (APES) pilot study in the waters off the coast of North Carolina represents a first step toward an operational, early-warning HABs monitoring capability for the United States eastern seaboard coastal and estuarine systems. Imagery of ocean color for APES is derived from the MERIS on the ESA's ENVISAT satellite (Environmental Protection Agency [EPA] 2009; Sokoletsky et al., 2009; Lunetta et al., 2009). The objective is to provide North Carolina agencies with near-real-time spatially explicit information for making decisions about safe water use for recreation, fishing, and shellfish harvesting. Federal agencies including NOAA and EPA will also use the information for management and regulatory oversight of nitrogen loading into estuaries and coastal wetlands.

Improved understanding of the factors causing HABs is a scientific priority that could lead to better forecasting as well as ways to reduce their incidence and severity. Some evidence suggests linkage to wind and water current conditions, excess nutrients from runoff, unusually high water temperatures, and sluggish water circulation (NOAA, 2008b). The Intergovernmental Panel on Climate Change (IPCC) includes HABs as an increased threat to food safety under possible impacts of global climate change on contaminated seafood (Confalonieri et al., 2007).

8.3.4.2 Cholera and Other Illnesses Caused by Vibrio Species of Bacteria

Cholera is a potentially fatal gastrointestinal illness caused by the pathogen *Vibrio cholerae* that is transmitted in water and food. Exposure to *V. cholerae* is high in parts of the world where people rely on untreated drinking water. Studies in recent decades have found an environmental reservoir of *V. cholerae* bacteria in rivers, estuaries, and coastal waters in both tropical and temperate regions (Constantin de Magny et al., 2008).

Remote-sensing tools are useful to detect environmental conditions that support the dissemination of *V. cholerae*. These bacteria attach to copepods—a major component of zooplankton whose numbers typically grow after algal blooms—due to this increase in the copepod food supply. The level of chlorophyll A is one indicator of the potential for cholera transmission after algal blooms (Figure 8.15); monitoring ocean color can signal potential risk. The relationships among chlorophyll, SST, rainfall, and cholera outbreaks have been studied in the Bay of Bengal as part of the foundation for an early warning system for cholera outbreaks (Constantin de Magny et al., 2008).

In the United States, *Vibrio parahaemolyticus* and *V. vulnificus* cause the majority of vibrio-related gastrointestinal infections (NRC, 2007b). Eating raw seafood, especially oysters, is a major mode of transmission; infections may also occur through open wounds exposed to contaminated seawater (Centers for Disease Control and Prevention [CDC] 2008 a,b). Severe disease is much more likely in people with weakened immune systems (CDC, 2008 a,b).

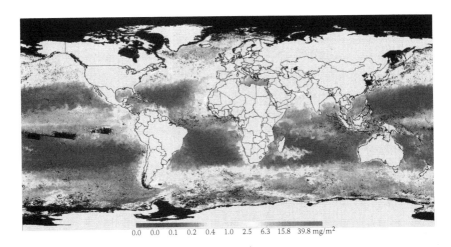

FIGURE 8.15
(See color insert following page 144.) SeaWiFS-derived chlorophyll concentration for March 1999. (Adapted from NASA DAAC.)

Studies are exploring the co-occurrence of pathogenic *Vibrio* and algal blooms with the aim of extending the scientific framework beyond *V. cholerae* to *V. parahaemolyticus* and *V. vulnificus*. By analogy with *V. cholerae*, remote-sensing tools offer the possibility of contributing to an early warning system for these harmful pathogens as well.

8.3.5 Conclusion

Environmental threats to public health are diverse. The discussion of heat waves, Lyme disease, and HABs provides a starting point for examining the multifaceted relationship between environment and public health. The GEOSS is leading to improved use of EO to better understand environmental influences on health and to improve decisions to benefit health; however, many applications are in relatively early stages and will grow as research expands and new technologies emerge. The EO are expected to contribute to basic research as well as programs to track pollution events, forecast health emergencies, construct risk maps, and target interventions.

8.4 The Roles of GIS in Studying Global Climate and Environmental Change*

Global environmental change is an interdisciplinary field relying on the intersection of science, technology, policy, and international affairs. From the

---

* This section is contributed by Sheryl Luzzadder-Beach and David Wong.

science and technology perspective, GIS and remote sensing may serve as the foundations of decision support systems to assist in detecting and measuring global and environmental change and also in modeling and understanding its mechanisms, its effects on, and linkages to the Earth and human systems. Decision support systems built on GIS and remote sensing can also serve in the policy, international affairs, and education arenas, by educating students, policymakers, and the public about global environmental change, its causes, its dimensions, its consequences, and human responses.

The domains of global and environmental change include the Atmosphere, Hydrosphere and Cryosphere, Lithosphere, Ecosphere and Sociosphere, and their linkages. For example, the sociosphere links to other domains by issues such as anthropogenic impacts, emergency response to short-term events, and long- term adjustment to change, among others. The GIS is among the data organizing and Earth observing tools for research that documents and models global change domains and their linkages. Global change modeling with GIS includes past conditions and future projections of the global environment and the human condition. This section will focus on GIS applications to global and environmental change, using the hydrosphere and cryosphere as primary examples.

As discussed elsewhere in this chapter, there are different levels of applications for decision support and education, which we introduce briefly here. This includes education for the public either directly or through policymaking, and student education, both involving research centers and institutional delivery. Global and environmental change has been largely represented in research programs and centers, including the USGCRP, also known from 2002 to 2008 as the U.S. Climate Change Science Program (USGCRP, 2009a). This program was launched in 1990 by the Global Change Research Act [Public Law 101–606 (11/16/90) 104 Stat. 3096–3104, USGCRP 2009b]. The program linked thirteen federal agencies (USGCRP, 2009c), mostly with GIS and remote-sensing capabilities. A long-running public education program is led by the U.S. Department of Energy (USDOE, 2009). Agencies and university research centers alike engage in providing data, models, and tools for researchers and policy makers and education for the broader public as part of their missions.

Formal degrees in global change have emerged very recently in the United States higher education. These efforts include, for example, the minor in global change, consisting of three core global change courses, launched at the University of Michigan in 2006 (van der Pluijm, 2006), and the major in global and environmental change, first offered at George Mason University in 2007 (GMU, 2009). In both of these degrees, dynamic Earth systems, their linkages to the anthrosphere, and use of spatial analysis tools are emphasized. The GIS and remote sensing are among the research tools presented to train students in detecting, measuring, documenting, and modeling global and environmental change, including response scenarios (van der Pluijm, 2006; GMU, 2009).

Education programs will be addressed further in the next chapter and in the final chapter of this edited volume; however, in the following sections, we review examples of applications of GIS to global and environmental change research in two closely related spheres, the hydrosphere and the cryosphere. The research reviewed in these sections links these spheres, respectively, through GIS to the atmosphere, lithosphere, ecosphere, and sociosphere. This section provides only a brief overview of an extensive literature.

8.4.1 Hydrosphere

The call for using GIS in global-change related hydrologic research reaches back nearly two decades in the formal literature, far longer if hydrologic research basic data storage and modeling functions for hydrology are considered, outside of the specific context for global change. A call to research by Din (1992) saw this practicality in GIS at various spatial scales, but it should be noted at the time that GIS did not have the more dynamic modeling capabilities required for decision support and response tools (Din, 1992). In a short time, scientists answered this call with more sophisticated GIS incorporating data tools and modeling functions (Hastings and Di, 1994a,b). Schultz (2000) addressed potential spatial data integration for global change studies in the context of water resources and European river basin management. In the area of policy and international affairs, Yoffee and Ward (1999) presented a GIS and spatial analysis methodology to be applied to river basins in transboundary situations, in an effort to model and mitigate international conflict.

Within the decade, GIS models for hydrologic resources had grown in sophistication and applications as computing, data storage, and transformation capabilities grew. Prominent issues addressed in this body of research include global, regional, and local scale models and impacts; also included are linkages of the hydrosphere to all domains of global change research. A sampling of this variable hydrologic spatial scale includes using GIS to study the effects on the atmosphere from irrigated rice farming and methane production in five Asian countries (Knox et al., 2000); assessing soil salination in a river basin in Cuba under irrigation and global warming scenarios (Utset and Borroto, 2001); and individual effects of climate change on the habitat and well-being of Appalachian trout (Clark et al., 2001). Besides questions of regional scale, systems are also frequent themes of GIS, hydrology, and global change research. The following sections briefly review examples of GIS research in sea level rise, water quality and public health, water supply issues, and the cryosphere, linked to global and environmental change.

8.4.1.1 Sea Level Rise Impacts

A leading global change GIS application in the hydrosphere is that of modeling sea level rise scenarios and impacts on human and natural systems.

Simas et al. (2001) presented a case study in estuarine salt marshes in Portugal, where GIS and IPCC data (Simas et al., 2001) were coupled with ecological models to estimate the impact of sea level rise in this coastal zone. Interestingly, they found the salt marsh ecosystems to be resilient in most scenarios (Simas et al., 2001) due to the natural fluctuation in tidal range in these ecosystems. Mean sea level rise models were applied to the Northern Adriatic Sea by Gambolati et al. (2002) to produce coastal management risk maps. They considered multiple hazards scenarios of land subsidence, extreme storm events and tides, and sea level rise and used GIS to manage the data sets generated by the models. Similar to these earlier works, a wealth of research has emerged in the last 2 years on establishing databases and conducting studies for coastal vulnerability. Examples include a GIS database for segments of all of the world's coasts outside of Antarctica (Vafeidis et al., 2008). This extensive database supports a modeling tool that operates at a variety of spatial scales, called Dynamic Interactive Vulnerability Assessment (DINAS) (Vafeidis et al., 2008). Torresan et al. (2009) conducted a similar study assessing coastal vulnerability to climate change and sea level rise, choosing to focus on one region, the Veneto region, and comparing 1 m global and 25 m regional DEM data. They found the latter data set to provide better representation of inundation risk (Torresan et al., 2009). Lam et al. (2009) focused on identifying the states with the most vulnerable populations along the United States coast, defining those living at elevations below 3 m as the most vulnerable. Florida and Louisiana had the highest percentage of residents vulnerable to sea level rise and hurricane scenarios associated with climate change, for a combined total of more than seven million residents (Lam, 2009, 1522). Dasgupta et al. (2009) used GIS to model the displacement of human populations in coastal zones due to sea level rise scenarios of 1–5 m and found that "tens of millions of people" concentrated in a small number of developing countries will be most impacted (Dasgupta, 2009, 379). Finally, Li et al. (2009) applied 1–6 m sea level rise scenarios in GIS to global data sets to allow visualization of these scenarios. They produced estimates from over 100 million to more than 430 million persons displaced worldwide (Li et al., 2009, 807), and they found that the most impacted ecosystems were forests and grasslands.

8.4.1.2 Water Quality and Public Health

Kistermann et al. (2001) integrated GIS and spatial analysis to create a tool to model drinking water quality and public health risks in the Rhein-Berg District, Germany. They related these risks to global climate change and responses in waterborne microbial communities. Their project successfully integrated existing drinking water databases into a tool for visualization, monitoring, and analysis. At the watershed scale, Tong and Chen (2002) used GIS to link land use and land cover scenarios to fecal, phosphorous, and nitrogen contamination of surface waters in the Little Miami River Basin in Ohio. They used the Better Assessment Science Integrated Point and Nonpoint

Sources (BASINS) model to establish significant statistical relationships between anthropogenic land use change and water quality, and they used GIS to spatially identify watersheds with compromised water quality and to quantify urban or agricultural lands. They concluded that the resulting models could be used to study and respond to global environmental change impacts (Tong and Chen, 2002). In one case, dissolved organic carbon (DOC) concentration in alpine lakes in Montana and Wyoming in the United States was used to develop baselines for proxy models of long-term climate change (Winn et al., 2009). The GIS was combined with remote sensing to model the relationships between physical watershed conditions influencing DOC and lake color. The authors concluded that the models were "promising" predictors of DOC.

8.4.1.3 Water Supply

In the arena of water supply, climate change is closely linked to the hydrosphere. An early work by Yates (1997) modeled the effects of climate change on runoff at the continental scale in South America. Yates used GIS to run annual models of runoff and evapotranspiration versus temperature and precipitation and found spatially varying increases and decreases in runoff. In a study organized by the USGRP, Rosenberg et al. (2003) combined Global Circulation model (GCM, U.K. Hadley Centre HadCM2 model, Rosenberg et al. 2003) climate change scenarios with GIS and the Soil Water Assessment Tool (SWAT model) to assess vulnerability of water yields in U.S. river basins to climate change, using decadal study periods. They included current and elevated CO_2 scenarios in their models. The study identified climate change as a major driver for decreased water yields in the Lower Mississippi River and Texas Gulf watersheds and for increased water yields in the west, and it found significant seasonal differences, affecting early snowmelt, thus increasing runoff (Rosenberg et al., 2003). Climate change impacts on groundwater recharge rates were modeled with a methodology using GIS and the Canadian Global Coupled Model 1 (CGCM1, Scibek et al., 2006). The methodology was tested in an unconfined aquifer in south central British Columbia, Canada. The study found that spatial variation in precipitation infiltration and recharge was a more important influence on groundwater levels than was temporal variation in recharge (Scibek et al., 2006). At the river basin scale, Graves and Chang (2007: 143) applied a "GIS based, distributed hydrologic model" and parameters from GCMs, including the U.K. Hadley Centre's Hadley Circulation Model (HadCM2) and the CGCM1, to predict changes in runoff in the upper Clackamas River basin in Oregon, the United States. The models predicted a large drop in snowpack in the area, smaller spring and summer flows, and a higher mean winter runoff. They found precipitation change as the dominant independent variable (Graves and Chang, 2007). The GIS was applied to forecast changes in river runoff in the Tien Shan alpine basins of Central Asia due to global and regional climate change (Aizen et al.,

2007). The study combined a DEM of the Tien Shan with a GIS-based river runoff model. It considered evapotranspiration and runoff and found that precipitation rather than air temperature was the best predictor of change in river runoff for this region. Numerous recent studies in Asia and in the Pacific northwest of North America used GIS to investigate the impacts of climate change on water supply and irrigation (e.g., Neilsen et al., 2006; Fu et al., 2007; Quilbe et al., 2008; Thomas, 2008).

8.4.1.4 Hydrosphere Conclusions

In the last two decades, GIS has evolved into a useful tool to assist in modeling hydrologic data for responses to global and environmental change. It has been successfully coupled with GCM, including the CGCM1 and the U.K. Hadley Centre's HadCM2, and has been coupled with surface and ground water hydrologic models. The GIS has also been used to combine hydrologic data sets to model scenarios of climate change impacts on water quality and human health and, in turn, for using water chemistry as an indicator of climate change. Finally, a dominant use of GIS in hydrologic response to global change is in modeling sea level change scenarios at local, regional, and global scales, for forecasting the displacement of coastal human populations and the impacts on coastal ecological systems. A major contribution of GIS has also been to reveal spatial variation in causes and consequences of global environmental change on hydrologic systems, in addition to temporal variations revealed by traditional statistical models.

8.4.2 Cryosphere and Sea Ice Research

Different spheres of the Earth system, regardless of whether they are the physical spheres or part of the bio-sociosphere, are all interconnected in respect to climate and global change (Steffen et al., 2005). One of the spheres that has attracted significant attention in recent decades due to its severe changes is the cryosphere. *Cyrosphere* may refer to different frozen parts of the Earth's surface. Apparently, the two largest components of this sphere are the Arctic and Antarctica. Although Antarctica, as a frozen continent, provides its records of climate change from its ice cores (Petit et al., 1999), effects of global change on the Arctic has been very much reflected by changes in sea ice. Relationships between sea ice characteristics in the Arctic and global change have already been established. Long-term climate modeling and recent global climate research suggest that the Arctic sea ice extent and thickness are sensitive indicators of climate variability (e.g., Wadhams, 1994; Vinnikov et al., 1999; Sturm et al., 2003). Studies have also detected significant decreases in sea ice extent over the past two to three decades (e.g., Johannessen et al., 1995; Parkinson and Cavalieri, 2002; Kerr, 2007; Meier et al., 2007). Effects of changes in the sea ice cover are not limited to the two poles but, in fact, have many far-reaching effects at the global scale through the interconnectedness of the climate systems extending beyond the polar region.

8.4.2.1 GIS Data on Sea Ice Characteristics

Sea ice is an integrated component of the Arctic climate system, particularly interfacing the ocean and atmosphere and moderating solar radiation upon the ocean. Several agencies and organizations are providing sea ice data. The National Ice Center (NIC, http://www.natice.noaa.gov), a tri-agency operational center involving the U.S. Navy, the NOAA, and the U.S. Coast Guard, is one of the major producers of sea ice data. A major mission of the NIC is to compile ice charts for the Arctic and Northern Hemisphere to support missions and the navigation of ships and submarines in the Arctic water. The ice chart data provide several important properties of sea ice, including the spatial extent and the concentration. Stage of development information can be used to derive additional sea ice characteristics such as ice thickness (Partington et al., 2003).

An ice chart consists of polygons delineating areas with homogeneous sea ice conditions (Figure 8.16). The ice chart also includes a set of symbols and codes ("egg codes") used by the World Meteorology Organization (WMO) to describe various sea ice characteristics (http://www.natice.noaa.gov/egg_code/index.html). To derive these ice charts, NIC has adopted a multi-sensor, multi-source approach to compile the ice charts (Dedrick et al., 2001). These sensors include passive microwave (PM) instruments such as Scanning Multi-frequency Microwave Radiometer (SMMR) and SSM/I (Cavalieri, 1994). Visible and infrared (IR) imagery data are also used to overcome some of the limitations of PM data. Newer remote-sensing data such as the synthetic aperture radar (SAR) data are also incorporated (Bertoia et al., 1998). Integrating these data sources, data from meteorological and oceanographic models and the analyst's expertise, composite pictures of ice extent, concentration, ice stage, and ice form are derived. Dedrick et al. (2001) provides a detailed explanation of the process used by NIC analysts. These ice charts and associated egg codes information are disseminated through the Web as images or can be downloaded as shapefiles or layers for use in GIS. Currently, NIC has archived ice data for the Arctic back from 1972.

Besides the ice chart data produced by NIC, the NSIDC in Colorado (http://nsidc.org/) also provides many data products of sea ice, including all satellite data mentioned earlier and related data from aircrafts, ships, and buoys. These data sets capture different aspects of sea ice. The NSIDC disseminates many of its sea ice data collections through the Web. Besides typical graphic formats, data are also distributed in GIS-compatible formats. Another important organization that disseminates sea ice related data is the Arctic Research Consortium of the United States (ARCUS), which consolidates a number of Web sites distributing sea ice related data into the Arctic GIS link (http://www.arcus.org/gis/data_links.html).

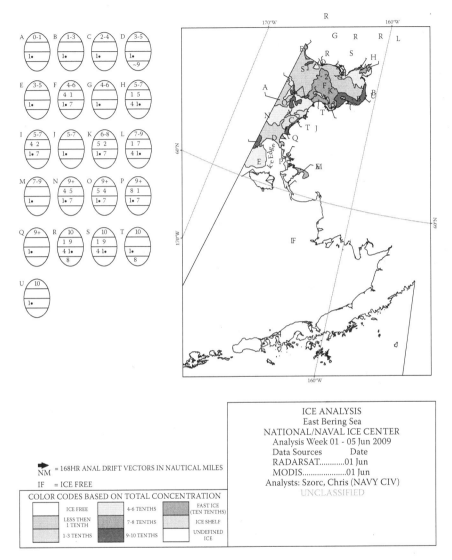

FIGURE 8.16
An ice chart for the East Bering Sea region, June 1, 2009, with egg codes on the upper left. (Adapted from National Ice Center [NIC, http://www.natice.noaa.gov].)

8.4.2.2 Sea Ice Data Used in GIS

Apparently, sea ice information can be derived from various sources of data from remote sensing to buoy data. Many scientists have been using data from these original data sources for sea ice research, and such research is abundant in the literature. This type of research can shed light on various aspects of climate change and the interaction between the hydrosphere and atmosphere. Some results are implicitly spatial, such the shrinking of sea ice coverage. On the other hand, many sea ice data sets are disseminated in GIS-compatible formats, and many systems built on GIS have been developed to process sea ice data into GIS formats in both the United States and other countries (e.g., Ramsay, 1998; Crooker and Carrieres, 2000; Helfrich et al., 2007). However, the use of sea ice data in GIS formats has been relatively limited (e.g., Howell and Yackel, 2004). Using GIS sea ice data, especially the ice charts, has several methodological challenges.

We have to understand the information captured in the ice chart first in order to deal with these challenges. There are three dimensions of information captured by the ice chart data. First, attributes in the ice chart data are a mix of nominal-categorical and interval-ratio scale data. Each polygon in an ice chart carries a set of sea ice characteristics reflected in the egg code: concentration levels in interval scale for different levels of ice thickness (C_a, C_b, and C_c); stages of development in ordered categories, where each category corresponds to a range of ice thickness (S_a, S_b, S_c); and forms of ice, each of which also corresponds to a range of ice thickness (F_a, F_b, F_c). Since most of these polygon attributes are not in interval-ratio scale, their measurement scales post some constraints on how these sea ice data can be manipulated and analyzed by GIS.

Second, the ice chart data are spatial in nature. Boundaries in polygons are drawn around areas such that locations within an area are believed to have homogeneous characteristics. In other words, polygons in the ice chart show the locations and spatial extents of given types of sea ice. However, the polygon data are somewhat different from other polygon data representing boundaries that are relatively persistent over time. Polygons are expected to change across ice charts to depict the changes in the spatial patterns of ice characteristics. Also, the number of features are expected to be different across ice charts. Therefore, using typical GIS functions such as overlay to compare multiple ice charts will not be effective and efficient.

Third, the ice chart data form a spatiotemporal series. Depending on the agencies or organizations generating the ice charts, the frequency of ice chart production is relatively frequent. Weekly seems to be a typical frequency, but more than once a week is possible. The implication is that within 1 year, fifty some charts are available for the same region assuming that the production frequency is weekly. From a time series analysis perspective, the ice chart data set is a very good set of time series data. As a result, many studies have mined the original data sources creating the ice charts to produce time-series analyses of sea ice coverage changes (e.g., Vinnikov, 1999). Unfortunately, space or location was not focused explicitly in many of these studies. On the other hand, GIS is quite incompetent to deal with the spatiotemporal series. The data for just 1 year will involve more than 50 layers, and such data volume overwhelms most GIS environments.

Given that sea ice data in GIS formats are very time-specific, usually have non-interval-ratio scale attributes and non-persistent polygon boundaries, therefore, numerous layers are needed for studies covering even a short period. As a result, most studies using the sea ice GIS data converted the polygon data into raster layers in order to derive changes in areas for certain sea ice characteristics. In such analysis, locations of change or the detailed geography of sea ice changes are often ignored or cannot be easily identified. Whether the detailed geography is important or not is difficult to determine in sea ice and cryosphere research, as such studies are often at a regional scale. Nevertheless, the current standard functions in GIS are not very effective in analyzing sea ice data, either for confirmatory or exploratory analyses. One specific attempt by Tang and Wong (2006) was to develop exploratory modules specifically designed to analyze and visualize sea ice data. Different modules focus on exploring changes in sea ice coverage by ice characteristics, identifying areas in which changes of ice characteristics have occurred, and animating the sea ice changes over time. These modules have the potential to be deployed as Web applications such that both the public and scientists can explore the valuable ice chart archives.

8.4.2.3 Cryosphere Conclusion

Numerous research results focusing on sea ice changes have pointed to the same conclusion that sea ice is melting at an unprecedented fast rate and therefore global warming is evidenced. Despite such a pessimistic conclusion, the voluminous ice chart archives likely have captured valuable or critical information, but these data have not yet been examined with regard to the outlook of global change. More effective and efficient tools are needed to mine these gigantic spatiotemporal data sets. The roles of GIS in producing the ice charts in GIS-compatible formats are apparent, but due to the specific attribute and spatial characteristics of the ice chart data, GIS have not been making significant contributions in mining and analyzing the spatiotemporal dataset. Innovative analytical tools that can handle long temporal series of spatial data in ordinal and nominal scales are needed. Other snow and ice data products may not be subject to the same limitations. Nevertheless, they have not been extensively exploited or utilized by GIS.

8.5 Other Applications in Social Science*

Applications of GIS in social sciences (including applied social sciences such as planning and public policy) have flourished in major university campuses and research institutes. Many conferences have been organized around this theme. The GIS have become the common thread among social sciences due

---

* This section is contributed by Fahui Wang.

to its capability of integrating and analyzing various data sets, in particular spatial data. The Center for Spatially-Integrated Social Science at UC Santa Barbara, funded by the National Science Foundation, has been an important force in promoting the usage of GIS technologies in various social sciences. As Michael Batty (2006) put it, "to do good social science that is policy relevant, ... methods, and the theory behind their practice, must be spatial." The GIS enrich social science research by providing a set of convenient spatial analysis tools, thus making public policy adaptable for various locations.

Several recent trends have contributed to the wide adoption of GIS in social science research. The first is the trend of increasing usage of scientific methods in social sciences, including computational and statistical methods, simulation models, and so on. The second is the emphasis on spatial issues in social sciences. The GIS have proved their value in tackling major issues such as globalization and intensified spatial interaction between regions, human-environment relations, sustainable development, and complexity of geopolitical conflicts. The third is attributable to the development of applied social sciences toward public policy relevance. Effective public policy needs to recognize regional differences and becomes adaptable to various locations. Finally, the advancement of GIS technology over the years has increased its capacity while making it more user friendly. All these factors create the "perfect storm" for the widespread of GIS applications across social sciences.

This section briefly discusses GIS applications in two of the most active fields closely related to public policy. One is *crime analysis and public safety* and the other is *public health*, particularly in health service access and provision. Both can be considered applied social sciences.

8.5.1 Crime Analysis and Public Safety

The GIS and related technologies have turned crime mapping and spatial analysis into powerful decision-making tools for law enforcement agencies (Wang, 2005a). The first use of computerized crime mapping in applied crime analysis occurred as early as in the mid-1960s in St. Louis (Weisburd and McEwen, 1997), at about the same time that the first GIS (i.e., the Canada Geographic Information System) was developed (Clarke, 1999, 9). A survey of police departments in 1997–1998 showed that an increasing number of departments used GIS regularly to analyze their crime problems (Mamalian et al., 1999), and the trend has continued (Weisburd and Lum, 2005). Earlier, GIS applications in crime studies were limited to archiving data, automated pin mapping, and cluster and hot spot analysis (Harries, 1999, 94), and they were gradually extended to applications of spatial statistics in testing spatial autocorrelation (e.g., Wasserman and Stack, 1993) and constructing spatial-lag variables in regression (e.g., Roncek and Montgomery, 1995).

The National Institute of Justice (NIJ) has been instrumental in promoting the use of GIS and spatial analysis in law enforcement and related research. The efforts are exemplified in funding the development of CrimeStat by Ned Levine and Associates (currently Version 3.1, see www.icpsr.umich.edu/CRIMESTAT/) and the Crime Analysis Application Extension for ArcView by the ESRI (see www.esri.com/industries/lawenforce/resources/crime_analysis.html) in the late 1990s. The NIJ also organizes an annual Mapping and Analysis for Public Safety (MAPS) conference (www.ojp.usdoj.gov/nij/maps/); the most recent was the tenth in New Orleans in August 2009. Recent collaborations between criminologists, geographers, statisticians, and others led to the applications of more advanced GIS and spatial statistics techniques in crime analysis. For example, Messner et al. (1999) used exploratory spatial data analysis to examine the homicide patterns in the United States; Baller et al. (2001) and Morenoff et al. (2001) used spatial regression models to explain homicide patterns across the United States and in Chicago, respectively; and Wang (2005b) used the scale-space clustering method to construct larger geographic areas for analyzing homicide rates in Chicago. One of the new trends in crime studies is the use of GIS-based computer simulation models such as the agent-based model in modeling individual crime behavior (Liu and Eck, 2008).

Here, we use one case study to illustrate the value of GIS in crime analysis. The case study concerns the linkage between crime rates and job market conditions. Most theories of crime suggest an inverse relationship between legal and illegal employment. Most research along this line has focused on the association between unemployment and crime rates (e.g., Chiricos, 1987). One danger of linking the rates of unemployment to crime is the implication that crimes were committed by the unemployed alone. Although crime rates among the unemployed were high, only one-third of prison inmates were unemployed at the time of their arrests (Beck et al., 1993). A study by Wang and Minor (2002) argues that not every job is an economic opportunity for all, and only an accessible job is meaningful. In addition to the adverse effect on employment prospects (and thus unemployment rate), poor job access incurs high monetary and psychological costs for workers already in the labor force and, as such, increases their willingness to risk losing their jobs through involvement in deviant or criminal behavior. The GIS are used to define the job market for a certain location as all jobs within a reasonable commuting time. Beyond the limit, a job is no longer accessible for a resident due to both time and monetary costs. In practice, a *Jobs to Resident Workers (JR) Ratio* is computed for each residential area within the commuting range. The higher the JR Ratio, the more jobs are available for resident workers in an area and, therefore, the more favorable the job market is. In summary, GIS enable a "localized" job market measure and permit researchers to examine the relationship between crime rates and job market conditions across neighborhoods within a city, instead of across cities as is the case in traditional crime studies.

8.5.2 Public Health

Public health is another major application area of GIS. The year 2001 marked the first ESRI Health GIS Conference in Washington, DC, which has continued ever since (www.esri.com/events/health). The Urban and Regional Information Systems Association (URISA) organizes an annual conference on a similar theme (www.urisa.org/conferences/health). Both have attracted hundreds of attendees from both academia and practitioners who are interested in public health applications of GIS. The GIS provide modern tools for exploring the dynamic connections between people, their health and well-being, and changing physical and social environments (Cromley & Mclafferty, 2002, 1). Major federal funding agencies related to health such as the National Institutes of Health (NIH) and the Centers for Disease Control and Prevention (CDC) have supported increasing numbers of GIS-related projects over the years.

One particular GIS application area in public health is to measure accessibility to healthcare services (e.g., physicians). Despite spending more per capita on medical care than any other nation, the United States ranks behind other industrialized nations in some key health performance measures (WHO, 2000). One major factor is the large disparities in access to care and health outcomes. The United States federal government has implemented various programs to alleviate health care access problems, including providing incentives or awarding financial assistance to providers serving designated physician shortage areas through the National Health Service Corps Program, the Medicare Incentive Program, and the J-1 visa waiver program, among others (GAO, 1995). These designations include the health professional shortage areas (HPSA) and the medically underserved areas or populations (MUA/MUP). The effectiveness of such programs relies primarily on appropriate and accurate *measures of accessibility* so that resources can be allocated to those needy areas.

Access may be affected by both spatial and nonspatial factors. *Spatial access* emphasizes the importance of spatial separation between supply and demand as a barrier or a facilitator, whereas *nonspatial access* stresses nongeographic barriers or facilitators (Joseph and Phillips, 1984). Nonspatial access is related to many demographic and socioeconomic variables such as social class, income, age, sex, race, and so on, and it also interacts with spatial access (Meade and Earickson, 2000, 389). Our focus here is on measuring spatial accessibility, where GIS are used most often. Measures of *spatial access to healthcare* (say, primary care physicians) need to account for the match between supply and demand within a region and the complex interaction between different regions. A study by Luo and Wang (2003) compared different methods for measuring spatial accessibility and recommended the two-step floating catchment area (2SFCA) method due to its intuitive interpretation and convenience of implementation in a GIS environment. In essence, the 2SFCA method measures spatial accessibility as a ratio of primary-care physicians to population: It first assesses physician availability at the physician's (supply) location as the ratio of physicians to their surrounding population (i.e., within a threshold travel time from the physicians) and then sums up the ratios (i.e., physician availability) around (i.e., within the same threshold travel time from) each residential (demand) location. Some recent developments related to the 2SFCA share a common objective: to seek the best way to capture actual physician-patient interactions, either by a function or by variable catchment areas (e.g., Guagliardo, 2004; Yang et al., 2006; 2008; Luo and Qi, 2009). For additional readings on GIS applications in social sciences, refer to Goodchild and Janelle (2004), Okabe (2005), and Wang (2006).

References

Aizen, V.B., E.M. Aizen, and V.A. Kuzmichenok. 2007. Geo-Informational simulation of possible changes in Central Asian water resources. *Global and Planetary Change*, 56(3–4):341–358.

Allan, B.F., R.B. Langerhans, and W.A. Ryberg. 2009. Ecological correlates of risk and incidence of West Nile virus in the United States. *Oecologia*, 158:699–708.

Allan, B.F., F. Keesing, and R.S. Ostfeld. 2003. Effect of forest fragmentation on Lyme disease risk. *Conservation Biology*, 17:267–272.

Allen, T.R. and D.W. Wong. 2006. Exploring GIS, spatial statistics, and remote sensing for risk assessment of vector-borne diseases: A West Nile Virus example. *The International Journal of Risk Assessment and Management*, 6(4/5/6):253–275.

Aron, J.L. and J.A. Patz. (eds.) 2001. *Ecosystem Change and Public Health: A Global Perspective*, 504p. The Johns Hopkins University Press, Baltimore, MD.

Baller, R.D., L. Anselin, S.F. Messner, G. Deane, and D.F. Hawkins. 2001. Structural covariates of U.S. county homicide rates: Incorporating spatial effects. *Criminology*, 39:561–590.

Batty, M. 2006. Foreword, in *Quantitative Methods and Applications in GIS*, ed. F. Wang. CRC Press, Boca Raton, FL.

Beck, A., D. Gilliard, L. Greenfeld, C. Harlow, T. Hester, L. Jankowski, T. Snell, J. Stephen, and D. Morton. 1993. *Survey of State Prison Inmates, 1991*. Report no. NCJ-136949. Bureau of Justice Statistics, U.S. Department of Justice, Washington, DC.

Bertoia C., J. Falkingham, and F. Fetterer. 1998. Polar SAR data for operational sea ice mapping. In *Recent Advances in the Analysis of SAR Data of the Polar Oceans*, eds. R. Kwok and C. Tsatsoulis, pp. 201–234. Springer-Verlag, Berlin, Germany.

Boone, J.D., K.C. McGwire, E.W. Otteson. 2000. Remote sensing and geographic information systems: Charting Sin Nombre virus infections in deer mice. *Emerging Infectious Diseases*, 6:248–258.

Brownstein, J.S., D.K. Skelly, and T.R. Holford. 2005. Forest fragmentation predicts local scale heterogeneity of Lyme disease risk. *Oecologia*, 146:469–475.

Brucker, G. 2005. Vulnerable populations: Lessons learnt from the 2003 summer heat waves in Europe. *Eurosurveillance*, 10(7):pii=551. Available online: http://www.eurosurveillance.org/ViewArticle.aspx?ArticleId=551

Cavalieri D.J. 1994. A microwave technique for mapping thin sea ice. *Journal of Geophysical Research*, 99(C6):12561–12572.

Centers for Disease Control and Prevention (CDC). 2008a. *Vibrio parahaemolyticus*. Division of Foodborne, Bacterial and Mycotic Diseases. http://www.cdc.gov/nczved/dfbmd/disease_listing/vibriop_gi.html

Centers for Disease Control and Prevention (CDC). 2008b. *Vibrio vulnificus*. Division of Foodborne, Bacterial and Mycotic Diseases. http://www.cdc.gov/nczved/dfbmd/disease_listing/vibriov_gi.html

Chiricos, T.G. 1987. Rates of crime and unemployment: An analysis of aggregate research evidence. *Social Problems*, 34:187–211.

Clarke, K.C. 1999. *Getting Started With Geographic Information Systems*. Prentice-Hall, Upper Saddle River, NJ.

Confalonieri, U., B. Menne, R. Akhtar, K.L. Ebi, M. Hauengue, R.S. Kovats, B. Revich, and A. Woodward. 2007. Human health. In *Climate Change 2007: Impacts, Adaptation and Vulnerability. Contribution of Working Group II to the Fourth Assessment Report of the Intergovernmental Panel on Climate Change*, eds. M.L. Parry, O.F. Canziani, J.P. Palutikof, P.J. van der Linden, and C.E. Hanson, pp. 391–431. Cambridge University Press, Cambridge, UK.

Connor. S.J., P. Ceccato, T. Dinku. 2006. Using climate information for improved health in Africa: Relevance, constraints and opportunities. *Geospatial Health*, 1:17–31.

Constantin de Magny G., R. Murtugudde, M.R.P. Sapiano, A. Nizam, C.W. Brown, A.J. Busalacchi, M. Yunus, et al. 2008. Environmental signatures associated with cholera epidemics. *Proceedings of the National Academy of Science*, USA, November 18, 105(46):17676–17681.

Crocker, G. and T. Carrieres. 2000. *Documentation for the Canadian Ice Service Digital Sea Ice Database*. Ballicater Consulting Ltd., Kingston, Ontario, Canada. Contract Report 00–02. 45pp.

Cromley, E. and S. McLafferty. 2002. *GIS and Public Health*. Guilford Press, New York.

Daniel, M., J. Kolar, and P. Zeman. 1999. Tick-borne encephalitis and Lyme borreliosis: Comparison of habitat risk assessments using satellite data. *Central European Journal of Public Health*, 7:35–39.

Das, A., S.R. Lele, G.E. Glass, et al. 2002. Modelling a discrete spatial response using generalized linear mixed models: Application to Lyme disease vectors. *International Journal of Geographical Information Science*, 16:151–166.

Dasgupta, S., B. Laplante, C. Meisner, D. Wheeler, and J. Yan. 2009. The impact of sea level rise on developing countries: A comparative analysis. *Climatic Change*, 93(3–4):379–388.

Dedrick, K.R., K. Partington, M. Van Woert, C.A. Bertoia, and D. Benner. 2001. U.S. National/Naval Ice Center Digital Sea Ice Data and Climatology. *Canadian Journal of Remote Sensing*, 27(5):457–475.

Delfino, R.J., S. Brummel, J. Wu, H. Stern, B. Ostro, M. Lipsett, A. Winer, et al. 2009. The relationship of respiratory and cardiovascular hospital admissions to the southern California wildfires of 2003. *Occupational Environment and Medicine* 66(3):189–197.

Din, A. 1992. Global environmental change data and modeling. *IFIP Transactions A-Computer Science and Technology*, 13:625–634.

Dister, S., L. Beck, B. Wood et al. 1993. The use of GIS and remote sensing technologies in a landscape approach to the study of Lyme disease transmission risk. In *Proceedings of GIS '93: Seventh Annual Symposium, Geographic Information Systems in Forestry*, Environmental and Natural Resource Management, pp. 1149–1155. Vancouver, BC, Canada, February 15–18.

Dister, S.W., D. Fish, S. Bros, D.H. Frank, and B.L. Wood. 1997. Landscape characterization of peridomestic risk for Lyme disease using satellite imagery. *American Journal of Tropical Medicine*, 57:687–692.

Environmental Protection Agency. 2009. Water Quality Parameters Derived From MERIS Imagery Data. http://maps6.epa.gov/aptw/viewer.htm

Epstein, P.R. 2005. Climate change and human health. *New England Journal of Medicine*, 353:1433–1436.

Estrada-Pena, A. 2002. Increasing habitat suitability in the United States for the tick that transmits Lyme disease: A remote sensing approach. *Environmental Health Perspectives*, 110:635–640.

Estrada-Pena, A. 2003. The relationships between habitat topology, critical scales of connectivity and tick abundance *Ixodes ricinus* in a heterogeneous landscape in northern Spain. *Ecography*, 26:661–671.

Ezenwa, V.O., M.S. Godsey, R.J. King et al. 2006. Avian diversity and West Nile virus: Testing associations between biodiversity and infectious disease risk. *Proceedings of the Royal Society B*, 273:109–117.

Frank, D.H., D. Fish, and F.H. Moy. 1998. Landscape features associated with Lyme disease risk in a suburban residential environment. *Landscape Ecology*, 13:27–36.

Frank, L.D., J.F. Sallis, T.L. Conway, J.E. Chapman, B.E. Saelens, and W. Bachman. 2006. Many pathways from land use to health: Associations between neighborhood walkability and active transportation, body mass index, and air quality. *Journal of the American Planning Association*, 72:75–81.

Fu, G., M.E. Barber, and S. Chen. 2007. Impacts of climate change on regional hydrological regi mes in the Spokane River watershed. *Journal of Hydrologic Engineering*, 12(5):452–461.

Gambolati, G., P. Teatini, and M. Gonella. 2002. GIS simulations of the inundation risk in the coastal lowlands of the Northern Adriatic Sea. *Mathematical and Computer Modelling*, 35(9–10):963–972.

General Accounting Office (GAO), 1995. Health care shortage areas: Designation not a useful tool for directing resources to the underserved, (GAO/HEHS-95-2000). General Accounting Office, Washington, DC.

George Mason University (GMU). 2009. BS Degree in Global and Environmental Change. Department of Geography and Geoinformation Science. Fairfax, VA. http://ggs.gmu.edu/page2.cfm?menu_id=14&sub_menu_id=39&sec_sub_menu_id=24 . Web site last accessed November 27, 2009.

Glass, G.E., B.S. Schwartz, J.M. Morgan III et al. 1995. Environmental risk factors for Lyme disease identified with geographic information systems. *American Journal of Public Health*, 85:944–948.

Glass G.E., J.E. Cheek, J.A Patz et al. 2000. Using remotely sensed data to identify areas at risk for Hantavirus pulmonary syndrome. *Emerging Infectious Diseases*, 6:238–247.

Goodchild, M.F. and D.G. Janelle (eds). 2004. *Spatially Integrated Social Science (Spatial Information Systems)*. Oxford University Press, New York.

Graves, D. and H. Chang. 2007. Hydrologic impacts of climate change in the upper Clackamas River basin, Oregon, USA. *Climate Research*, 33(2):143–157.

Group on Earth Observations. 2009. Group on Earth Observations. http://earthobservations.org

Guagliardo M.F. 2004. Spatial accessibility of primary care: Concepts, methods and challenges. *International Journal of Health Geography*, 3(1):3.

Gupta, P., S.A. Christopher, J. Wang, R. Gehrig, Y. Lee, and N. Kumar. Satellite remote sensing of particulate matter and air quality assessment over global cities. *Atmospheric Environment*, 40(30):5880–5892.

Hales, S., P. Weinstein, and A. Woodward. 1997. Public health impacts of global climate change. *Reviews on Environmental Health*, 12:191–199.

Harries, K. 1999. *Mapping Crime: Principle and Practice*. National Institute of Justice, U.S. Department of Justice, Washington, DC.

Hasting, D.A. and L.P. Di. 1994a. Modeling of global change phenomena with GIS using the global change data base. 1. Modeling with GIS. *Remote Sensing of Environment*, 49(1):1–12.

Hasting, D.A. and L.P. Di. 1994b. Modeling of global change phenomena with GIS using the global change data base. 2. Prototype synthesis of the AVHRR-based vegetation index from terrestrial data. *Remote Sensing of Environment*, 49 (1):13–24.

Helfrich, S.R., D. McNamara, B.H. Ramsey, T. Baldwin, and T. Kasheta. 2007. Enhancements to, and forthcoming developments in the interactive multisensor snow and ice mapping system (IMS). *Hydrological Processes*, 21:1576–1586.

Homer, C., L. Huang, L. Yang et al. 2004. Development of a 2001 National Land-Cover Database for the United States. *Photogrammetric Engineering and Remote Sensing*, 70:829–840.

Howell, S.E.L. and J.J. Yackel. 2004. A vessel transit assessment of sea ice variability in the Western Arctic, 1969–2002: Implications for ship navigation. *Canadian Journal of Remote Sensing*, 30(2):205–215.

Jackson, L.E., E.D. Hilborn, and J.C. Thomas, 2006a. Towards landscape design guidelines for reducing Lyme disease risk. *International Journal of Epidemiology*, 35:315–322.

Jackson, L.E., J.F. Levine, and E.D. Hilborn. 2006b. A comparison of analysis units for associating Lyme disease with forest-edge habitat. *Community Ecology*, 7: 189–197.

Johannessen, O.M., M.W. Miles, and E. Bjørgo. 1995. The Arctic's shrinking sea ice. *Nature*, 376(6536):126–127.

Joseph, A.E. and D.R. Phillips. 1984. *Accessibility and Utilization–Geographical Perspectives on Health Care Delivery*. Harper & Row Publishers, New York.

Kerr, R.A. 2007. Is battered Arctic sea ice down for the count? *Science*, 318:33–34.

Kistemann, T., S. Herbst, F. Dangendorf, and M. Exner. 2001. GIS-based analysis of drinking-water supply structures: A module for microbial risk assessment. *International Journal of Hygiene and Environmental Health*, 203(4):301–310.

Kitron, U. and J.J. Kazmierczak, 1997. Spatial analysis of the distribution of Lyme disease in Wisconsin. *American Journal of Epidemiology*, 145:558–566.

Knox, J.W., R.B. Matthews, and R. Wassmann. 2000. Using a crop/soil simulation model and GIS techniques to assess methane emissions from rice fields in Asia. III. Databases. *Nutrient Cycling in Agroecosystems*, 58(1–3):179–199.

Kovats, S.R. and K.L. Ebi. 2006. Heatwaves and public health in Europe. *European Journal of Public Health*, 16(6):592–599.

Kovats, S. R. and C. Koppe. 2005. Heat waves: Past and future impacts on health In *Integration of Public Health with Adaptation to Climate Change: Lessons Learned and New Directions*, eds. K.L. Ebi, J.B. Smith, and I. Burton, Chapter 8. Taylor & Francis, London.

Kunzli, N., E. Avol, J.Wu, W.J. Gauderman, E. Rappaport, J. Millstein, J. Bennion, et al. 2006. Health effects of the 2003 southern California wildfires on children. *American Journal of Respiratory and Critical Care Medicine*, 174:1221–1228.

Lam, N.S.-N., H. Arenas, Z. Li, and K. Liu. 2009. An estimate of population impacted by climate change along the U.S. coast. *Journal of Coastal Research*, 2(56):1522–1526.

Levin, M., J.F. Levine, C.S. Apperson et al. 1995. Reservoir competence of the Rice Rat (Rodentia: Cricetidae) for *Borrelia burgdorferi*. *Journal of Medical Entomology*, 32: 138–142.

Li, X., R.J. Rowley, J.C. Kostelnick, D. Braaten, J. Meisel, and K. Hulbutta. 2009. GIS analysis of global impacts from sea level rise. *Photogrammetric Engineering and Remote Sensing*, 75(7):807–818.

Liu, L. and J. Eck (eds). 2008. *Artificial Crime Analysis Systems: Using Computer Simulations and Geographic Information Systems*. Idea Group Publishing, Hershey, PA.

Lo, C.P. and D.A. Quattrochi 2003. Land-use and land-cover change, urban heat island phenomenon, and health implications: A remote sensing approach. *Photogrammetric Engineering and Remote Sensing*, 69(9):1053–1063.

Luber, G. and M. McGeehin. 2008. Climate change and extreme heat events. *American Journal of Preventive Medicine*, 35(5):429–435.

Lunetta, R.S., J.F. Knight, H.W. Paerl, J.J. Streicher, B.L. Peierls, T. Gallo, J.G. Lyon, T.H. Mace, and C.P. Bozzelli, 2009. Measurement of water color using AVIRIS imagery to assess the potential for an operational monitoring capability in the Pamlico Sound Estuary, USA. *International Journal of Remote Sensing*, 30(13):3291–3314.

Luo W. and Y. Qi. 2009, An enhanced two-step floating catchment area (E2SFCA) method for measuring spatial accessibility to primary care physicians. *Health and Place*, 15:1100–1107.

Luo, W. and F. Wang. 2003. Measure of spatial accessibility to health care in a GIS environment: Synthesis and a case study in the Chicago region. *Environmental and Planning B: Planning and Design*, 30:865–884.

Mamalian, C.A., N.G. La Vigne, and the staff of the Crime Mapping Research Center. 1999. *The Use of Computerized Crime Mapping by Law Enforcement: Survey Results*. National Institute of Justice, U.S. Department of Justice, Washington, DC, FS 000237.

Mather, T.N., M.L. Wilson, S.I. Moore et al. 1989. Comparing the relative potential of rodents as reservoirs of the Lyme disease spirochete (*Borrelia burgdorferi*). *American Journal of Epidemiology*, 130:143–150.

McCarthy, J.J., O.F. Canziani, N.A. Leary et al. 2001. Climate change 2001: Impacts, adaptation and vulnerability. *Contribution of Working Group II to the Third Assessment Report of the Intergovernmental Panel on Climate Change*. Cambridge University Press, Cambridge, UK.

Meade, S.M. and R.J. Earickson. 2000. *Medical Geography* (2nd ed.). The Guilford Press, New York.

Meier, W., J. Stroeve, and F. Fetterer. 2007. Whither Arctic sea ice? A clear signal of decline regionally, seasonally and extending beyond the satellite record. *Annals of Glaciology*, 46:428–434.

Messner, S.F., L. Anselin, R.D. Baller, D.F. Hawkins, G. Deane, and S.E. Tolnay, 1999. The spatial patterning of county homicide rates: An application of exploratory spatial data analysis. *Journal of Quantitative Criminology*, 15:423–450.

Morenoff, J.D., R.J. Sampson, and S.W. Raudenbush. 2001. Neighborhood inequality, collective efficacy, and the spatial dynamics of urban violence. *Criminology*, 39:517–559.

National Interagency Fire Center. 2009. National Interagency Coordination Center. http://www.nifc.gov/nicc/index.htm

National Oceanic and Atmospheric Administration. 2008a. Harmful Algal Bloom Forecasting System. August 27, 2008. http://tidesandcurrents.noaa.gov/hab

National Oceanic and Atmospheric Administration. 2008b. Harmful Algal Blooms. http://oceanservice.noaa.gov/topics/coasts/hab/. Date last updated November 26, 2008.

National Oceanic and Atmospheric Administration. 2009. Fire Weather. http://www.noaawatch.gov/themes/fire.php

National Research Council. 2007a. Human health and security. In *Earth Science and Applications from Space: National Imperatives for the Next Decade and Beyond*, pp. 152–189. National Academies Press, Washington, DC.

National Research Council. 2007b. Oceans and *Human Health—Highlights of National Academies Reports*. National Academies Press, Washington, DC.

Neilsen, D., C.A.S. Smith, G. Frank, W. Koch, Y. Alila, W.S. Merritt, W.G. Taylor, M. Barton, J.W. Hall, and S.J. Cohen. 2006. Potential impacts of climate change on water availability for crops in the Okanagan Basin, British Columbia. *Canadian Journal of Soil Science*, 86(5):921–936.

Nicholson, M.C. and T.N. Mather. 1996. Methods for evaluating Lyme disease risks using geographic information systems and geospatial analysis. *Journal of Medical Entomology*, 33:711–720.

O'Connell, T.J., L.E. Jackson, and R.P. Brooks 2000. Bird guilds as indicators of ecological condition in the central appalachians. *Ecological Applications*, 10: 1706–1721.

Okabe, A. 2005. *GIS-based Studies in the Humanities and Social Sciences*. Taylor & Francis, Boca Raton, FL.

Ostfeld, R.S. and F. Keesing. 2000. The function of biodiversity in the ecology of vector-borne zoonotic diseases. *Canadian Journal of Zoology*, 78:2061–2078.

Parkinson, C.L. and D.J. Cavalieri. 2002. A 21 year record of Arctic sea ice extents and their regional, seasonal and monthly variability and trends. *Annals of Glaciology*, 34:441–446.

Parmenter, R.R., E.P. Yadav, C.A. Parmenter et al. 1999. Incidence of plague associated with increased winter-spring precipitation in New Mexico. *American Journal of Tropical Medicine and Hygiene*, 61:814–821.

Parmeson, C. and Yohe, G. 2003. A globally coherent fingerprint of climate change impacts across natural systems. *Nature*, 421:37–42.

Parry, M.L., O.F. Canziani, J.P. Palutikof et al. 2007. Climate change 2007: Impacts, adaptation and vulnerability. *Contribution of Working Group II to the Fourth Assessment Report of the Intergovernmental Panel on Climate Change*, 976pp. Cambridge University Press, Cambridge, UK.

Partington, K., T. Flynn, D. Lamb, C. Bertoia, and K. Dedrick. 2003. Late twentieth century northern hemisphere sea ice record from U.S. National Ice Center ice charts. *Journal of Geophysical Research-Oceans*, 108(C11), doi:10.1029/2002JC001623.

Patz, J.A., P.R. Epstein, T.A. Burke et al. 1996. Global climate change and emerging infectious diseases. *Journal of the American Medical Association*, 275:217–223.

Patz, J.A., J.M. Martens, D.A. Focks et al. 1998. Dengue fever epidemic potential as projected by general circulation models of global climate change. *Environmental Health Perspectives*, 106:147–153.

Patz, J.A., M.A. McGeehin, S.M. Bernard et al. 2000. The potential health impacts of climate variability and change for the United States: Executive Summary of the Report of the Health Sector of the U.S. National Assessment. *Environmental Health Perspectives*, 108:367–376.

Petit, J.R., J. Jouzel, D. Raynaud, N.I. Barkov, J.-M. Barnola, I. Basile, M. Bender, et al. 1999. Climate and atmospheric history of the past 420,000 years from the Vostok ice core, Antarctica. *Nature*, 399:413–429.

Phillips, M.L. 2008. Dengue reborn: widespread resurgence of a resilient vector. *Environmental Health Perspectives*, 116:A382–A388.

Quilbe, R., A.N. Rousseau, J. Moquet, N.B. Trinh, Y. Dibike, P. Gachon, and D. Chaumont. 2008. Assessing the effect of climate change on river flow using general circulation models and hydrological modelling—Application to the Chaudiere River, Quebec, Canada. *Canadian Water Resources Journal*, 33(1):73–93.

Ramsey, B.H. 1998. The interactive multisensor snow and ice mapping system. *Hydrological Processes*, 12:1537–1546.

Reeves, W.C., J.L. Hardy, W.K. Reisen et al. 1994. Potential effect of global warming on mosquito-borne arboviruses. *Journal of Medical Entomology*, 31:323–332.

Riitters, K.H., J.D. Wickham, R.V. O'Neill et al. 2002. Fragmentation of continental United States forests. *Ecosystems*, 5:815–822.

Roncek, D.W. and A. Montgomery. 1995. Spatial autocorrelation revisited: Conceptual underpinnings and practical guidelines for the use of the generalized potential as a remedy for spatial autocorrelation in large samples. In *Crime Analysis Through Computer Mapping*, eds. C.R. Block, M. Dabdoub, and S. Fregly, pp. 99–110. Police Executive Research Forum, Washington, DC.

Rosenberg, N.J., R.A. Brown, R.C. Izaurralde, and A.M. Thomson. 2003. Integrated assessment of Hadley Centre (HadCM2) climate change projections on agricultural productivity and irrigation water supply in the conterminous United States—I. Climate change scenarios and impacts on irrigation water supply simulated with the HUMUS Model. *Agricultural and Forest Meteorology*, 117(1–2):73–96.

Schmidt, Charles W. 2005. Terra Cognita: Using Earth observing systems to understand our world. *Environmental Health Perspectives*, 113(2):A99–A104.

Schmidt, K.A. and R.S. Ostfeld. 2001. Biodiversity and the dilution effect in disease ecology. *Ecology*, 82:609–619.

Schultz, G.A. 2000. Potential of modern data types for future water resources management. *Water International*, 25(1):96–109.

Schumaker, N. 1996. Using landscape indices to predict habitat connectivity. *Ecology*, 77:1210–1225.

Scibek, J. and D.M. Allen. 2006. Modeled impacts of predicted climate change on recharge and groundwater levels. *Water Resources Research*, 42(11):AR W11405 1–18, doi:10.1029/2005WR004742.

Sheridan, Scott C. 2007. A survey of public perception and response to heat warnings across four North American cities: an evaluation of municipal effectiveness. *International Journal of Biometeorology*, 52:3–15.

Simas, T., J.P. Nunes, and J.G. Ferreira. 2001. Effects of global climate change on coastal salt marshes. *Ecological Modelling*, 139(1):1–15.

Sokoletsky, L.G., O.V. Nikolaeva, V.P. Budak, L.P. Bass, R.S. Lunetta, V.S. Kuznetsov, and A.K. Kokhanovsky. 2009. A comparison of numerical and analytical radiative-transfer solutions for plane albedo of natural waters. *Journal of Quantitative Spectroscopy & Radiative Transfer*, 110(13):1132–1146.

Steffen, W., A. Sanderson, P. Tyson, J. Jager, P. Matson, B. Moore III, F. OldField, et al. 2005. *Global Change and the Earth System: A Planet under Pressure*. Springer, Berlin, Germany.

Sturm, M., D.K. Perovich, and M.C. Serreze. 2003. Meltdown in the North. *Scientific American*, 289(4):60–67.

Swaddle, J.P. and S.E. Calos. 2008. Increased avian diversity is associated with lower incidence of human West Nile infection: Observation of the dilution effect. *PLoS One*, 3:e2488.

Tang, Y. and D.W. Wong. 2006. Exploring and visualizing sea ice chart data using Java-based GIS tools. *Computers and Geosciences*, 32(6):846–858.

Thomas, A. 2008. Agricultural irrigation demand under present and future climate scenarios in China. *Global and Planetary Change*, 60(3–4):306–326.

Tong, S.T. and W.L. Chen. 2002. Modeling the relationship between land use and surface water quality. *Journal of Environmental Management*, 66(4):377–393.

Torresan, S., A. Critto, M.D. Valle, N. Harvey, and A. Marcomini. 2008. Assessing coastal vulnerability to climate change: Comparing segmentation at global and regional scales. *Sustainability Science*, 3(1):45–65.

U.S. Department of Energy (U.S. DOE). 2009. Global Change Education Program. Office of Biological and Environmental Research, Washington DC. http://www.atmos.anl.gov/GCEP/. Web site last accessed November 27, 2009.

U.S. Global Change Research Program (U.S. GCRP). 2009a. Washington, DC. http://www.globalchange.gov/. Web site last accessed November 27, 2009.

U.S. Global Change Research Program (U.S. GCRP). 2009b. Washington, DC. http://www.globalchange.gov/about/program-structure/global-change-research-act. Web site last accessed November 27, 2009.

U.S. Global Change Research Program (U.S. GCRP). 2009c. Washington, DC. http://www.globalchange.gov/about/. Web site last accessed November 27, 2009.

U.S. Environmental Protection Agency (U.S. EPA). 2006. *2006–2011 EPA Strategic Plan: Charting Our Course*. http://www.epa.gov/ocfo/plan/2006/entire_report.pdf Web site last accessed January 20, 2010.

Utset, A. and M. Borroto. 2001. A modeling-GIS approach for assessing irrigation effects on soil salinisation under global warming conditions. *Agricultural Water Management*, 50(1):53–63.

Vafeidis, A.T., R.J. Nicholls, L. McFadden, R.S.J. Tol, J. Hinkel, T. Spencer, P.S. Grashoff, G. Boot, and R.J.T. Klein. 2008. A new global coastal database for impact and vulnerability analysis to sea-level rise. *Journal of Coastal Research*, 24(4):917–924.

van der Pluijm, B.A. 2006. The global change curriculum and minor at the University of Michigan. *Journal of Geoscience Education*, 54(3):249–254.

Vinnikov, K., A. Robock, R.J. Stouffer, J.E. Walsh, C.L. Parkinson, D.J. Cavalieri, J.F.B. Mitchell, D. Garrett, and V.F. Zakharov. 1999. Global warming and northern hemisphere sea ice extent. *Science Magazine Reprint Series*, 286:1934–1937.

Wadhams, P. 1994. Sea ice thickness changes and their relation to climate. *Polar Oceans and their Role in Shaping the Global Enviroment, The Nansen Centennial Volume* (Geophysical Monography 85), pp. 337–362. American Geophysical Union, Washington, DC.

Wang, F. 2005a. Preface, in *Geographic Information Systems and Crime Analysis*, ed. F. Wang. Idea Group Publishing, Hershey, PA.

Wang, F. 2005b. Job access and homicide patterns in Chicago: An analysis at multiple geographic levels based on scale-space theory. *Journal of Quantitative Criminology*, 21:195–217.

Wang, F. 2006. *Quantitative Methods and Applications in GIS*. Taylor & Francis, Boca Raton, FL.

Wang, F. and W.W. Minor. 2002. Where the jobs are: Employment access and crime patterns in Cleveland. *Annals of the Association of American Geographers*, 92:435–450.

Wasserman, I.M. and S. Stack. 1993. Spatial autocorrelation patterns with regard to suicide in the United States. In *Workshop on Crime Analysis through Computer Mapping Proceedings*, C.R. Block and D. Margaret, pp. 69–75. Illinois Criminal Justice Information Authority, Chicago, IL.

Watts, R.D., R.W. Compton, J.H. McCammon et al. 2007. Roadless space of the conterminous United States. *Science*, 316:736–738.

Weisburd, D. and C. Lum. 2005. The diffusion of computerized crime mapping in policing: Linking research and practice. *Police Practice and Research*, 6:419–434.

Weisburd, D. and J.T. McEwen (eds). 1997. *Crime Mapping and Crime Prevention*. Criminal Justice Press, Monsey, NY.

Weng, Q. and S. Yang. 2004. Managing the adverse thermal effects of urban development in a densely populated Chinese city. *Journal of Environmental Management*, 70(2):145–156.

White, J.E. 2009. AIRNow-International: The future of the United States real-time air quality reporting and forecasting program and GEOSS participation. 33rd ISRSE—Session: Air Quality and GEOSS—TS34–1, May 6, 2009. http://wiki. esipfed.org/images/2/21/TS34–1_White.ppt. Web site last accessed January 21, 2009.

Winn, N., C.E. Williamson, R. Abbitt, K. Rose, W. Renwick, M. Henry, and J. Saros. 2009. Modeling dissolved organic carbon in subalpine and alpine lakes with GIS and remote sensing. *Landscape Ecology*, 24(6):807–816.

World Health Organization (WHO). 2000. The world health report 2000—Health systems: improving performance. http://www.who.int/whr/2000/en/index. html

Woods Hole Oceanographic Institute. 2008. Harmful algae. http://www.whoi.edu/ redtide. Date last updated May 7, 2008.

Wu, J., A.M. Winer, and R.J. Delfino. 2006. Exposure assessment of particulate matter air pollution before, during, and after the 2003 Southern California wildfires. *Atmospheric Environment*, 40(18):3333–3348.

Xu, H.Q. and B.Q. Chen. 2004. Remote sensing of the urban heat island and its changes in Xiamen City of SE China. *Journal of Environmental Science (China)*, 16(2): 276–281.

Yang, D., R. Goerge, and R. Mullner. 2006. Comparing GIS-based methods of measuring spatial accessibility to health services. *Journal of Medical Systems*, 30:23–32.

Yates, D.N. 1997. Climate change impacts on the hydrologic resources of South America: An annual, continental scale assessment. *Climate Research*, 9(1–2):147–155.

Yoffe, S.B. and B.S. Ward. 1999. Water resources and indicators of conflict—a proposed spatial analysis. *Water International*, 24(4):377–384.

9

Other Applications

Meixia Deng, Zhong Liu, Qianjun Miao, Qisheng Pan, Ming Zhang,
Long S. Chiu, Liping Di, Steven Kempler, Lenard Milich, Hualan Rui,
Jiqiang Tan, Yanli Tang, William Teng, Ruixin Yang, Hongwei Yu,
and Chaowei Yang

CONTENTS

This chapter follows the previous chapter on the specific usage of geoinformation and geoinformation science and technology beyond environmental and related applications. First, an application example of EOSDIS (refer Chapter 4) for a global agriculture information system with online real-time precipitation data is introduced. Next, Geographic Information Systems for Transportation (GIS-T), which was developed for solving transportation related problems reflecting federal legislations, was discussed, and the GIS-T applications in various projects and applications including location-based services (LBS) were reviewed. After the GIS-T section, GIS as a decision support tool in an urban planning was reviewed with case studies for the advanced storm water treatment plans in Los Angeles. Then, GI Science, in particular, GIS, and remote-sensing applications in emergency response cases were summarized and demonstrated with application to the 2008 8.0 Ms Wenchuan earthquake. Finally, the focus of the chapter is moved to data-intensive Earth Science (ES) education, issues associated with it, and a solution, the GeoBrain system, another NASA sponsored project.

9.1 Global Agriculture Information System and Its Applications*

9.1.1 Introduction

Weather is important to agriculture (Special Issue of Agricultural and Forest Meteorology 2000). For example, prolonged droughts can inhibit crop growth and reduce yield. Monitoring global agricultural crop conditions during the growing season and estimating potential seasonal production are critically important for market development of U.S. agricultural products and for global food security. Meteorological information is one key component in monitoring crop conditions. It is essential to have such information in a timely and accurate manner to ensure successful monitoring activities (Special Issue of Agricultural and Forest Meteorology 2000). In developed countries, people rely on a network of ground observational stations to provide meteorological information, such as rainfall, temperature, and so on, along with other observational networks. However, in developing countries, few ground stations are available, therefore hindering agriculture monitoring activities.

Satellite remote-sensing technology (e.g., Liu et al. 2007) provides a unique way to monitor agricultural conditions from space, especially in data-sparse regions. Multi-satellites provide more frequent measurements than a single satellite, making global monitoring activities a reality. In particular, the TRMM is a joint mission between NASA and the Japan Aerospace Exploration Agency (JAXA) designed to monitor and study tropical rainfall (Special Issue on the TRMM 2000). Combining with other satellites

* This section is contributed by Zhong Liu, Steve Kempler, William Teng, Hualan Rui, Long S. Chiu, and Lenard Milich. This work and the Agriculture DISC are a contribution of the GES DISC, via NASA Grant (REASoN CAN-02-OES-01).

(e.g., geostationary satellites), TRMM products can be greatly improved in terms of their temporal resolution, such as the 3-hourly TMPA (Huffman et al. 2007). However, data access can be a challenging task for developing countries and non data experts (Liu et al. 2007). For example, some products require special software and computer platforms for processing and visualization, which could require a significant investment from the user side (Liu et al. 2007).

NASA GES DISC (http://disc.gsfc.nasa.gov/) is the home of TRMM data archives. It distributes TRMM data and provides value-added data services. Its Agriculture DISC (http://disc.gsfc.nasa.gov/agriculture/index.shtml) is a part of NASA Earth Science Division Applied Sciences Program, Agricultural Efficiency Program, aiming at improving agricultural competitiveness through a better understanding of weather and climate, especially in the prediction of events with an increasing accuracy, longer lead times, and the integration of predictions and observations into local and regional decision support systems used in agricultural management. The Agriculture DISC has been established to facilitate data access and allow users to focus on their intended activities; it has developed a Web-based Agriculture Information System (AIS). The AIS offers two major services: (1) Agriculture Online Visualization and Analysis System (AOVAS) and (2) Analysis maps for current conditions.

9.1.2 AOVAS Architecture

The development of AOVAS (Liu et al. 2007) is based on the TRMM Online Visualization and Analysis (TOVAS, http://disc2.nascom.nasa.gov/Giovanni/tovas/). The TOVAS primarily consists of TRMM data and has been in operation since March 2000 to support various research and application efforts (Liu et al. 2007). Although TOVAS offers quite a number of rainfall products, additional rainfall products specifically targeted for agriculture, such as 10-day, need to be derived.

The AOVAS (Figure 9.1) and TOVAS are a part of the GES-DISC Interactive Online Visualization and Analysis Infrastructure (Giovanni) (Acker and Leptoukh 2007; Berrick et al. 2009). The principle design objective for Giovanni is to provide a quick and simple interactive means for science data users to study various phenomena by trying various combinations of parameters measured by different instruments to arrive at a conclusion, and then generate graphs suitable for publication. Alternatively, Giovanni would provide means to ask relevant what-if questions and get answers that would stimulate further investigations. This would all be done without having to download and preprocess large amounts of data.

Application users can quickly and easily access the latest information regarding current environmental conditions through AOVAS. For example, by accessing near-real-time rainfall products, they can learn the amount and duration of rainfall that an area has received. By accessing historical data,

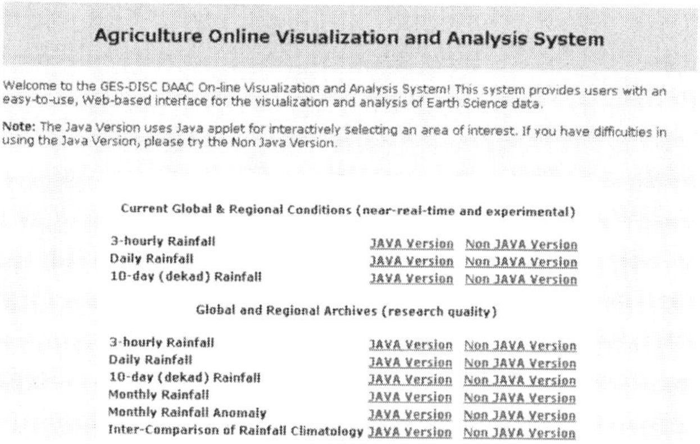

FIGURE 9.1
AOVAS home page. (Adapted from http://agdisc.gsfc.nasa.gov/Giovanni/aovas/.)

they can study past events and their impacts on agricultural production. Users can also obtain data through AOVAS for further analysis and applications. In short, AOVAS provides users a simple and easy way to access global satellite data information.

The Giovanni AOVAS consists of HTML and CGI (Common Gateway Interface) scripts written in Perl, Grid Analysis and Display System (GrADS, http://grads.iges.org/grads/) scripts, and one or more GDS running on remote machines that have GrADS readable data. In addition, there is an image map Java applet through which a user can select a bounding box area to define an area of interest. The data flow is illustrated in Figure 9.2. The AOVAS currently runs on Linux platforms.

A user selects one or more data sets (see Table 9.1), the spatial area, the temporal extent, and the type of output with the Giovanni Web interface (Figure 9.2). Supported output types or functions are listed in Table 9.2. The selection criteria are passed to the CGI scripts for processing.

9.1.2.1 Agriculture Information System

The Agriculture Information System (AIS) provides environmental data information to support agricultural activities; specifically, it provides access to NASA Earth System Science and other data products (e.g., rainfall, crop model outputs) and services in operation mode. Currently, AIS consists of TRMM and other rainfall products:

Analysis of Current Conditions: Contains links to rainfall product analysis maps for selected regions worldwide and the conterminous United States (http://disc.sci.gsfc.nasa.gov/agriculture/additional/tools/current_conditions.shtml). The maps are updated daily, and they include accumulated rainfall, anomaly, and normalized anomaly analysis for the following time periods (Figure 9.3):

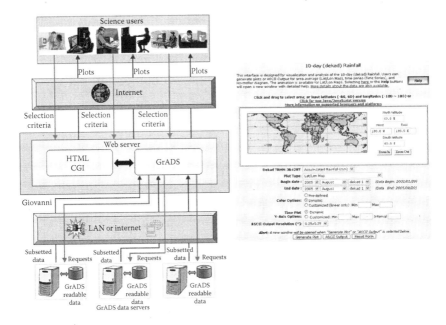

FIGURE 9.2
Left: AOVAS system diagram; Right: A sample Web interface.

- 3-hourly Global and Regional Rainfall Maps
- 24-hour Global and Regional Rainfall Maps
- 10-day Global and Regional Rainfall Maps
- 30-day Global and Regional Rainfall Maps

TABLE 9.1

Major Rainfall Products for AOVAS

Product	Description	Usage
3B42RT (experimental)	Near-real-time, 3-hourly, daily, 10-day, 0.25 deg., 60°S–60°N, multi-satellite precipitation analysis. February 2002—present	Flood monitoring and other applications
3B42 (research)	3-hourly, daily, 10-day, 0.25 deg., 60°S–60°N, multi-satellite precipitation analysis. February 2002—present	Historical case studies
TRMM 3B43 (research)	Monthly, 0.25 deg., 50°S–50°N, TRMM, and other data sources rainfall estimate. January 1998—present	Drought monitoring and crop yield estimate, deriving climatology and anomaly. Historical case studies

TABLE 9.2

AOVAS Functions

Output Type	Description
Area plot	Area plot averaged or accumulated over any available data period within any rectangular area
Time Plot	Time series averaged over any rectangular area or point
Hovmoller plots	Longitude–time and latitude–time plots
Animations	Animations available for area plots
ASCII output	ASCII output available for all plot types, suitable feeding GIS, or other applications

- 60-day Global and Regional Rainfall Maps
- 90-day Global and Regional Rainfall Maps

Agriculture Online Visualization and Analysis System (AOVAS): Allows to access global, current, and historical data and generate customized maps, time series, and digital data.
Current (near-real-time and experimental):

- 3-hourly Global and Regional Rainfall
- Daily Global and Regional Rainfall
- 10-day (dekad) Global and Regional Rainfall

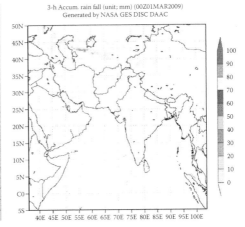

FIGURE 9.3

(See color insert following page 144.) Left: List of maps of current conditions, providing a quick and easy access to current global and regional rainfall conditions. Rainfall anomaly and normalized anomaly can be used to identify flood or drought events. Right: A sample of the maps.

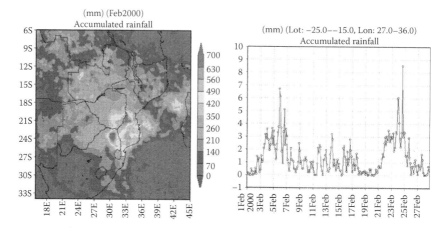

FIGURE 9.4
(See color insert following page 144.) Left: Regional accumulated rainfall (TRMM 3B43) for February 2000. Parts of Mozambique received rainfall of over 300 mm in 1 day. Over 300 people died and 2 million had been displaced or affected according to news reports. Right: Time series of 3-hourly rainfall (TRMM 3B42) for the southern part of Mozambique (see Left). Two major rain events are identified in this plot.

Archives (research quality):

- 3-hourly Global and Regional Rainfall
- Daily Global and Regional Rainfall
- 10-day (dekad) Global and Regional Rainfall
- Monthly Global and Regional Rainfall
- Monthly Global and Regional Rainfall Anomaly and Climatology
- Inter-comparison of Rainfall Climatological Data Products

Figure 9.4 is an example that shows that AOVAS is a powerful tool that provides customized analysis and visualization.

9.1.3 AOVAS Applications

Two major operational users who utilize satellite remote sensing for global crop monitoring are the USDA Foreign Agricultural Service (FAS) and the UN's World Food Program (WFP). The primary goal of FAS is to improve foreign market access for U.S. agricultural products. The WFP uses food to meet emergency needs and to support economic and social development. Both use global agricultural decision support systems that can integrate and synthesize a variety of data sources to provide accurate and timely information on global crop conditions.

One of the most prominent additions to the FAS Web site over the last couple of years has been the development of a Web-based analytical tool called

Crop Explorer (http://www.pecad.fas.usda.gov/cropexplorer/). Crop Explorer provides customers with timely and accurate crop condition information on a global scale unavailable from any other source (Kanarek 2005).

Developed and managed by FAS' Production Estimates and Crop Assessment Division (PECAD), the Crop Explorer Web site features near-real-time global crop condition information based on satellite imagery and weather data (Kanarek 2005). Thematic maps of major crop growing regions depict vegetative vigor, precipitation, temperature, and soil moisture. Time-series charts show growing season data for specific agro-meteorological zones. Regional crop calendars and crop area maps are also available for selected regions of major agricultural significance.

As a part of AOVAS, a CGI script provides near-real-time TMPA maps to Crop Explorer. The maps include 10-day accumulated rainfall and its anomaly (percent normal). Users can also access the maps through a dedicated page (http://www.pecad.fas.usda.gov/cropexplorer/mpa_maps.cfm) in Crop Explorer.

Satellite precipitation products are produced by NASA via a semi-automated process and made accessible from this Web site for USDA and public viewing. Monitoring precipitation for agriculturally important areas around the world will greatly assist USDA FAS to quickly locate regional weather events as well as to improve crop production estimates.

Set-up in 1963, WFP is the UN's frontline agency in the fight against global hunger. The WFP depends on donors worldwide although the United States has been the largest donor. Natural disasters, such as floods and droughts, occur every year in third-world countries, and emergency food aids are often required. Moving large quantities of foods over a long distance is not an easy task. Time and planning are required. Accurate and timely environmental information will facilitate better decision making on the food distribution and maximize the use of contributions.

The data services provided to WFP regional offices in supporting its operations in Indonesia, East Timor, and Africa are near-real-time rainfall maps, monthly rainfall, anomaly and normalized anomaly maps, time series plots, and more. These services are used to address global and regional water-related issues such as floods and droughts. During the 2002–2003 growing season, bulletins (http://disc.sci.gsfc.nasa.gov/agriculture/additional/applications) were compiled with data and information provided by the GES DISC and other agencies to closely monitor El Nino's impacts on agriculture in Southern African countries and provide important information for decision making.

The International Society for Agricultural Meteorology (INSAM, http://www.agrometeorology.org/) consists of nearly 1000 operational members around the world. Recent studies conducted by INSAM have shown that ES information services can be very useful to many developing countries (e.g., in Southern Africa, South America, etc.). However, due to a number of difficulties in data accessing, which often involves a high level of expertise, maximizing the rich NASA's ES research results in agriculture still remains a major challenge to global farmers. The data and services at the GES DISC provide some important environmental information to members around the world though the services are limited and additional products are needed (http://www.agrometeorology.org/topics/online-weather-and-climate-information-for-agrometeorologists/).

9.1.4 Conclusions and Future Directions

The AIS was developed at the GES Agriculture DISC, and it provides a simple and easy way for agricultural users around the world to access NASA's near-real-time and historical remote sensing products over the Internet, making monitoring global agriculture a reality. Unlike many existing similar systems that provide only graphic products, AIS can provide not only customized maps but also customized data for further applications and analysis. Despite limited products and services, the concept of AIS seems to be working well in supporting several domestic and international operational organizations. Since AIS focuses on rainfall products that are usually not enough for decision making, more data products and additional functionalities are needed. The details are discussed below.

Surface observations, such as temperature, wind, soil moisture, and so on, are all important, and adding these products will provide additional information to monitoring activities. Agricultural activities are usually carried out on a small scale, so improving spatial and temporal resolutions of meteorological products is also very important, especially for variables with high spatial variations such as rainfall.

Climatic information is important to agriculture. Currently, there are very limited services (such as anomaly analysis) available in AIS. Additional data and services are needed, especially information related to short-term events, such as ENSO, and long-term events, such as global warming.

Biosphere data products, such as NDVI and its departure from normal, provide valuable information for monitoring crop conditions. Ground imagery can provide not only current conditions but also important information regarding the extent of damages, such as floods, fires, and so on.

Numerical model predictions for both the weather and the climate are also important. Timely and accurate short- and medium-range numerical weather forecast will allow users to plan ahead to minimize potential weather-related damages. Accurate long-range forecasts will greatly help users to further plan ahead though the forecasting skill needs to be improved. Model reanalysis data are important for conducting case studies, NASA Modern Era Retrospective-Analysis for Research and Applications (MERRA) (Bosilovich 2008; Bosilovich et al. 2008) has a special focus on the hydrological cycle; therefore, adding it to AIS will benefit agriculture users.

Additional functionalities include inter-comparison, histogram, trend detection, and so on. Capabilities such as the ability to integrate heterogeneous data as well as to support GIS are needed. Continuing to work with major clients and ordinary users to collect and analyze their feedback and suggestions is a must.

9.2 Transportation and LBS*

9.2.1 Introduction

The GIS-T refers to the principles and applications of geographic information technologies for problem solving in the transportation field (Miller and Shaw 2001). Traditionally, transportation deals with people and the movement of goods. Increasingly, the field has been paying more attention to mobility-related economic, social, and environmental issues. The broadening scope of transportation is reflected in (and motivated by) a series of federal legislation over the past 20 years, for example, the Clean Air Act Amendments (CAAAs 1990), the Americans with Disabilities Act (ADA 1990), the Intermodal Surface Transportation Efficiency Act (ISTEA 1991), the Transportation Equity Act for the twenty-first century (TEA-21 1998), and, most recently, the Safe, Accountable, Flexible, Efficient Transportation Equity Act: A Legacy for Users (SAFETEA-LU 2005). These laws mandate a systematic and comprehensive approach to transportation planning, policy-making, and investment decisions that must address broad societal concerns. The development of GIS-T gains momentum from these laws. The GIS-T offers an effective way to integrate information needed for the analysis and presentation of growingly complex transportation challenges.

9.2.2 Components of GIS-T

Similar in structure to the standard GIS (USGS 2009), GIS-T consists of four basic components, namely, (1) information assembling, (2) storage, (3) analysis, and (4) reporting. For transportation applications, GIS-T often requires particular consideration.

9.2.2.1 Information Assembling for Transportation

There are traditionally four major modes of transportation: ground (road and rail), air, water, and pipeline. The GIS-T applies to all of these technologies and requires data input from a variety of sources. Depending on application purposes, some input data may be highly local, for instance, the site condition of a transit station. Others may come at the regional or global scale, for example, expressway network and airline service routes. The U.S. Bureau of the Census provides, free of charge, the Topologically Integrated Geographically Encoded and Referenced (TIGER) Line files (U.S. Census

---

* This section is contributed by Ming Zhang.

Bureau 2009), which are widely used base files, to create roadway networks in urbanized areas in the United States. Additional information about the network should be collected and coded for transportation analysis and management; it includes roadway width, number of lanes, direction, design capacity, speed, and so on. For urban transportation planning, data on the socioeconomic and demographic characteristics of individuals and households are essential to understand travel demand. The U.S. population and business censuses provide base information in various levels of geographies. These base data, however, are insufficient for transportation planning. State departments of transportation (DOTs) and metropolitan planning organizations (MPOs) conduct activity-travel surveys to gain further knowledge on local travel for the study area. The GIS-T must assemble and integrate these spatial and nonspatial data needed for transportation applications from different sources and times.

9.2.2.2 Storage of Transportation-Specific Datasets

Similar to other GIS systems, information in GIS-T can be stored by spatial features (e.g., point, line, and polygon), by thematic attributes (e.g., trip end locations, neighborhood income, and roadway network), or by temporal dimensions (e.g., year, month, week, and time of day). In practice, transportation professionals have adopted a conventional geographic unit for area data called the "traffic analysis zone" (TAZ). Information needed for travel demand forecasting or other policy analysis purposes is commonly represented at the TAZ level. Decennial census data are now available for general transportation applications by TAZ in Census Transportation Planning Products (CTPP 2009).

In many cases, however, transportation analysis and management require data input in certain formats, which distinguish GIS-T from most other GIS applications. For example, travel demand analysis estimates existing origin-destination (O-D) flows and forecasts future O-D trip distributions. The O-D flow data are conventionally and conveniently represented in matrix form. Most commercial GIS software using relational database structure have limited capabilities in handling matrix data. A GIS-T will need to provide special handling functions for users to work with matrix data.

Network analysis is a common task for managing and monitoring roadway performance and traffic conditions. Generic GIS does have the capacity to represent a network as a set of links interconnected with a set of nodes. Transportation applications, however, often involve linear measurements of a segment of the link with respect to a pre-specified starting point. One typical example is location of a traffic accident from the nearest milepost. This one-dimensional linear referencing approach usually cannot be efficiently handled by the two-dimensional Cartesian coordinate system used in a generic GIS system. The GIS-T applies a dynamic segmentation data model developed for the need of the transportation community.

9.2.2.3 Transportation Analysis

The transportation field has developed numerous analytical methods and models such as the shortest path algorithm to solve traveling salesman problems, routing algorithm for vehicle routing problems, and facility location problems. Most commercial GIS software supports these common methods. There are other important methods that are not supported by generic GIS, such as travel demand models.

For decades, transportation professionals have applied the four-step models for long-range travel demand forecasting in metropolitan areas. The four steps are trip generation, trip distribution, modal split, and traffic assignment. In the first step, trip generation, the number of trips is estimated through cross-tabulation, regression, or other advanced statistical approaches in all the TAZs of a region. Each TAZ produces trips with people in the TAZ traveling to the rest of the region. It also attracts trips with people traveling from the rest of the region to the TAZ. Next, the trips produced and attracted are distributed between each pair of TAZs in the region to produce O-D matrices of trip flows. Trip distribution often applies the gravity model developed based on Newton's Law of Gravitation. The third step involves estimation of the percentages of trips made by various travel means, for instance, by driving alone, shared ride, bus and/or rail transit, and walking or biking. Multinomial logic and nested logic are widely used choice models. Finally, traffic assignment estimates the amount of traffic on highway links and transit routes, indicating the performance levels of the transportation services and hence suggesting the places that demand investments for improvement. Many algorithms have been developed for this last step.

Traditionally, these analytical tasks are carried out in a command-based computing environment. Generic GIS software does not support these highly specific analytical procedures. The GIS-T will have these methods and routines built in and will improve analytical efficiency significantly. Examples of commercial GIS-T software include TransCAD developed by Caliper Corp., PTV by PTV Inc., and Cube by Citilabs Inc.

9.2.2.4 Reporting on Transportation Information and Analysis

The broadening scope of transportation concerns has increased the complexity of understanding and analyzing transportation problems. The visualization capability of GIS-T provides transportation professionals with a more user-friendly interactive map environment for data input and analysis than the traditional computing packages. Further, transportation policy making and investment decisions involve intensive public participation. The GIS-T provides a useful tool to help break the *black-box* of transportation modeling and better inform people who otherwise may not be able to visualize the relationships among study variables and the likely consequences.

The LBS rely on telecommunications, which is popularly considered as the fifth mode of transportation. An LBS is an information service accessible with IP-capable mobile devices through the mobile network. An LBS typically includes five components: the service provider's software application, a mobile network to transmit data, a content provider to supply the end user with geo-specific information, a positioning component, and the end user's mobile device. An LBS can be query based and provides the end user with location information such as "Where is the nearest gas station?" It can be push based, delivering e-coupons or other marketing information to customers who are in a specific geographical area.

9.2.3 Applications of GIS-T and LBS

The GIS-T has a wide range of applications in transportation for public services and private business. The applications are growing in all modes of passenger and goods movement, including highway, aviation, rail, water and maritime, and pipeline transportation. Telecommunication as an integrated element of GIS-T lies at the core of intelligent transportation systems (ITS). Each of these applications requires a customized design for data input and maintenance as well as a calibration of application-specific analytical procedures. Common application examples of GIS-T include transportation network management, asset management, travel demand forecasting, transit planning and operation, traffic monitoring and control, traffic safety analysis, routing and scheduling, site selection, and service area analysis.

The U.S. Federal Highway Administration (FHWA) maintains an online repository listing an extensive array of GIS-T projects carried out by transportation professionals and GIS-T services offered by transportation agencies at the federal, state, and local level (FHWA 2009a). The FHWA's GIS practices range from Web-based mapping to data collection, sharing, and analysis. For example, the Highway Performance Monitoring System (HPMS) Viewer allows users to map, view, and compare HPMS data for areas throughout the nation. The Freight Analysis Framework (FAF) provides a policy and systems analysis tool to analyze freight demands, assess implications for the surface transportation system, and improve freight mobility. Analysts can use the FAF database to understand the geographic relationships between local trade flow and the nation's overall transportation system.

All of the state DOTs have employed GIS-T (FHWA 2009a). In the state of New York, for example, a customized GIS called Over Size or Over Weight Pre-Screening Tool (OS/OW PST) has been developed to help the transportation industry and the general traveling public with choices in both OS or OW and legal vehicle routing. Bridge closure or weight restrictions, highway maintenance or construction, and other permanent or temporary conditions impose travel restrictions on OS or OW vehicles. The OS or OW PST, a Web-based map viewer, helps visualize travel restrictions and identify appropriate routes of OS or OW travel. The Orange County GIS for Transit Planning is a local GIS practice for transit planning by the Orange County Transportation Authority (OCTA) in the state of California. It provides the

OCTA with detailed information on the demographic and land characteristics of all locations throughout its service area.

The GIS-T provides an essential platform for MPOs to perform urban transportation planning and travel demand analyses. Figure 9.5 illustrates an application of GIS to visualize forecasted roadway condition in the Austin, Texas area. First, the four-step models are estimated to obtain traffic volumes on highway links for the forecasting year of 2030. Next, a congestion index is calculated based on the ratio of volume over capacity, along with consideration of other contextual factors (Zhang and Lomax 2007). Maps of the

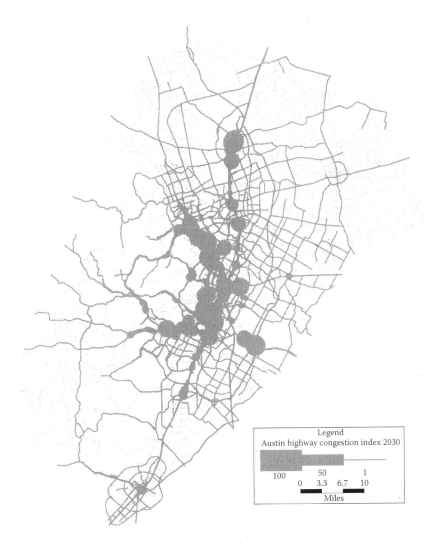

FIGURE 9.5
Visualization of forecasted roadway congestion index (CI) in the Austin, Texas Area.

congestion indices clearly show the congestion *hot spots* at different times of day and under different development scenarios.

The GIS-T is applied widely for the analysis of nonmotorized travel as well. Figure 9.6 shows an example of applying a common spatial analysis technique of GIS buffering to examine pedestrian walkability in the metro station area of San Juan, Puerto Rico (FHWA 2009b). Each circle indicates a buffer in 0.25-mile Euclidian distance from the station. The irregular-shaped gray area shows the walking path-based buffer in the same distance from the station. Pedestrian walkability in each station area is clearly shown by the size of the walking buffer. The area ratio of the path distance buffer over the Euclidian distance buffer gives a quantification of the walking condition, which can be used as input variable for further modeling analyses.

The U.S. Federal Rail Administration (FRA) offers Web-based GIS services that provide the user with information about FRA's rail lines, rail crossings, freight stations, and milepostings (FRA 2009). Private railroad companies have also employed GIS-T to enhance asset management, minimize disruptions and delays, and maximize throughput and safety. The Union Pacific Railroad, the largest and oldest operating railroad company in the United States, has created an ArcIMS-based network surveillance tool. The tool provides immediate notification to all related parties when weather conditions exceed defined tolerance levels. The BNSF Railway is the second largest freight railroad company in North America. It has developed a GIS-T that provides rapid access to information about the rail infrastructure, chemical spill handling procedures, environmental risks, and community demographics. The

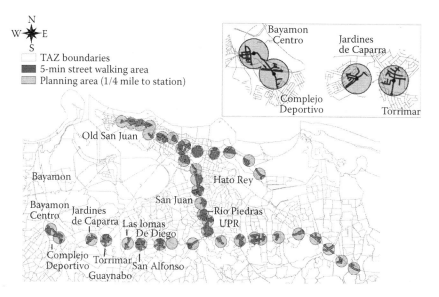

FIGURE 9.6
GIS for pedestrian walkability assessment in the metro station area in San Juan, Puerto Rico.

BNSF GIS allows point-and-click emergency response information about locations across over 32,000 miles of the BNSF rail network.

In the aviation industry, GIS-T finds a wide range of applications, for example, managing airport facilities, airspace usage, and noise modeling. The GIS adds spatial information and 3D visualization to support efficient aviation operations, and, in turn, the aviation industry offers rich data on air traffic (BTS 2009). With GIS, the data can be better utilized for research and practice within and outside the aviation industry. Figure 9.7 shows air passenger flows among major cities in the United States in 2005. It is created from the national transportation databases that provide annual O-D traffic counts of air passengers in the United States since 1990. With more than 800 airports nationwide, the raw O-D tables contain an overwhelming amount of data that is not informative until it is reorganized and represented. TransCAD GIS offers a built-in procedure to map the matrix data by creating desired lines among origins and destinations. The map illustrates patterns of spatial interactions among cities, as indicated by air travel. Apart from better understanding air travel, the mapping exercise also helps inform other inquiries such as the campaign for national spatial development strategy in megaregions (RPA 2006).

Pipeline transportation and the water and maritime industry have utilized GIS-T widely. The National Pipeline Mapping System (NPMS) is a GIS created by the U.S. DOT's Pipeline and Hazardous Materials Safety Administration (PHMSA) in cooperation with other federal and state government agencies and the pipeline industry. The NPMS consists of a single National Repository

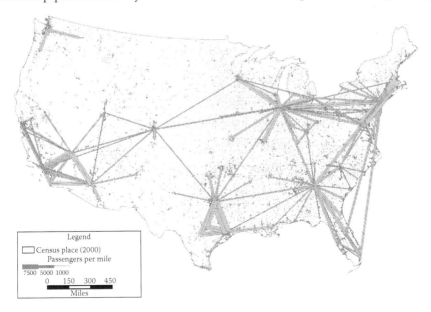

FIGURE 9.7
Air passenger flows among major cities in the United States in 2005.

(NR) that serves as the final processing and storage facility for all pipeline data submitted by pipeline, LNG plant, and breakout tank facility operators in the nation. The NR staff converts the submitted data to append to the nationwide GIS layer and facilitates all of the Web-based and stand-alone applications associated with the NPMS. The PHMSA also uses the NPMS as a tool to support various regulatory programs, pipeline inspections, and authorized external customers.

Today, port operators face an increasing demand for operational efficiency, comprehensive security, and sensitive environmental management. The GIS helps port operators to access, update, and visualize detailed information to meet this demand. The Port of Houston, which is the second largest in shipping tonnage in the United States, uses GIS to map all the Port's property leases. The Web-based lease management system links a Microsoft Access database of relevant leasehold information for port employees to view and perform lease management.

Rapid growth of the Internet and wireless communications in recent years has enabled GIS-T for various LBS services. The GPS, once reserved for military applications only, is now widely available for civilian use as a built-in device in vehicles or as a portable device. With GPS, LBS guides the traveler to the nearest business or service, such as an ATM or a restaurant. It can provide the driver with turn-by-turn navigation to any destination. It can also locate the mobile phone caller in response to emergencies. Additional examples of LBS include personalized weather services and even location-based games. Parcel tracking and vehicle tracking services are common LBS for business. The first global LBS services for commercial purposes were launched in 2001 in Japan by DoCoMo for pre-GPS handsets and by KDDI for the first mobile phones equipped with GPS.

9.2.4 Conclusion

The GIS has been applied in various forms and will continue to expand its applications in transportation. Transportation practitioners and scholars are finding GIS-T essential for transportation planning, management, education, and research. Increasingly, individuals and business owners are seeing GIS-T and LBS as indispensable elements of their daily life or business operations.

9.3 Applications of GIS in Urban Planning*

9.3.1 Introduction

As a computer system designated for analyzing and managing spatial data, GIS has been widely implemented in urban planning research and practice.

---

* This section is contributed by Qisheng Pan.

In addition to the descriptions of GIS concepts and methodologies, Easa and Chan (1999) also listed a wide range of applications of GIS in urban planning research and development, including regional planning, transportation, public utilities, storm water and waste management, cultural and natural resources, emergency response and disaster management, environment assessment, program evaluation and policy analysis, and so on.

The increasing demands of spatial data by planning professionals have extended the applications of GIS in urban planning practice. The ESRI (2006), one of the leading GIS companies, summarized the GIS solutions for professionals in urban and regional planning, including the integration of Web services and GIS for e-government, the enhancement of business workflow with GIS in the enterprise, the facilitation of planning, and public participation with GIS-based planning support system. It also provided numerous case studies to demonstrate the implementation of GIS in professional planning practice.

Some recent publications have summarized various applications of GIS, ranging from modeling to managing urban environments. Berke et al. (2006) discussed the GIS elements and functions to develop planning support systems for urban land use planning. Scally (2006) introduced the application of GIS to build an efficient model for turning brownfields into productive property in a major seaport city in England. Maantay and Ziegler (2006) described a variety of GIS applications in urban environment, such as urban environmental planning, health planning, crime analysis, emergency management, community-based planning, historic preservation, and so on, Maguire et al. (2005) summarized GIS tools and techniques for spatial analysis and modeling and also introduced the applications of GIS to study various socioeconomic and environmental issues in urban systems.

This section will review the major applications of GIS in various issues of urban planning. It will also demonstrate the GIS applications with some case studies in urban planning.

9.3.2 Applications of GIS in Urban Planning

9.3.2.1 Applications of GIS in Analyzing Planning Policies and Developing Planning Support Systems

Powered by its spatial and statistical data analysis functions, GIS has also been widely applied as a decision support platform in urban planning to assist numerous local government plans for housing and community development, economic development, transportation and land use planning, public health, law enforcement, environmental management, citizen participation, and so on.

O'Looney (2000) claims that the values of GIS for public decision making lie in its capability of illustrating the results of complex statistical and spatial analysis in a simple and straightforward way to facilitate the early identification of issues, efficient communication, fast response, and intervention in planning decision processes. More importantly, GIS technology can help local government to confine and resolve policy conflicts identified at their early stages and balance or enhance several key values, including efficiency, equity, community visibility, and environment quality. Greene (2000) also discussed the applications of GIS in public policy and addressed policy issues such as education, public health and safety, public services, environment, social services, and international relationships.

Klosterman (1997, 2000) reviewed the planning support system and explained the evolution of planning processes and GIS-based information systems. In the 1960s, planning was considered an applied science approach, whereas IT focused on system optimization and provided necessary information for rational planning. In the 1970s, planning became a political process. During the same period, IT was applied in planning to enhance political structure and transform policy making processes. In the 1980s, planning emphasized communication, and IT was implemented to facilitate information transmission from planners to others. In the 1990s, planning became a process to combine the reasoning procedures, and IT offered information infrastructure to assist communication, social interaction, and decision making. Klosterman (2000) did not give an overview for the changes in the 2000s. Based on our observation, however, planning has continued to be a multi-disciplinary practice, whereas a planning support system has become a more distributed system than before. The GIS-based IT has provided essential tools to assemble the distributed planning components in a more smooth and efficient way.

Batty et al. (2000) presented newly developed technologies in remote-sensing survey and 3D modeling that can be integrated into GIS spatial database to construct 3D city models to support planning decision making. Kwartler and Bernard (2000) also described an Arcview GIS-based decision support system, communityViz, for community design and planning.

By recognizing the advantage of case-based reasoning (CBR) over the traditional rule-based reasoning, Yeh (2005) showed how to integrate the CBR and GIS in a planning support system for development control, which is an important component of urban planning.

9.3.2.2 Application of GIS in Constructing Planning Data Models

Planning data models are needed to encode the content of urban development projects and store the plans and processes to the database of planning support systems. The GIS data models designated for spatial and statistical analysis provide a solid base for developing more specific data models in urban planning. For example, *The Transport Data Model* developed by ESRI has been applied in transportation research and industry to satisfy the dramatically increased GIS data demands in transportation planning.

Besides the general data model released by leading GIS companies, there are also particular data models designated for urban planning research and practice. Hopkins et al. (2005) highlighted the benefits of planning data models to users in specifying scenarios, working with multiple urban development models for comparison purposes, and bridging the models in different specializations, such as land use, transportation, environment, and socioeconomic analysis. They proposed a planning data model (PDM) to represent plans and regulations in urban simulation models and described a planning geo-database derived from the PDM and designated for urban planning projects. In its application to support the simulations of land use, transportation, and environmental impact analysis by Waddell (2002)'s UrbanSim modeling system, the PDM collects and stores data of households and jobs by a grid and validates the control totals with external macroeconomic models. The PDM also provides data support to other planning models, such as travel demand, urban growth boundary policy, infrastructure investment, and land development models through a model coordinator in the modeling system.

9.3.2.3 Application of GIS in Transportation and Land Use Modeling

Traditionally, the primary concern of transportation planning is to improve the efficiency of transportation networks, that is, to maximize vehicular mobility and reduce traffic congestion. However, in recent years, the major interest of transportation planners has been shifted to provide better service for users within the existing transportation facilities. The GIS technology has been widely implemented to assist transportation planning to confront the challenge and make the transition smoother. A new concept of GIS for Transportation, or so-called GIS-T, has been proposed by researchers and professionals to apply GIS to understanding and describing complex transportation systems and solving the issues in transportation related to construction, operation, management, and so on.

Miller (2009) proposed an idea of Transport 2.0 to incorporate GIS, geovisualization, agent-based model, virtual reality, and Web 2.0 tools and their underlying science to meet the challenges and achieve the levels of sophistication and interoperability beyond what is currently available. He also emphasized the major research directions, including the advanced technologies to process spatial and temporal transportation data collected by new location-aware technologies and geosensor networks, new knowledge and methods for performing better collaborative decisions in transportation, simulation and evaluation of transportation projects for various purposes and at different scales, and the building of spatial infrastructure for real-time transportation systems.

The interaction between land use and transportation is a major issue addressed in urban planning research and practice. Wegener (2004) reviewed twelve land use and transportation models that are selected as most representative from large pools of models. The characteristics of the models, including structure, equilibrium, prices, and agents, were described. Wegener (2005) summarized the interactions between land use and transportation.

He emphasized that high standard spatial data and powerful GIS functions can be applied to enhance the state of urban land use and transportation modeling.

Some scholars pointed out that there is no real full integration between urban land-use transportation models and the GIS system, because specific data organizations, complex analytical methods, and high computational efficiency required by urban land-use transportation models cannot be provided by current GIS. In the context of urban planning, various approaches, such as stand-alone and loose or tight coupling models, have been adopted to integrate urban models with GIS (Goodchild et al. 1992). All the integrated urban models are classified into four types: (1) embedded GIS-like functionalities into urban modeling packages; (2) embedded urban modeling into GIS by software vendors; (3) loose coupling for two software packages; and (4) tight coupling by embedding urban models as GIS macros or programs (Sui 1998). Wegener (2005) stated that a full integration may not be feasible, but the loose coupling mechanism has been implemented to integrate a GIS system with the twelve models he has reviewed. It is obvious that loose coupling is a practical mechanism to integrate land-use transportation models with GIS.

9.3.2.4 Applications of GIS in Environmental Planning

As Goodchild (2005) pointed out, some of the earliest GIS activities were rooted in environmental studies, and environmental applications have continued to drive the development of functions and tools of GIS systems. The first GIS system using digital data on a computer system developed by Alan Tomlinson in 1962 covered the environmental elements, including agriculture, forestry, wildlife, and recreation in Canada, which was called the *Canadian Geographic Information System.*

Today, enormous amounts of environmental data are available in GIS formats. There are also a tremendous number of tools available at different GIS platforms to support a variety of environmental studies, including water management, disaster monitoring, hazard mitigation, air pollution and controls, and so on. There are some barriers that obstruct the applications of GIS in environmental planning, including the lack of suitable data, the expense of GIS technology, and the limited knowledge of professionals in GIS (Maidment and Djokic 2000). However, these barriers have been broken down quickly with the development of GIS data and technologies.

In water management and modeling, many GIS data models and tools have been built up, for instance, the DEMs reviewed by Garbrecht and Martz (2000), the GIS-based hydrologic modeling system (HMS) developed by the Hydrologic Engineering Center (HEC) of the U.S. Army Corps of Engineers, and the Arc Hydro data model developed by Maidment (2002) at the Center for Research in Water Resources in the University of Texas at Austin. Maidment et al. (2005) also developed a hydrologic information system that integrates space and time information with hydrologic models and supports management of water resources using Arc Hydro data model in ArcGIS.

The GIS has also been applied to study the change of land cover using an empirical approach of modeling transition likelihood (Eastman et al. 2005) or agent-based models (Parker 2005), to examine the relationship between human and animal activities (Ahearn and Smith 2005), and to simulate landscape design in GIS (Duh and Brown 2005).

9.3.2.5 Other Applications

The GIS has also been extensively applied in urban planning education. Many urban planning programs in the United States provide GIS classes to both undergraduate and graduate students. Some offer GIS certificate programs or continuous education credits (CEUs) to planning professionals.

New technologies have been a driving force in GIS applications. The integration of GIS with new technologies in LIDER in remote sensing, GPS, wireless communication, and Internet has enhanced its application in urban planning. Peng and Tsou (2003) introduced the distributed GIS techniques and services for Internet and wireless network. They reviewed the applications of Internet GIS in urban planning and resource management, including some case studies in community planning, emergency management, and infrastructure planning and management. They also presented several examples of Internet GIS applications in intelligent transportation systems.

9.3.3 Case Studies

9.3.3.1 Applications of GIS in the Cost–Benefit Analysis of Advanced Storm water Treatment in Los Angeles

According to the Water Quality Act of 1987, the USEPA was required to set the criteria for the National Pollutant Discharge Elimination System (NPDES). State agencies have been authorized by the Federal Clean Water Act to serve as the NPDES permitting authority in lieu of the USEPA. County water quality control committees are required by USEPA to reapply for an NPDES permit every 5 years. It is critical for state and local water quality control units to preserve and enhance water quality and protect the beneficial uses of waters. However, many state bodies of water do not yet meet applicable water quality standards. In California, the California State Water Resources Control Board and nine Regional Water Quality Control Boards are in charge of monitoring and controlling water quality in the state. The Los Angeles Regional Water Quality Control Board (LARWQCB) is responsible for managing water resources and protecting water quality in Los Angeles.

In Los Angeles, the regulations and standards set by the LARWQCB for the NPDES require three levels of treatment of storm water before it is discharged into a public body of water. Level I is a physical treatment that concentrates on settling and removing suspended solids and particulates.

Level II is a disinfectant treatment that focuses on filtering and disinfecting to remove biological contaminants. Level III is an advanced treatment that removes small concentrations of priority toxics and heavy metals.

A previous study estimated that 10-year costs for storm water treatment in the Los Angeles County is in the range of US \$53–65 billion for 480 new storm water treatment plants in 13,950 acres (Brown and Caldwell 1998). Based on the existing studies and additional empirical data, Moore et al. (2004) examined the costs for storm water treatment associated with nine combinations of construction cases and rainfall scenarios. Their study intended to address multiple research questions:

- Are there alternative treatment plans?
- What are the annual capital cost equivalents for various alternatives?
- What are the annual operating and maintenance costs for these alternatives?
- What are the economic impacts of the storm water treatment plants for the region and its subareas?

A key issue in this project is how to apply GIS to identify construction locations for treatment plans.

To define alternative treatment plans, three rainfall scenarios were proposed, including (1) 1.25-in rainfall in 24 h examined by Brown and Caldwell (1998); (2) 0.5-in rainfall in 24 h basing on the assumption of 70% of all wet days; and (3) 2.25-in rainfall in 3 days basing on the assumption of 97% of all wet days.

It also specified three construction cases, including (1) 480 plants, 45.2 MG, and 11.1 acres per plant, defined by Brown and Caldwell (1998); (2) 65 regional plants sited based on geography and one large treatment plant in each of Los Angeles County's sub-basins; and (3) 130 plants sited based on political equity and at least one treatment plant located in each of the cities in Los Angeles County. It includes 123 basin-CDP (Census Designated Places) combinations and seven residual basin areas that are neither incorporated nor designated as a CDP. It is obvious that GIS will play an important role in developing the three construction cases. The identification of the location of the storm water treatment plants by TAZ for the three construction cases using GIS is demonstrated in Figures 9.8 through 9.10. The zones highlighted (in yellow) are the TAZs identified using the GIS spatial analysis tools for the location sites of 480 plants in Figure 9.8; 65 regional plants in Figure 9.9; and 130 plants in Figure 9.10. In the figures, the TAZs are classified by elevation.

After the locations of the treatment plants are identified, the construction costs, collection system costs, operation and management (O&M) costs, and the economic impacts are calculated. The direct impacts of construction, collection system, and O&M costs are located at the plant site. Indirect impacts on industrial sectors triggered by the direct impacts are allocated

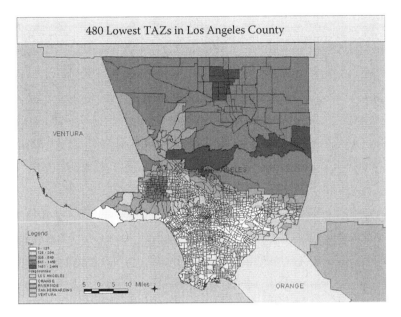

FIGURE 9.8
(See color insert following page 144.) Identifies the locations of 480 lowest TAZs for storm water treatment plants.

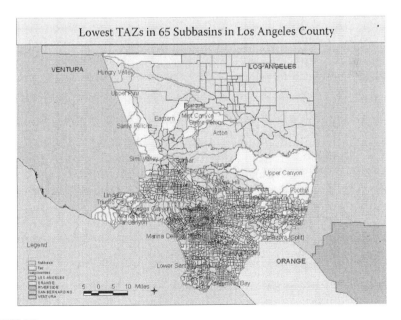

FIGURE 9.9
(See color insert following page 144.) Identifies the locations of lowest TAZs in the 65 subbasins for storm water treatment plants.

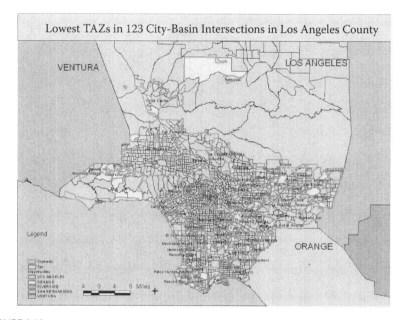

FIGURE 9.10
(See color insert following page 144.) Identifies the locations of lowest TAZs in the 123 city–basin interactions for storm water treatment plants.

in terms of the spatial distribution of employment. Induced impacts on household expenditure are allocated using the work-to-shop O-D matrix. The allocation of the impacts is completed with a GIS-based modeling procedure.

9.3.3.2 Applications of GIS in the Impact Analysis of Highway Expansion Projects in Los Angeles

Transportation planning has always been a major field for the GIS applications. The traditional spatial and statistical analysis functions and the advanced Web-based techniques in GIS have enhanced transportation research and planning practice. Nyerges (2004) discussed why GIS should be used in transportation planning and how to use GIS to support urban and regional transportation analysis. He also provided some cases to demonstrate the application of GIS in transportation studies. In the transportation industry, most of the leading transportation analysis packages, including Caliper's TransCAD, Citilabs' Cube system, and INRO's Emme, are GIS-based transportation modeling software.

The California Department of Transportation (Caltrans) has proposed a project to widen the Interstate 5 freeway in the Los Angeles five-county area to cope with the growth in regional transportation demand. In 1998, the

Caltrans initiated an I-5 corridor Major Investment Study (MIS). The MIS study found traffic problems within the I-5 corridor, assessed some alternatives, and encouraged public involvement (Caltrans 1998).

Funded by the Caltrans, a group of researchers at the University of Southern California conducted research to estimate the economic impacts of the I-5 major improvement project. They developed a regional input-output model, or the so-called Southern California Planning Model (SCPM), to reports results in considerable spatial detail. In the modeling procedures of the SCPM shown in Figure 9.11, GIS has played a significant role in estimating and allocating the economic impacts for the highway expansion project.

The important inputs to the model are the estimates of the number of households and businesses that can be expected to leave their original sites and relocate elsewhere in the region as a result of highway expansion. The Geocoding technique, a common GIS spatial analysis tool, was employed to identify the location of the displaced land parcels or households by their street addresses (Figure 9.12).

According to various empirical findings (Quigley and Weinberg 1977; Clark and Burt 1980; Clark et al. 2003), intra-regional residential move distances are distributed exponentially. The probability distribution function (PDF) is:

$$F_x(x) = 1 - e^{-\lambda x}, \quad x \geq 0 \tag{9.1}$$

where x is the average residential move distance. According to the study by Clark et al. (2003) on the relationship between residential changes and commuting behavior in the greater Seattle area, the mean moving distance is 6.28 miles.

To relocate the displaced households, two GIS programs were developed. One was a statistical module used to generate household moving distances based on the above PDF function. The other was a GIS-based spatial analysis tool used to find a plausible move-in place, given the distance and the background information, for each move-out household. The GIS program draws a circle with the distance calculated for a given move-out household and pinpoints the location on the circle with the neighborhood characteristics fitting to the household background. An additional GIS program was developed to relocate the displaced business establishment from the origin to the destination TAZs according to the business type and their regional distribution.

After the displaced households and businesses were relocated, the direct economic impact of highway expansion including the change of household expenditure, tax revenue, and job loss or gain in small areas was calculated on a GIS platform. The direct impact was the input to the IMPLAN I-O model to report the indirect and induced effects, which do not have spatial details. The spatial distribution of the effects were estimated and reported using the SCPM model shown in Figures 9.12 and 9.13.

378 *Advanced Geoinformation Science*

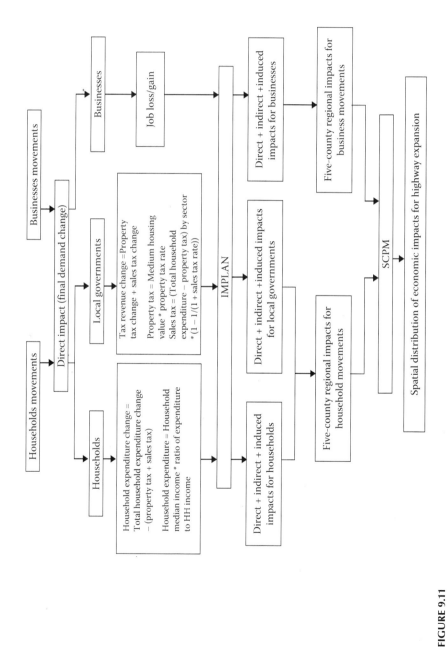

FIGURE 9.11
Flowchart of economic impact analysis for highway expansion.

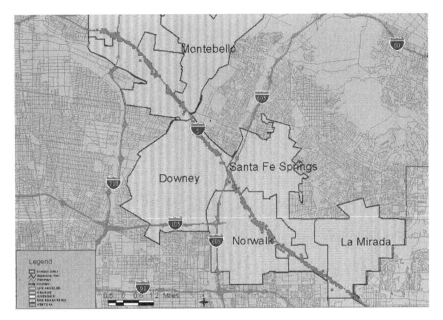

FIGURE 9.12
(**See color insert following page 144.**) The identification of the displaced land parcels by geocoding.

9.3.4 Conclusions

This section focused on GIS in an urban planning context. It reviewed the applications of GIS in various urban planning aspects, including the analysis of planning policies and the development of planning support systems, describing the construction of planning data models, transportation and land use modeling, and environmental planning.

This section also provided a case study on the advanced storm water treatment plans in Los Angeles to demonstrate the applications of GIS in environmental planning. It also addresses the applications of GIS in transportation planning and economic analysis using a case study on the Caltrans' I-5 highway expansion project.

The introduction section addressed some important urban planning issues related to GIS. There is no doubt about the limitation of GIS applications. Currently, most urban models only have a loose coupling with GIS. In many cases, GIS has only been applied as a visualization tool or a spatial data storage platform. Maguire et al. (2005) pointed out that the limitation of the author's knowledge and capabilities would be the major cause. The applications of GIS on urban planning research and professional practice are continuing to mature with the further advancement of GIS technology and the rapid spread of GIS knowledge.

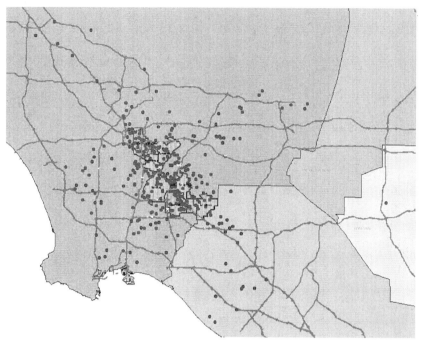

FIGURE 9.13
(**See color insert following page 144.**) The distribution of the relocated (move-in) households.

9.4 Emergency Response*

9.4.1 Overview

Spatial information is one of the most important factors for disaster relief. Information Services may be implemented through the Geomatics emergency response system. The service may be applied in the whole course of pre-disaster warning, post-disaster emergency relief, and disaster recovery. Geomatics emergency response system has played an important role in the 2008 Wenchuan earthquake relief in the Sichuan province of China.

9.4.2 Application of Geoinformation Technologies in Emergency Response

9.4.2.1 Acquisition, Processing, and Extraction of Emergency Response Data

The key time period for rescue efforts after a disaster is limited to just a few days. With the delay of rescue, the effect of rescuing becomes smaller. It is

---

* This section is contributed by Qianjun Miao, Yanli Tang, Jiqiang Tan, and Hongwei Yu.

necessary to acquire the high spatial resolution remote-sensing images, which are continuously updating and analyzing the possible casualty of people, the damage situation of basic establishment, and other detailed information. Therefore, the requirement of the temporal and spatial resolution of remote-sensing images is relative high (Tronin 2006).

9.4.2.1.1 Acquisition Methods of Satellite Remote-Sensing Data

Remote sensing satellites acquire data according to their fixed orbits. They have the following advantages: stable operation, well-rounded processing methods, cost-effective data acquisition, and so on.

Certainly, it could not meet the requirement of disaster monitoring depending on the satellites of a country. Moreover, the spatial resolution of the remote sensing satellites of a country could not achieve the objective of disaster relief easily. Therefore, it is necessary to use international remote sensing satellite resources for disaster relief.

Taking account of the problems of ownership and control rights of the satellites, the country, which is in a disaster, may not acquire the relevant remote-sensing data of the disaster area. Therefore, spatial information resources sharing mechanism was built on an international level—International Charter "Space and Major Disasters." The mechanism was initiated by ESA and Centre National d'Etudes Spatiales (CNES). The International Charter aims at providing a unified system of space data acquisition and delivery to those affected by natural or man-made disasters through AUs. Each member of the agency has committed resources to support the provisions of the Charter and thus, is helping to mitigate the effects of disasters on human life and property (Li et al. 2008).

The only bodies authorized to request the services of the Charter are the authorized users, who have been given a confidential phone number. An authorized user (AU) is a civil protection, rescue, defense, or security body from the country of a Charter member. The sequence of events that occur once the Charter has been activated are as follows: The AU provides the basic information of the disaster area to the On-Duty Operator; then, the On-Duty Operator transmits the information to the Emergency On-Call Officer; the Emergency On-Call Officer verifies the validity of the disaster relief data request, identifies the most timely and appropriate satellite resource, prepares a draft plan, gets the Space Agency user's approval, gathers all relevant information in the dossier, transfers the dossier to the Project Manager, and informs Space Agencies of the status of their space resources through the Project Manager; once the data is acquired, it is processed into images by Value Added (VA) Reseller; and VA further processes and interprets the data acquired over the area affected by the disaster and delivers the images to the End User (Figure 9.14).

The international Charter can provide the resources of RADARSAT, ERS, ENVISAT, SPOT, Formosat, IRS, SAC-C, NOAA satellites, POES, GOES, Landsat, Quickbird, GeoEye-1, ALOS, DMC satellites, ALSAT-1, NigeriaSat,

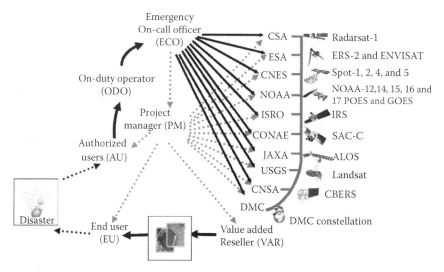

FIGURE 9.14
Activation procedures of Charter.

BILSAT-1, UK-DMC, TopSat, FY, SJ, ZY satellite series, and others to obtain data and information on a disaster occurrence.

9.4.2.1.2 Acquisition Methods of Airborne Remote-Sensing Data

Although the satellites can extend the acquiring area by orbit maneuver and position adjusting, the satellites may not locate on the disaster area on a disaster occurrence, so it is still difficult to meet the practical requirements of some disasters. Therefore, under the permissible condition, the acquisition of airborne remote-sensing data is one of the essential technical methods.

The sensor types of an airborne remote-sensing platform can be divided into an airborne optical image sensor, an airborne LIDAR, and an airborne SAR.

The airborne optical image sensor mainly includes Leica ADS, Z/I Imaging DMC, and SWDC-4.

Siwei Digital Camera-4 (SWDC-4) is a large format aerial camera developed by Beijing Geo-Vision Tech. Co., Ltd. in 2006 to meet the need of topographical mapping. The SWDC-4 is composed of four nonmetric unit cameras or camera heads. It has the following main characteristics: changeable lens, large field of view, large base-to-height ratio, and high height accuracy.

As a kind of active and direct georeferencing technique with high accuracy, airborne LIDAR has gained more and more attention by researchers and users, which is being applied in many areas such as large- scale DEM generation, true orthorectified images generation, 3D city modeling, topographic mapping, forest resource management, risk assessment, and so on.

Airborne SAR provides high-resolution images, which can be used for identifying individual buildings. The airborne SAR is able to provide full-polarization information. The complete polarization characteristics are obtained by the analysis of polarimetoric SAR images. The polarization characteristics are highly suitable for the identification of detailed surface conditions of objects, because they differ according to factors such as building materials and the density of city blocks. Due to the fact that SAR data acquisition is suitable for all types of weather, SAR data can be easily acquired during the process of disaster relief. Single-polarization SAR images can show limited information of ground features. Therefore, the application of multi-polarization SAR images is widely used in emergency response (Liu et al. 2008).

9.4.2.1.3 *Ultra-Light Aviation Low-Altitude Digital Remote-Sensing System*

The digital remote-sensing system, based on an ultra-light aviation platform, uses a $4K \times 5K$ plane array camera to obtain large-scale true color aerial imagery of a disaster region in low altitude, with the advantage of high resolution and high quality. It can make up the disadvantage of traditional aerial photography technique and dependence on airport and weather. It has the characteristics of management convenience, fast processing, and low costs. It is an important means of technology in low altitude and small areas, for emergency services and security of surveying and mapping. This system is extensively used on images of the affected small towns and villages that are acquired quickly, mainly takes on the role of DOM (Digital Ortho Map) production faced with disaster and 1:2000 mapping of reconstruction after the disaster. The production quality is good, resolution is high, image information is rich, and interpretation of disaster is more accurate and detailed. We could view collapsed housing, and even tile damage, road subsidence, and so on. Meanwhile, it has clear performance on mud-rock flow, landslides, landslide lakes, and so on, and meets the requirements of surveying, mapping, and disaster management (Zhong and Ding 2008).

9.4.2.1.4 *Quick Processing of Emergency Response Data*

The most important technical problem in the application of remote-sensing data for disaster reduction is quick processing. It mainly includes satellite image rectification with or without a few ground control points, quick rectification and mosaic of aerial data, and image dodging processing.

The rectified remote-sensing image is positionable and measurable, and it can be widely applied in disaster rescue. The common methods for image orthorectification normally need whole fundamental geographic data and more control points. However, under general conditions, it is very difficult to select ground control points after disaster occurrence.

A new type of remote-sensing satellite sensor imaging model, rational polynomial coefficients (RPC), can acquire accurate and simple conceptual models that are approximately consistent with the rigorous imaging model of a satellite remote-sensing image. Rational polynomial satellite sensor model is a simpler empirical mathematical model relating image space (line and column position) to latitude, longitude, and surface elevation. It can acquire the rectified remote-sensing image model at high positioning accuracy with a few or without ground control points. The accuracy meets the requirements for emergency response (Li et al. 2008).

After a disaster, the quick rectification of an aerial photo is very difficult. Due to the sudden occurrence of a disaster, it is impossible to design an aerial photogrammetry course according to a common specification. The aerial photogrammetry task should be implemented along seriously damaged residential area and transportation lines. The disorderly and unsystematic aerial photogrammetry course and position causes the problems of a big rotation angle and lean angle of image pair. It increases the difficulties of image matching and influences the image overlap degree. The products of Pixel Factory and Digital Photogrammetric Grid (DPGrid) provide the possibility to process massive remote-sensing data in short time and can solve the problem of rectification and mosaic caused by emergency response aerial photogrammetry. In addition, both products are designed based on parallel processing to improve the processing speed.

The acquisition time, external illumination, and other factors of optical remote-sensing images result in the color difference of acquired images. Therefore in the process of mosaic it is difficult to ensure the continuity of images and it brings some difficulties for the interpretation and analysis of ground features. The images need a dodging process. With the assistance of image automatic dodging processing software, the customers can set parameters to implement the dodging process for a single image and several images in the disaster area to effectively decrease work load and improve the production efficiency.

9.4.2.1.5 Changes Detection of Disaster Information

The visual interpretation needs professional interpretation experts. Since a large amount of data needs to be processed in a disaster occurrence, the interpretation only depends on manual work that restricts the speed and amount of disaster information extraction. Therefore, automatic and semi-automatic disaster information changes detection technology is the key to improving the speed and amount of disaster information extraction.

The changes detection technology mainly applies the difference of ground features in shape and spectrum between the information before and after the disaster occurrence to extract the information with serious damage and obvious change. The following two kinds of change detection technology can be used: the change detection technology based on different sources' remote-sensing images and the change detection technology based on a single image.

The car-borne road information collecting and updating system (LD2000) developed by Wuhan University can acquire measurable stereo images with geo-coding of a street and lane one by one in the disaster area. The information can be used in disaster evaluation and reason analysis for house damage (Li et al. 2008).

9.4.2.2 Emergency Response Database Construction

Different kinds of natural disasters are different in the mechanism of occurrence and extension, the intensity of disasters, the disaster causing modes and processes, and so on. For example, a drought normally lasts for a long time and influences a large area; but landslides, mud-rock flow, and earthquakes have the characteristics of sudden occurrence. Therefore, in the process of an emergency response, different kinds of disaster need different geographical information. From the point of view of disaster occurrence and the process of disaster relief, an emergency response database should include the following kinds of geographical information.

9.4.2.2.1 Basic Geographical Data

Fundamental geographical data is composed of administrative divisions (including national, provincial, municipal, and county levels), boundaries (including national, provincial, municipal, county, and town levels), hydrological systems (rivers and lakes), highways (national and provincial levels), railways, residential areas, and so on. Each kind of feature is divided according to a classification code. Each kind of feature is stored respectively in a point, line, and polygon geometry.

In order to meet the requirement of multilevel analysis from macro, middle, and micro views by customers in different stages of disaster management, the basic geographical data includes multiple kinds of basic scale data.

9.4.2.2.2 Image Data

The image data acquired by remote-sensing technology has the advantages of speed, real time, visual, multi resolution, multi temporal, and so on. In an emergency response database, the images of different remote-sensing platforms before and after a disaster should be stored.

9.4.2.2.3 Thematic Data

The professional data of different fields, such as GIS technology and spatial data, should be integrated to form the various thematic data for describing special geographical phenomena. The GIS thematic data includes spatial, attribute, temporal, and other types of information related to an application theme or field. Thematic data is the collection of geographical data for describing special themes. In the process of monitoring and evaluation of disaster and disaster area reconstruction, it involves the thematic data of ecological environment, earthquake, geological, social and human environment, planning, construction, and so on.

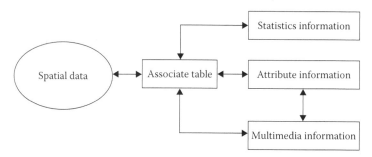

FIGURE 9.15
Data relating method.

9.4.2.2.4 Digital Elevation Model (DEM)

In order to construct the 3D landscape for disaster area, DEM data, combined with the remote sensing image acquire in real time, should be included in the database.

9.4.2.2.5 Other Data

In order to realize the comprehensive situation of the area that suffered from the disaster, except for the spatial information of the location and region, the database should include attribute information, statistical information, photos, pictures, videos, and other multimedia information.

9.4.2.2.6 Organization of Data

To manage the massive spatial database, spatial data is organized by blocks according to regions and by levels according to contents. Under several kinds of basic scales, each scale of spatial data was divided into blocks according to the cover area. Each block data under a different scale is managed by a level according to the contents of features. The number of levels under different scales is different as well. Generally, the larger the scale, the more detailed the data is and the greater the level number is.

The spatial, attributes, statistics, and relevant multimedia information are all integrated using an associate table. Each feature is given only the ID code in the vector data; the association relationship among feature identification codes, attribute information, statistics information, and multimedia information in the database is built (Figure 9.15) (Han et al. 2008).

9.4.2.3 Emergency Response Geoinformation Public Services System

9.4.2.3.1 Integrated Fundamental Geographical Information Emergency Services

The emergency response services mainly include (1) quick integration, basic surveying, and providing mapping products and the most recent remote-sensing data for the disaster area; (2) quick construction of the thematic GI system, which integrates the management of massive data; has the functions of 3D image browsing, comparison, analysis, and provides the spatial data integration demonstration and analysis platform for disaster situation assessment, analysis, and reconstruction planning; and (3) thematic map and atlas compilation based on acquirements to demonstrate disaster-affected areas, disaster-affected degrees, response of disaster relief, planning ideas, and so on, to effectively represent the spatial and temporal distribution of disaster, reconstruction planning arrangement, and so on (Chen et al. 2008a).

After the occurrence of a natural disaster, it is necessary to design, organize, and implement the integrated emergency response services, which include data and result services, thematic system services, and thematic mapping services.

9.4.2.3.2 *Emergency Response Geographical Information System*

The GIS is the tool that acquires, integrates, analyzes, stores, and displays the multi-source massive spatial data. In recent years, with the development of information, aerospace, and digital integration technology, GIS has been widely applied in the fields of forestry, agriculture, prospection, planning, and so on, especially in disaster management. The application of GIS in disaster management mainly includes the following stages:

1. *Disaster prevention:* The various measurements are adopted for disaster-causing factors [landslide, SARS (Severe Acute Respiratory Syndromes), dangerous chemicals, etc.] to avoid the occurrence of a disaster.

2. *Disaster reduction:* Various cautionary measurements are adopted before the disaster can cause serious damage. The main objective is to reduce the disaster influence to the smallest possible extent, for example, stabilizing the river bank to prevent flooding.

3. *Disaster preparedness:* Various measurements are adopted before the occurrence of a disaster to ensure the successful implementation of suitable and effective activation after disaster.

4. *Emergency response and rescue:* Actions are immediately taken after the occurrence of a disaster for search and rescue and to meet the requirements of the victims for food, temporary living place, and health care.

5. *Restoration and reconstruction:* Completely restore the function of the disaster-affected area to the level before the disaster, include assisting the reconstruction of homestead of victims, reconstructing essential public establishments, and restoring the main economic and social activities.

The GIS can provide information and technical support for establishing a disaster emergency response plan, evaluating disaster-affected situations, sharing disaster information, and so on. The information is not only important at the beginning of disaster relief but also indispensable for planning, coordinating, and supervising work in the whole process of disaster rescue. Through applying GIS technology, the scientific decision, information exchange, and collaboration can be realized.

9.4.2.4 Compilation of Emergency Response Thematic Map

9.4.2.4.1 Compilation of Image Map of Disaster Area

Image map is the combination of remote-sensing image and map. It integrates the advantages of image and line symbol map and is the composite representation of multi-source spatial data. As a kind of representation form of map products, image map can accurately and comprehensively represent topographic and terrain information and overcome the limitation of information load and representation methods of a line map. Especially, aiming at the requirements of surveying and mapping data by different organizations in the period disaster emergency response, the application of image map products shorten the mapping time and reduce the mapping cost. It can provide working map and Geoinformation timely.

According to the task requirements for damage assessments after a disaster, disaster investigation and monitoring, and reconstruction plans after a disaster by relevant departments, the image products in different levels and scales can satisfy the basic application (Chen et al. 2008b).

9.4.2.4.2 Disaster Relief Emergency Map

Disaster relief emergency map provides the geographical situation of a disaster area, guidance of the transportation of personnel and material in the disaster area for decision-making departments in the disaster rescue, disaster relief, reconstruction, and other emergency response activities. The emergency response thematic map guarantees that the regional disaster distribution, the water system, road transportation network, topographic situation, and the dynamic change of disaster situation are timely showed on the map in appropriate ways.

The producing time of disaster relief emergency map is very limited. In order to meet the time efficiency requirements of disaster rescue and relief, the disaster relief emergency map must be compiled in a very short time.

The most important characteristics and values for the emergency response map are that it is quick, accurate, and practical. The objective of mapping is to produce the map after the quick processing of symbolization and generalization using the accurate basic geographical information data in the shortest possible time. The design of the symbol should be simple and easy to understand. Meanwhile, the present situation of map and the dynamic development of disaster situation are the key points during the compilation of an emergency response map. It is essential to update changes to the situation of boundaries, residential areas, transportation, and water systems into the basic geographical information database and highlight the disaster information on the map (Wang and Yang 2008).

9.4.2.5 Management and Decision Support of Emergency Respond

9.4.2.5.1 Geomatics Assistant System for Disaster Relief

The Geomatics assistant system for disaster relief applies client or server technical framework and integrates database technology, spatial data engine technology, GIS, and 3D virtual visualization technology. Based on fundamental geographical information, integration of real-time image data and disaster situation data of the disaster area, and, additionally, through the abundant representation methods of GIS, the system realizes the function of browsing, querying, and analyzing statistics under the 2D and 3D environment and provides assistant information services using the method of integrating graphics and text. It demonstrates the disaster and geographical situation of the disaster area for the decision makers and provides support for disaster relief.

The system is constructed on a multi-level system structure. The system has four levels: data level, services level, application level, and representation level. The data level stores the fundamental spatial data, disaster thematic data, and related accessorial data using database and spatial database management software. The service level publishes the data using the methods of map services. The application level encapsulates the functional features of related tasks to realize the various functions. The representation level mainly includes the display of the client server, that is, the users acquire the required information sources using the system function (Chen et al. 2008a).

9.4.2.5.2 Geoinformation Integrated Platform for Emergency Response

An emergency response platform is composed of data resources inputting level, data integrating and managing level, data publishing and services level, and application level. Based on the requirements of the diversity of the emergency response business and emergency response information amounts, emergency response information services involve the emergency response data of different departments. Due to the different data types and formats of different departments, the emergency response platform of different organizations unifies data formats to construct the emergency response geographical information database through the data exchange platform. With the support of the sharing database, through the integration and publishing of emergency response geographical information, it realizes the integrated visualization of the geographic information of the meteorological, hydrological, and seismic departments to provide powerful assistance tools for emergency response services' decision making (Pu and Wang 2008).

9.4.2.6 Emergency Response and Guarantee

9.4.2.6.1 Emergency Response and Guarantee of Surveying and Mapping

The objective of emergency response, surveying, and mapping is to provide support for the following aspects: understanding the disaster situation, decision making and commanding, disaster rescue and disaster relief, restoration and reconstruction after disaster, and disaster relief propagation.

The contents of guarantee services mainly include

1. *Basic surveying and mapping products emergency response services:* provides basic surveying and mapping products for related national departments, the government of disaster area timely.

2. *Emergency response spatial data acquisition services:* urgently assembles unmanned aerial vehicles, helicopters, and other transportation vehicles, equipped with digital aerial camera that implements aerial photogrammetry for the disaster area, coordinates the relevant remote-sensing satellite, both domestics and abroad, and continuously acquires images of disaster areas to provide to related organizations.

3. *The construction of geographical information services system:* constructs the thematic systems, such as disaster rescue integrated services geographical information system, disaster integrated evaluation geographical information system, and reconstruction planning information system. It provides decision making and surveying and mapping technical support for disaster relief work and restoration and reconstruction work after disaster.

4. *Disaster thematic information extraction from remote-sensing information:* implements the remote-sensing interpretation and statistics analysis of the landslide, landslip, mud-rock flow, and barrier lake for the disaster area; constructs disaster situation monitoring and evaluation database; and assists the disaster situation monitoring, secondary disasters prevention, and related work for the land resources, hydrology, and earthquake departments.

It monitors and analyzes the topographical situation of the disaster area and implements the distortion monitoring and security analysis for the important basic establishments and main cultural treasures.

It produces the public version disaster map and provides free downloads and online services for all circles of society through the Internet.

9.4.2.6.2 Evaluation of Restoration and Reconstruction

The evaluation of restoration and reconstruction after a disaster is one of the important contents of disaster evaluation. At present, evaluation of restoration and reconstruction mainly focuses on the evaluation of reconstruction of housing damages. Evaluation of restoration and reconstruction requires relative high-resolution remote-sensing image data, such as QUICKBIRD, SPOT, and so on.

9.4.2.6.3 *Post-Disaster Reconstruction*

Surveying and mapping provide a powerful guarantee for the scientific planning of reconstruction after a disaster. Surveying and mapping scientific planning is the prerequisite of restoration and reconstruction after a disaster; surveying and mapping are important for scientific planning. The following work needs scientific reference and technical support of surveying and mapping: the fundamental work of disaster evaluating; analyzing geological and geographical conditions and resource environment load support capability; integratively analyzing the economic, social, cultural, and natural factors; scientifically dividing suitable and unsuitable areas for reconstruction; and adjusting and optimizing urban and rural layout, population distribution, industrial structure, distribution of productive forces, and so on.

According to the requirements such as being "human oriented," "respecting nature," "making overall plans by taking all factors into consideration," and "scientific reconstruction," the surveying and mapping departments will actively work out the relevant planning for national government and local government of a disaster area. Therefore, we can provide topographic and thematic maps, remote-sensing images, and other basic geographical information data to analyze the assessment report of topographical change and geological situation for a disaster area.

The requirements of basic surveying and mapping guarantee services include acquiring the airborne and space-borne remote-sensing image after disaster in all of the disaster-affected areas; constructing the emergency response surveying and mapping datum system of the disaster area; compiling the series image map of the disaster area; implementing the surveying of large-scale topographic map of the urban area; establishing the disaster situation monitoring and evaluating GIS; providing timely, reliable, and applicable surveying and mapping guarantee for restoration and reconstruction; and constructing and updating the fundamental geographical information database for disaster area timely.

In order to improve the capability of surveying and mapping emergency response guarantee services, it is necessary to further consummate the emergency response plan of surveying and mapping guarantee; quicken the construction of surveying and mapping emergency response guarantee team; strengthen the integration of power and resources of surveying and mapping emergency response; make overall plan on the acquisition and publishing of basic aerial photogrammetry and high-resolution satellite image for surveying and mapping; construct multiple navigation satellite integrated application and services system, high-resolution stereo mapping satellite application system and advanced aerial remote-sensing platform; promote the emergency response guarantee capability of remote-sensing image; assist in constructing the national integrated disaster reduction and risk management information sharing platform, national integrated disaster reduction and risk management GIS; and fully play the important, fundamental, and advanced role of basic surveying and mapping in disaster reduction and risk management.

9.4.3 Application of Geoinformation Technologies in the Emergency Response of Wenchuan Earthquake

The 2008 Wenchuan earthquake of a magnitude 8.0 MS occurred at 14:28 (Beijing time) on May 12, 2008 in the Sichuan province of China. The seriously damaged region involves 51 counties, 1271 towns, and 14,565 administrative villages with a total population of 19.867 million and a total area of 132,596 km². After the earthquake, Geoinformation technologies have been fully used for China disaster-relief-related organizations and divisions to provide scientific and effective decision making on the disaster relief and post-disaster reconstruction. It has been playing an important role in the disaster emergency response (Li et al. 2008).

A spatial information–based comprehensive emergency services system was designed and completed by China State Surveying and Mapping Bureau, its GIS-related sections, and universities. The basic spatial information data services, specific system services, and thematic mapping services are taken as the core components of the system.

Based on the terrain feature of the disaster area and the requirement of disaster relief, the available fundamental surveying and mapping data and update imagery have been quickly integrated and provided. Meanwhile, geomatics information system has been rapidly built with the functionalities of massive data management and integration as well as 3D image viewing, compared analysis, to provide spatial data integration and an analysis platform for disaster estimation and post-disaster reconstruction planning. Several types of thematic maps and atlases including disaster hit area, disaster intensity, relief response, and reconstruction planning, have been made for responding to the earthquake.

9.4.3.1 Provision of Fundamental Geographic Information Data

Immediately after the Wenchuan earthquake, the State Bureau of Surveying and Mapping of China started the Emergency guarantee plan and provided rapid surveying and mapping service for rescue and relief. From 12 May to June, the State Bureau of surveying and Mapping totally provided for the General Office of the State Council, the emergency response office of the State Council, National Committee for Disaster Reduction, The State Flood Control and Drought Relief Headquarters, China Seismological Bureau, Ministry of Environmental Protection, Ministry of Housing and Urban-Rural Development, Ministry of Land and Resources, and Ministry of Transport and Peking University with 53 thousand map sheets of disaster area, including 31 thousand map sheets made just after the earthquake and 12,000 GB digital imagery data.

Meanwhile, 1200 control points were provided for the rescue and relief paradrop. The map sheets provided include the topographic map at the scale of 1:50,000, 1:100,000, 1:200,000, 1:250,000, and 1:1,000,000; Digital Linear Graph; DEM; DOM; and navigation geographic infrastructure data (Li et al. 2008).

9.4.3.2 Real-Time Imagery Data Acquisition, Processing, Information Abstraction, and Thematic Map Making

9.4.3.2.1 Acquisition of High Temporal and Spatial Resolution Remote-Sensing Data

The post-disaster images covering 770,000 km² are acquired and used during the disaster rescue and relief. China and Brazil Earth Resource Satellite (CBERS) and Beijing-1 satellite owned by China are fully and continuously observing the disaster area. Since the resolutions of Chinese satellites are relatively lower, they cannot completely meet the high requirement of disaster monitoring on surface details. This is a limitation of only applying Chinese-owned satellites to monitor the disaster area. Therefore, it is necessary to request international remote-sensing satellite resources for the disaster rescues and relief.

The space agencies of more than ten countries, including ESA, French Space Agency, Canadian Space Agency (CSA), NOAA, and China Space Agency, have joined the International "Space and Major Disasters" Charter, and China National Space Administration (CNSA) joined the Charter on 24 May 2007. On the Wenchuan earthquake disaster occurrence, National Committee for Disaster Reduction of China as the AU and Project Manager started an emergency call to the operators of Charter and requested the data and information, and on 13 May, the first SAR imagery was received by the National Committee for Disaster Reduction (CDRC) of China. Through the Charter, CDRC totally obtained more than 130 scene imageries of Topsat, Radarsat-1, SPOT2/4/5, ALOS, UK-DMC, NigeriaSat-1, Landsat-7, ENVISAT, TerraSAR, and EROS-B. It has made a great data support to the disaster rescue and relief.

At the moment that the earthquake occurred, the satellite might have just passed the disaster area. It may be able to expand the area of acquired data by means of changing orbits and poses of the satellites, but some requirements for the disaster relief still could not be met. Under the possible condition, airborne remote-sensing data acquisition is necessary. During the period of the disaster's occurrence, the State Bureau of surveying and Mapping of China, Chinese Academy of Sciences, Ministry of Land and Resources of China, Chinese Military Department of Surveying and Mapping, and Wuhan University delivered important spatial IT support by implementing aerial photography projects and acquired aerial photography data covering an area of 110,000 km².

Airborne sensors, including airborne optical, LIDAR, and SAR sensors, are the types of sensors that were used for the disaster monitoring. During

Wenchuan disaster relief, the airborne optical sensors used mainly include Leica ADS40, Z/I Imaging DMC, and SWDC-4.

The SAR has played a great role in the emergency response to the Wenchuan earthquake, because it has day and night operational capabilities. Most of remote-sensing data used for the disaster rescue is the airborne SAR data acquired by carrying out a number of airborne SAR flights in the phases of first rescue and handling secondary disasters. From 14 to 20 May, the average flight time is about 11 h per day. The total flight area covered 65,000 km^2, including the heavily hit area of 25,000 km^2. Since single-polarization SAR image has its limitation on the display of Earth surface features information, multi-polarization SAR images have been widely used. During the disaster rescue period, due to the high risk of Tangjiashan Barrier Lake, Leica ALS50 airborne Lidar data were used for the DEM extraction. Being a new type of Lidar equipment, its vertical and horizontal accuracy may reach up to 20–30 cm, with the precise ephemeris parameter available by using single point positioning algorithm without ground control points (no ground base station). This has provided precise data for the decision making of solving the problem of Tangjiashan Barrier Lake.

9.4.3.2.2 *Remote-Sensing Data Processing*

During the period of Wenchuan disaster relief, the integrated processing software developed by Wuhan University played a great role. The software is based on the RPC model, aiming at both optical sensors, such as SPOT-5, Worldview, P5, Quickbird, and IKONOS, and SAR sensors, such as TerraSAR-X and COSMO SkyMed, and it has produced high precision data with less control points to meet the requirement of disaster relief.

The rectification of images also is quite a difficult process. Due to the outbreak of the earthquake, during the disaster relief it is impossible to design flight routes of the aerial photogrammetry as usual. The flight can only be taken over the heavily damaged cities and towns as well as along the transportation routes.

The hovering flight results in a large relative rotation angle and big obliquity between adjacent images; therefore, it causes more difficulties for the image matching and as a consequence, overlap between adjacent images was heavily affected. Wuhan University has developed a rapid processing system called DPGrid for parallel processing of aerial images. The DPGrid with its powerful functionalities to match stereo pairs with a large relative rotation angle and under the hovering and broken flight condition has provided effective solutions for the image rectification and matching under the rapid response flight condition. As a result, parallel processing is more time efficient when compared with the traditional image by image processing strategy. Image maps at a resolution of 0.3 m were produced on 19 May, 4 days after the images were collected for the Yingxiu—Wenchuan—Maoxian flight strip. The processing of a total of 4,507 DMC images and rapid production of an orthophoto map were completed within 111 CPU hours. The DPGrid system can automatically process about 40 images per hour, which is much higher than the rate of a few images per hour in the conventional digital photogrammetry workstation (Li et al. 2008).

9.4.3.3 Geographic Information System Services for Emergency Response and Guarantee

As a basic platform for geographic information integrating and sharing, GIS has provided a great support for disaster monitoring, emergency respond, and post-disaster reconstruction via spatial visualization.

Three sets of thematic GIS including 3D GIS, disaster comprehensive evaluation system, and reconstruction planning information integrated system have been built by National Geomatics Center of China at the three phases of disaster rescue, disaster evaluation, and reconstruction planning, respectively. The 3D GIS for the disaster area was developed at the beginning of the quake occurrence to provide basic geographic information to experts and leaders to inspect the disaster. It integrated multi-scale basic geographic data (1:1,000,000; 1:250,000 and 1:50,000); multi-temporal and multi-resolution remote-sensing images before and after the earthquake; and a large amount of collected thematic data including earthquake intensity, earthquake fault belt, barrier lakes' distribution, and so on. It provides functionalities with real-time linkage of raster and vector data of disaster area, 3D image viewing, and image contrast analyses before and after the disaster as well as thematic information marking and drawing, quick search of geographic names, hot information marking, and so on (Han et al. 2008).

This system has been offered free of charge to a total of 41 divisions including the Wenchuan Seismological expert committee of the State Council, Ministry of Civil Affairs of China, Disaster rescue expert team of Ministry of Science and Technology of China, National Committee for Disaster Reduction of China, Ministry of Water resources of China, and the Ministry of Environmental Protection of China. It has provided an advanced tool for the 3D visualization and analysis of disaster area, and it has played a great role in disaster situation analysis and targeted assistance.

Disaster comprehensive evaluation system integrated a number of types of thematic geographic data including geographic or geologic environment data, disaster situation, damage loss evaluation, and disaster rapid response, thus providing tools for information marking, searching and viewing, and 3D display. It has been of great assistance to disaster evaluation experts to get to know the distribution of disaster, to make a contrast with multi-factors overlay, and to extend the quantity research on disaster loss.

In accordance with the requirement of the national reconstruction plan, the Reconstruction Planning Information Integrated System provided the functionality of a 3D based general layout plan markings and representations, and it has become an organic integration of disaster evaluation results, planning results, and basic geographic information. The system integrated reconstruction plan results covering an area of 51 counties extremely heavily hit by the earthquake based on latest remote-sensing image maps with scales of 1:25,000 and 1:5000.

Spatial IT has played an important role in geographic information comprehensive services for Wenchuan disaster rescue and relief, the design and making of disaster image maps, the development of GIS for the disaster relief, the acquisition of images, rapid rectification and processing of remote-sensing images, and the map editing of disaster relief.

It has indicated that China has made great progress in the airborne digital remote-sensing system, acquisition of space remote-sensing images, the rapid processing of images and geographic information emergency services, and decision making.

However, this has also unveiled some shortages in the capacities of real-time acquisition, rapid processing, and emergency integration of geographic information. It has put forward challenges and requirements for the future development and construction of emergency guarantee system of surveying and mapping and scientific innovation. It will further improve application of spatial IT in emergency guarantee for disasters.

9.5 GeoBrain for Data-Intensive Earth Science (ES) Education*

9.5.1 Introduction

Recent dramatic advances in remote-sensing capabilities and related data and IT have allowed increasingly detailed and comprehensive observations and a better understanding of the Earth and its atmosphere, the Earth system. *Earth* science—a term used in this chapter in a broad sense—means studies directed toward understanding, documenting, and explaining the ES. The ES has been undergoing significant changes. One such change is that ES has been transformed into a highly interdisciplinary and data-intensive field. Modern ES studies largely rely on computer-based data processing and model simulation capabilities.

In adapting to the changes and responding to the various challenges and opportunities of a data-intensive and, thus, computation-intensive world, modern ES research needs a large scientific workforce of individuals well trained not only in multiple related disciplines but also in mastering computational approaches, software tools, and essential computing and IT skills. In order to produce such a workforce, ES education needs to keep pace with advancing technologies, integrate state-of-the-art research, and explore innovative educational approaches. The importance of innovating educational

* This section is contributed by Meixia Deng and Liping Di, and research reported is supported by a NASA Grant (NNG04GE61A, PI: Dr. Liping Di).

approaches with different aspects and emphases has been widely recognized, in studies including the following:

1. Involving students in an inquiry process (National Research Council 1996; Barstow and Geary 2001)
2. Using extensive remote-sensing data in classroom teaching (Manduca and Mogk 2002; Ledley et al. 2008)
3. Cultivating technical competence and intellectual self-confidence in research (Marlino et al. 2004)
4. Addressing computational thinking and quantitative skills (Marlino, et al. 2004; Manduca et al. 2008)
5. Engaging student interest and creating understanding through exciting, real-world applications (Marlino et al. 2004)

Despite the different perspectives, their essence is to engage students with intensive, real-world scientific data. In comparison to traditional ES education, these modern ES educational concepts and approaches can be generally classified as *data-intensive ES education.*

9.5.2 Issues in Data-Intensive ES Education

Obviously, data-intensive ES education relies on data accessibility and usability and computing capability. Both the technological and educational communities have worked actively in bringing real scientific data and analysis tools into teaching and learning (DLESE 2001; Domenico et al., 2002; Wright and Summer 2003). However, due to the diversity and complexity of ES data (e.g., multiple sources, heterogeneous data structure and types, and different scales and projections) and the shortcomings of the current data and computing infrastructure, data accessibility and usability remain major problems. Educators still face obstacles when trying to access and use data and analysis tools in ES education (Marlino et al. 2004; Hanson and Carlson 2005; Ledley et al. 2008). In general, the major obstacles that hamper ES teaching and learning activities can be classified into two categories: (1) difficulty in finding, accessing, integrating, and using large quantities of ES data and (2) inadequate data processing and computing capability.

Due to the difficulty in finding, obtaining, and using ES data, the remote-sensing data used in the ES classroom are still largely sample data sets prepared by the educator of the class. Educators need to spend significant time in obtaining and preprocessing (e.g., subsetting, georectifying, reprojecting, and reformatting) the data to the form acceptable by the in-house analysis systems to form the "sample datasets." Students are rarely exposed to the richness of remote-sensing data that has been collected, and they will not be able to learn how to use the vast amount of remote sensing data available for real-world applications. Further, due to the computing capability required for dealing with the large quantities of ES data, universities may not have software systems and resources large enough for students for dynamic analysis of ES data. Students will not be able to cultivate the technical competence and intellectual self-confidence in research they need.

9.5.3 GeoBrain Solution

In order to be successful, data-intensive ES education needs considerable support from the broadest science, education, and technology communities. Many efforts have been and are being made to use innovative information technologies to remove obstacles from data-intensive ES education and to advance ES research and education. The GeoBrain project (http://geobrain. laits.gmu.edu) is one such effort. Funded by the insightful NASA ES REASoN program, the GeoBrain project aims at building a comprehensive data, information, and knowledge system named GeoBrain to facilitate modern ES higher education (Di 2004; Deng and Di 2006).

In order to meet the essential needs of current and future data-intensive ES education, a unique strategy, "by the community and for the community," which attempts to encourage the broadest participation, collaboration, and sharing, has been adopted in the development of GeoBrain. This strategy ensures that the GeoBrain system is built with flexibility, effectiveness, and responsiveness to the demands of data-intensive ES education.

For efficiency, the GeoBrain infrastructure and functionalities are built with a focus on removing the two major obstacles to data-intensive ES education by taking advantages of recent advances in CI (NSF CIC 2007). By adopting existing research results, such as NASA HDF-EOS Web GIS Software Suite (NWGISS) (Di et al. 2002) and developing state-of-the-art Web services, geospatial interoperability, and knowledge management technologies, GeoBrain comprehensively addresses the problems for data-intensive ES education . Deng and Di (2009) have described the problems and the approach to solving them.

9.5.3.1 Architecture

The GeoBrain system is an integrated, standards-compliant geospatial Web service system that provides users open access to spatial data, information, knowledge, and processing services. The system is a comprehensive and data-enhanced computing CI designed to make interoperability, scalability, reusability, and adaptability possible to meet a wide range of data-intensive ES education and research needs. Figure 9.16 shows the top-level system architecture.

The figure shows multiple components in each of the three tiers of the system: the Web portal and client tier, the Middleware service tier, and the interoperable Data provider tier. To users, only the components in the Web portal and client tier (front-end tier), which provide users dedicated, specialized

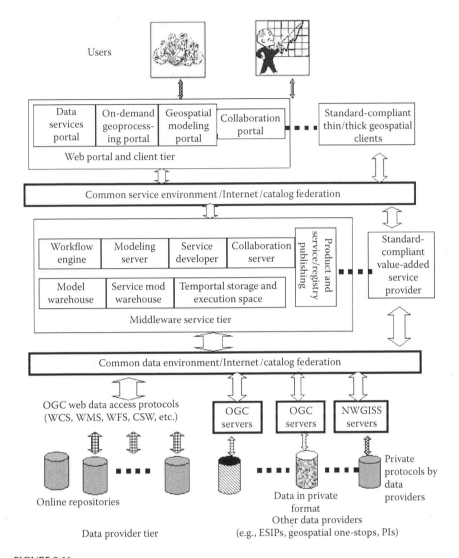

FIGURE 9.16
The GeoBrain system.

data and computing services, really matter. The important components in the front-end tier are the data portal, the geoprocessing portal, the spatial modeling portal, and geospatial clients.

9.5.3.2 Implementation

The primary mechanism for developing an open and interoperable GeoBrain system is to use open, consensus-based technology standards. Specifications and standards from the OGC, W3C, the Organization for the Advancement of Structured Information Standards (OASIS), and the International Organization for Standardization (ISO) have been followed due to the authority of the organizations and the popularity and applicability of the standards they promulgate. Many standards-compliant technologies have been developed in GeoBrain, including, but not limited to:

1. The OGC standards-based WCS (Evans et al. 2003), WMS (OGC WMS), WFS (OGC WFS), and CSW (Nebert 2003) servers and clients.

2. The OGC geoprocessing services, for example, the Web Coordinate Transformation Service (WCTS), the Web Image Classification Service (WICS), the Feature Cutting Service, and the Reformatting Service.

3. Many geospatial Web services, either converted from GRASS (Mitasova and Neteler 2002) or built on it with the Geospatial Data Abstraction Library (GDAL), and functioning as either OGC Web Processing Service (WPS) (OGC WPS) based services or SOAP-based services.

4. The Business Process Execution Language (BPEL) (OASIS BPEL) Workflow Engine extended for Geospatial Web Services, called BPELPOWER.

5. Geospatial Web service chaining and geoprocessing modeling tools.

GeoBrain enables better interoperability from data, function (service), and system levels by adopting standards and establishing common data and common service environments. All standards-compliant clients can access GeoBrain functionalities, and standards-compliant services or systems can be dynamically integrated into the GeoBrain system. The GeoBrain system and services can also be plugged into other larger infrastructures. These plug-in-and-play facts of GeoBrain are due to the advantage of using standards and enabling interoperability.

GeoBrain provides easy, convenient access from any Internet-connected computer to distributed data, geospatial Web services, and computing resources through developing specialized, dedicated, and extensible Web portal systems. There are four major types of Web portals in GeoBrain: Data services portal, On-demand geoprocessing portal, geospatial modeling Portal, and Collaboration Portal (the Collaboration Portal is not to be addressed, because it is beyond the scope of this section).

Two operational data services portals have been implemented and maintained in GeoBrain to enhance dynamic, on-demand data discovery, access, integration, and customized downloading services:

1. The "GeoDataDownload" data portal (http://geobrain.laits.gmu. edu/GeoDataDownload/) allows users to find the data they want

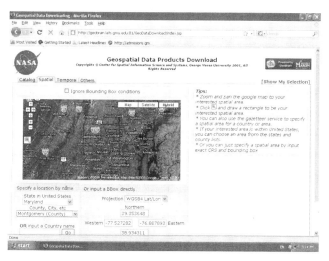

FIGURE 9.17
GeoDataDownload portal.

and to get the data from distributed data archives in a user-requested form (coverage, resolution, format, projection, etc.) regardless of how they are archived. Figure 9.17 shows the Web interface of the portal.

2. The "DEM Explorer" data portal (http://ws.csiss.gmu.edu/DEMExplorer/), shown in Figure 9.18, provides more specific data exploration and customized downloading services for DEM data.

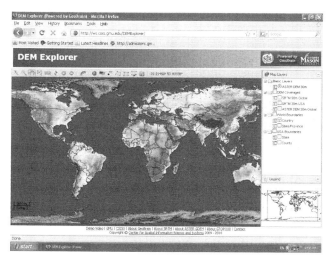

FIGURE 9.18
DEM explore.

FIGURE 9.19
GeOnAS homepage.

The GeoBrain online analysis portal (GeOnAS, http://geobrain.laits.gmu.edu/OnAS/) is an operational geoprocessing portal, providing users dynamic, on-demand data access, processing, integration, and analysis services. This portal provides advanced spatial analysis for any data available through the GeoBrain system, by dynamically invoking a geospatial Web service or chaining several Web services to perform a spatial computation task defined by a user. Figure 9.19 is the portal homepage, and Figure 9.20 shows the interface using GeOnAS to start a new project.

The Geoprocessing modeling portal (http://laits.gmu.edu/vdp/), also named the Abstract Model Designer (AMD), is a prototype portal that enables design, registration, execution, and sharing of geoprocessing models. Figure 9.21 shows the design of a geoprocessing model by constructing a service chain through AMD.

All these important portals in GeoBrain are implemented with and powered by

- Interoperable, personalized, and on-demand data access services (IPODAS)
- Ontology-augmented and federated catalog services
- A large number of interoperable, chainable, and value-added geospatial Web services
- Workflow-based spatial modeling and infusion technologies
- Machine-to-machine access to data cataloged by multiple archives, for example, NASA ECHO

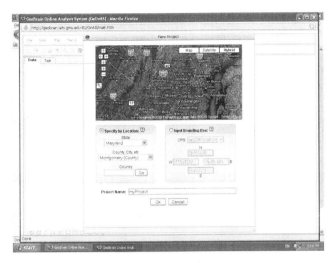

FIGURE 9.20
Using GeOnAS.

Besides the Web portals, a stand-alone client, the Multiple-Protocol Geospatial Client (MPGC) (http://geobrain.laits.gmu.edu/mpgc/), has been implemented for users to download and install into their local machine for access to all GeoBrain functionalities. The MPGC (Figure 9.22) has some improvements in efficiency and offline operation.

9.5.3.3 Capabilities

GeoBrain has been implemented as an interoperable, standard-compliant, and Web-service-based data, information, and knowledge building system

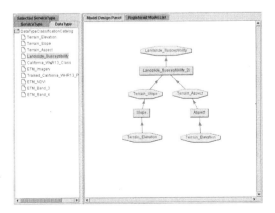

FIGURE 9.21
Designing a Geoprocessing model.

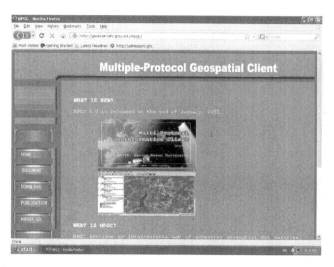

FIGURE 9.22
The MPGC client.

with dedicated web portal services. It has, among others, the following capabilities:

1. Making massive, distributed data, services, and computing resources available easily online from one single point entry
2. Customizing data at a user's request
3. Enabling automated or semiautomated multi-source spatial data integration
4. Providing online on-demand data mining, visualization, and analysis
5. Promoting spatial modeling, knowledge building, and sharing

GeoBrain online data resources are unlimited. As introduced earlier, machine-to-machine interfaces are built between GeoBrain and many large data archives. Thus, GeoBrain provides seamless and transparent access to distributed data, cataloged by NASA ECHO (about 4 PB EOS data) or in the NOAA's CLASS and the USGS Landsat. In fact, a great advantage of GeoBrain, as an interoperable, extensible, and standard-compliant system, is that any distributed data sources that are compliant with OGC WCS, WFS, WMS, or Sensor Observation Service (SOS, for access to live sensors) protocols can be easily accessed through GeoBrain. It is this advantage that enables GeoBrain unlimited data resources. In the meantime, an online data repository with about 20 TB typical ES remote-sensing data is embedded in the local GeoBrain system in order to provide users faster and easier data access than access to data in remote archives. Data in the GeoBrain online repository are selected based on their importance and popularity to ES studies or users' feedback.

The GeoBrain online service resources mainly include OGC standards-based data services (such as WCS, WMS, and WFS), OGC WPS services (e.g., WCTS and WICS), and more than 200 value-added geospatial Web services for data analysis and computation based on open source software programs or functions in legacy systems (e.g., GRASS). Any standard-based Web service can be invoked through GeoBrain to perform spatial computing or analysis. All services developed in GeoBrain are dynamically chainable and executable through GeoBrain BPELPower Workflow Engine to fulfill more complicated spatial computing tasks.

Through the workflow-based approach, GeoBrain recognizes different users' requirements for data and geoprocessing and enables users to dynamically examine, analyze, and explore any data available through the GeoBrain system.

GeoBrain has high-performance computing resources available for online users, mainly an Apple G5 Cluster Server, including a head node, multiple cluster nodes, multiple Xserver RAIDs, 2 Gbps Fiber Switch, 8 Port Gigabit Ethernet Switch, and lots of supporting software tools.

With the dedicated Web portals and MPGC client services, GeoBrain meets a wide range of computing needs of data-intensive ES research and education. GeoBrain requires few user-side computing resources, so that complicated geoprocessing and computing tasks can be achieved with a Web browser from any Internet-connected computer. Users can easily access and use the GeoBrain online data, geoprocessing services, and modeling software tools as if those data, analysis functions, and modeling tools were in users' local machines.

GeoBrain provides an integrated framework and dynamic online mechanism for easy access, integration, and analysis of distributed spatial data managed by worldwide spatial data providers. Users can easily obtain customized data and advanced information products as well as original data through GeoBrain. Multi-source data products obtained through GeoBrain can be easily integrated due to the automated data co-registration services powered by IPODAS and value-added geopspatial Web services. Using GeoBrain shortens the typical time required for ES data acquisition, processing, and analysis from weeks to just *minutes or seconds*. GeoBrain also allows automation of many steps in a typical ES research project, and it provides useful help to some other steps. For example, a typical data-intensive ES research project involves the following steps:

Step 1: Find a real-world problem to solve.

Step 2: Develop or modify a hypothesis or model.

Step 3: Implement the model or develop analysis procedure on computer systems and determine the data requirements.

Step4: Search, find, and order the data from data providers.

Step 5: Preprocess the data into the ready-to-analysis form (e.g., reprojection, reformatting, subsetting, subsampling, and geometric or radiometric correction).

Step 6: Execute the model or analysis procedure to obtain the results.

Step 7: Analyze and validate the results.

Step 8: Repeat steps 2–7 until the problem is solved.

Without using GeoBrain, a researcher may need to spend a lot of time and computing resources for steps 2–7. Therefore, many projects requiring real or near-real time data could not be conducted. With GeoBrain, steps 3–6 could be fully automated, saving significant amount of time and computing resources for users. In addition, the geoprocessing modeling and knowledge sharing capability of GeoBrain can provide useful help to users for steps 2–7.

9.5.3.4 Use in ES Higher Education

GeoBrain makes enormous NASA EOS remote-sensing data and other important global change data products in distributed locations readily available online and easily usable. These data are especially valuable to higher-education users (professors, undergraduates or graduates, and ES majors or nonmajors) due to a variety of data sources, multi-disciplinary coverage, science-oriented data collection, and no or low cost of the data. GeoBrain also makes adequate data services, geoprocessing and analysis functions, and modeling software tools freely available online. Thus, GeoBrain provides an ideal online learning and research environment to train students on how to deal with real-world problems that need a huge amount of Geoinformation resources to solve the data-intensive ES problems. This online environment effectively supports a wide range of teaching, learning, and discovery activities including classroom teaching, dynamic demonstration, distance learning, online courses, home exercises, and research projects (Deng et al. 2006 and http://geobrain.laits.gmu.edu/courselist.html).

A dozen of universities have been selected as funded educational partners to pilot the use of GeoBrain in their classroom teaching and research projects and provide feedback and student surveys. Based on their feedback, thousands of graduates, undergraduates, and other students have been benefited directly or indirectly from GeoBrain each year; numerous new courses have been developed with full or partial support from GeoBrain; a large number of existing courses have incorporated GeoBrain into classroom teaching and student learning activities; teaching and learning effectiveness has been improved; and most of research projects in the institutes have used GeoBrain as the research tool or data source or both (Deng et al. 2006; Prakash 2008; Colby et al. 2009; Gentry and Badurek 2009). Through the data-intensive

GeoBrain, students are easily trained with better understanding of scientific concepts, adequate global-scale, real-world application experience, and necessary computational skills. By tracking the educational partners, GeoBrain is reported as an ideal platform to support data-intensive ES education with many benefits, including

- Enhancing classroom teaching and laboratory exercises
- Improving teaching and learning effectiveness
- Supporting inquiry-based, problem-based, and authentic learning activities
- Supporting new curriculum development
- Providing free and open computational cyber-laboratory experiments

GeoBrain is suitable for any interested individual to conduct spatial learning or discovery activities from any Internet-connected computer at any time, despite its dedication to ES higher education. A log of online user access statistics reveals that about 2500 worldwide distinct users (distinct online users have different IP addresses; multiple users who share the same IP address are counted as one distinct user) use GeoBrain each month. GeoBrain is an effective and unlimited CI system for its open accessibility and dramatically relaxed constraints (or no constraints) of time and distance to its users. In general, GeoBrain can be used as

- An unlimited data, services, and computing resource
- An online on-demand data visualization, geoprocessing, and spatial analysis platform
- An online spatial modeling, knowledge building, and sharing platform
- An online collaboration, cooperation, and resource-sharing agent

9.5.4 Conclusions

The ES education faces both opportunities and challenges in an increasingly data-intensive world. Easy access, integration, and analysis of distributed ES data play essential roles in facilitating data-intensive ES education. The GeoBrain project has designed and implemented an effective computing infrastructure and a dynamic online mechanism to resolve the major issues in data-intensive ES education and meet a wide range of needs of current and future data-intensive ES education. It largely helps the new trend and paradigm shift from having all of the data and computing resources owned locally to having them shared over the web with the latest technologies in CI and Web services. It creatively builds an unprecedented, data-intensive, and CI-enabled online learning and research environment with adequate data services, geoprocessing and analysis functions, and modeling software tools that are freely available to worldwide users. The GeoBrain online environ-ment enables easy, dynamic, and seamless access; integration; and analysis of distributed ES data.

Practice with GeoBrain in data-intensive ES classroom teaching, learning, and discovery activities demonstrates that GeoBrain may have a significant impact on how scientific research, education, and applications can be con-ducted. Any person who has an Internet-connected desktop or laptop com-puter will be able to access or utilize a vast amount of spatial data, analytic functions, and other computing resources that used to be only available to a few privileged people such as researchers.

References

Acker, J.G. and G. Leptoukh. 2007. Online analysis enhances use of NASA earth science data. *EOS Transactions of American Geophysics Union*, 88(2):14–17.

Ahearn, S. and J.L.D. Smith. 2005. Modeling the interaction between humans and animals in multiple-use forests. In *GIS, Spatial Analysis, and Modeling*, eds. J.M. David, M. Batty, and M.F. Goodchild, pp. 387–402. ESRI Press, Redland, CA.

Barstow, D. and E. Geary. 2001. Blueprint for Change: Report from the National Conference on the Revolution in Earth and Space Science Education, Snowmass, Colorado, TERC: 100p. http://www.earthscienceedrevolution.org/acknowledge/acknowledge01.cfm (access March 7, 2010).

Batty, M., D. Chapman, S. Evans, M. Haklay, S. Kueppers, N. Shiode, A. Smith, and P.M. Torrens. 2000. Visualizing the city: Communicating urban design to plan-ners and decision makers. In *Planning Support Systems*, eds. R.K. Brail and R.E. Klosterman, pp. 405–442. ESRI Press, Redland, CA.

Berke, P.R., D.R. Godschalk, E.J. Kaiser, and D.A. Rodriguez. 2006. *Urban Land Use Planning*, 5th edn. University of Illinois Press, Chicago.

Berrick, S.W., G. Leptoukh, J.D. Farley, and H. Rui. 2009. Giovanni: A web service workflow-based data visualization and analysis system. *IEEE Transation of Geoscience and Remote Sensing*, 47(1):106–113.

Bosilovich, M. 2008. NASA's modern era retrospective-analysis for research and applications: Integrating earth observations. *Earthzine* http://www.earthzine.org/2008/09/26/nasas-modern-era-retrospective-analysis/.

Bosilovich, M., J. Chen, F.R. Robertson, and R.F. Adler. 2008. Evaluation of global precipitation in reanalyses. *Journal of Applied Meteorology and Climatology*, 47(9):2279–2299.

Brown and Caldwell. 1998. Costs of storm water treatment for Los Angeles NPDES permit area. Final Report prepared for the California Department of Transportation.

BTS. 2009. Airline Data and Statistics. http://www.bts.gov/programs/airline_information/ (accessed April 5, 2009).

Chen, J., C. He, W. Zhu, Z. Peng, and S. Jin. 2008a. Integrated geographical informa-tion emergency service for Wenchuan earthquake. *Geomatics World*, 6(6):7–11.

Chen, W., H. Wang, and Y. Zhuang. 2008b. A rapid method of producing 1:25,000 image map for earthquake disaster area. *Geomatics World*, 6(6):16–19.

Clark, W.A.V. and J. Burt. 1980. The impact of workplace on residential relocation, Annals. *Association of American Geographers*, 79:59–67.

Clark, W.A.V., Y. Huang, and S. D. Withers. 2003. Does commuting distance matter? Commuting tolerance and residential change. *Regional Science and Urban Economics*, 33:199–221.

CLASS, NOAA's Comprehensive Large Array-Data Stewardship Data System. http://www.class.ncdc.noaa.gov/saa/products/welcome (accessed March 7, 2010).

Colby, J., C.A. Badurek, and L.A. Edwards. 2009. Application of NASA EOS Data to research and education in mountain-focused geographic information science. *Annual Meeting of the Association of American Geographers*, Las Vegas, NV.

CTPP. 2009. Census Transportation Planning Products. http://www.fhwa.dot.gov/ctpp/ (accessed April 5, 2009).

Deng M., L. Di, M. Abolins, H. Gong, and G. Zhou. 2006. Making differences: Uses of GeoBrain in classroom teaching of earth system and geospatial information sciences at higher education institutes worldwide. *GSDI-9 International Conference for Global Spatial data Infrastructure*, Santiago, Chile.

Deng, M. and Di L. 2009, Building an online learning and research environment to enhance use of geospatial data. *International Journal of Spatial Data Infrastructures Research*, 4:77–95.

Deng, M. and L. Di. 2006. Utilization of latest geospatial web service technologies for remote sensing education through GeoBrain system. In *Proceedings of 2006 IEEE International Geoscience and Remote Sensing Symposium (IGARSS06)*, 4pp. IEEE Publisher, Denver.

Di, L. 2004. GeoBrain-A web services based geospatial knowledge building system. In *Proceedings of NASA Earth Science Technology Conference 2004*. Palo Alto, CA.

Di, L., W. Yang, M. Deng, D. Deng, and K. McDonald. 2002. Interoperable access of remote sensing data through NWGISS. In *Proceedings of IEEE International Geoscience and Remote Sensing Symposium (IGARSS 2002)*, pp. 255–257. Toronto, Canada.

DLESE. 2001. DLESE Strategic Plan. http://www.dlese.org/library/index.jsp (accessed March 7, 2010).

Domenico, B., J. Caron, E. Davis, R. Kambic, and S. Nativi. 2002. Thematic real-time environmental distributed data services (THREDDS): Incorporating interactive analysis tools into NSDL. *Journal of Digital Information*, 2(4). http://journals.tdl.org/jodi/article/viewArticle/51/54 (accessed March 7, 2010).

Duh J. and D.G. Brown. 2005. Generating prescribed patterns in landscape models. In *GIS, Spatial Analysis, and Modeling*, eds. D.J. Maguire, M. Batty, and M.F. Goodchild, pp. 423–444. ESRI Press, Redland, CA.

Easa, S. and Y. Chan. 1999. *Urban Planning and Development Applications of GIS*. American Society of Civil Engineers (ASCE), Washington DC.

Eastman, J.R., M.E. Van Fossen, and L.A. Solorzano. 2005. Transition potential modeling for land-cover change. In *GIS, Spatial Analysis, and Modeling*, eds. D.J. Maguire, M. Batty, and M.F. Goodchild, pp. 357–386. ESRI Press, Redland, CA.

ECHO, http://www.echo.nasa.gov/ (accessed March 7, 2010).

ESRI. 2006. *GIS solutions for Urban and Regional Planning; Designing and Mapping the Future of Your Community with GIS*. ESRI Press, Redland, CA.

Evans, J., B. Bernhard, L. Di. et al. 2003. *Web Coverage Service (WCS), Version 1.0.0*, OGC 03–065r6. OpenGIS© Implementation Specification. Open GIS Consortium. http://ww.opengis.org/docs/03-065r6.pdf (accessed March 7, 2010): 101p.

FHWA (U.S. Federal Highway Administration). 2009a. GIS in Transportation. http://www.gis.fhwa.dot.gov/ (accessed December 18, 2009).

FHWA. 2009b. *Toolbox for Regional Policy Analysis*. http://www.fhwa.dot.gov/Planning/toolbox/tren_overview.htm (accessed April 5, 2009).

FRA (U.S. Federal Rail Administration). 2009. FRA's GIS Web Application. http://fragis.frasafety.net/GISFRASafety/ (accessed December 18, 2009).

Garbrecht J. and L.W. Martz. 2000. Digital elevation model issues in water resources modeling. In *Hydrologic and Hydraulic Modeling Support with Geographic Information Systems*, eds. D. Maidment and D. Djokic, pp. 1–27. ESRI Press, Redland, CA.

Gentry, J. and Badurek, C.A. 2009. A spatial analysis of socioeconomic indicators and refugee camp environmental impacts in Western Tanzania. *Annual Meeting of the Association of American Geographers*, Las Vegas, NV, March 22–27, 2009.

Goodchild M.F. 2005. GIS, spatial analysis, and modeling overview. In *GIS, Spatial Analysis, and Modeling*, eds. D.J. Maguire, M. Batty, and M.F. Goodchild, pp. 1–18. ESRI Press, Redland, CA.

Goodchild, M.F., R. Haining, and S. Wise. 1992. Integrating GIS and spatial data analysis: Problems and probabilities. *International Journal of Geographical Information Systems*, 6:407–23.

Greene R.W. 2000. *GIS in Public Policy*. ESRI Press, Redlands, CA.

Han, G., S. Jin, F. Wang, and Z. Zhou. 2008, The design and realization of Wenchuan earthquake relief geographical information system. *Geomatics World*, 6(6):25–29.

Hanson, K. and B. Carlson. 2005. Effective access: Teachers' use of digital resources in STEM Teaching, Education Development Center, Inc: p97. http://www2.edc.org/GDI/publications_SR/EffectiveAccessReport.pdf (accessed March 7, 2010).

Hopkins, L.D., N. Kaza, and V.G. Pallathucherl. 2005. A data model to represent plans and regulations in urban simulation models. In *GIS, Spatial Analysis, and Modeling*, eds. D.J. Maguire, M. Batty, and M.F. Goodchild, pp. 173–202, ESRI Press, Redlands, CA.

Huffman, G.J., R.F. Adler, D.T. Bolvin, G. Gu, E.J. Nelkin, K.P. Bowman, Y. Hong, E.F. Stocker, and D.B. Wolff. 2007. The TRMM multi-satellite precipitation analysis: Quasi-global, multi-year, combined-sensor precipitation estimates at fine scale. *Journal of Hydrometeorology*, 8(1):38–55.

Kanarek, H. 2005. The FAS Crop Explorer: A Web Success Story. FAS *WorldWide*, USDA Foreign Agriculture Service. http://www.fas.usda.gov/info/fasworldwide/2005/06–2005/Cropexplorer.htm (accessed March 7, 2010).

Klosterman, R. 1997. Planning support systems: A new perspective on computer-aided planning. *Journal of Planning Education and Research*, 17(1):45–54.

Klosterman, R. 2000. Planning support systems: A new perspective on computer-aided planning. In *Planning Support Systems*, eds. R.K. Brail and R.E. Klosterman, pp. 1–23. ESRI Press, Redlands, CA.

Kwartler M. and R.N. Bernard. 2000. Communityviz: An integrated planning support system. In *Planning Support Systems*, eds. R.K. Brail and R.E. Klosterman, pp. 285–308. ESRI Press, Redlands, CA.

Ledley, T.S., A. Prakash, C. Manduca, and S. Fox. 2008. Recommendations for making geoscience data accessible and usable in education. *EOS*, 89(32):291.

Li, D., X. Chen, and X. Cai. 2008. Spatial information techniques in rapid response to Wenchuan earthquake. *Journal of Remote Sensing*, 12(6):841–851.

Liu, B., H. Tao, J. Fan, B. Tian, J. Zhang, and D. Yan. 2008, Application of high-resolution SAR images in Wenchuan earthquake hazard monitoring and assessment. *Journal of Mountain Science*, 26:267–271.

Liu, Z., H. Rui, W. Teng, L. Chiu, G. Leptoukh, and G. Vicente. 2007. Online visualization and analysis: A new avenue to use satellite data for weather, climate and interdisciplinary research and applications. *Measuring Precipitation from Space— EURAINSAT and the future, Advances in Global Change Research*, 28:549–558.

Maantay J. and J. Ziegler. 2006. *GIS for the Urban Environment*, ESRI Press, Redlands, CA.

Maguire, D.J., M. Batty, and M.F. Goodchild. 2005. GIS, spatial analysis, and modeling: Current status and future prospects. In *GIS, Spatial Analysis, and Modeling*, eds. D.J. Maguire, M. Batty, and M. F. Goodchild, pp. 445–455. ESRI Press Redland, CA.

Maidment, D.R. 2002. *Arc Hydro: GIS for Water Resources*, ESRI Press, Redland, CA.

Maidment, D.R. and D. Djokic. 2000. *Hydrologic and Hydraulic Modeling Support with Geographic Information Systems*, ESRI Press, Redland, CA.

Maidment, D.R., O. Robayo, and V. Merwade. 2005. Hydrologic modeling. In *GIS, Spatial Analysis, and Modeling*, eds. D. J. Maguire, M. Batty, and M.F. Goodchild, 319–332. ESRI Press, Redland, CA.

Manduca, C.A. and D. Mogk. 2002. Using data in undergraduate science classroom: Report from an Interdisciplinary Workshop, Carleton College. http://dlesecommunity.carleton.edu/research_education/usingdata/ (accessed March 7, 2010).

Manduca, C.A., E. Baer, G. Hancock, R.H. Macdonald, S. Patterson, M. Savina, and J. Wenner. 2008. Making undergraduate geoscience quantitative. *EOS*, 89(16):149–150. http://serc.carleton.edu/serc/EOS-89–16–2008.html (accessed March 7, 2010).

Marlino, M., T. Sumner, and M. Wright. 2004. Geoscience education and cyberinfrastructure. Report from a workshop held in Boulder, Colorado. http://www.dlese.org/documents/reports/GeoEd-CI.pdf (accessed March 7, 2010).

Miller, H.J. 2009. Transport 2.0: Meeting grand transportation challenges with geographic information science. *ArcNews*, 30(4).

Miller, H.J. and S.L. Shaw. 2001. *Geographic Information Systems for Transportation: Principles and Applications*. Oxford University Press, New York.

Mitasova, H. and M. Neteler. 2002. *Open Source GIS: A GRASS GIS Approach*. Kluwer Academic Press, Dordrecht.

Moore, J.E., J. Kuprenas, J.J. Lee, P. Gordon, H. Richardson, and Q. Pan. 2004. Cost analysis methodology for advanced treatment of stormwater: The Los Angeles Case. *Journal of Construction Research*, 5(2):149–170.

National Research Council. 1996. *National Science Education Standards*, 262pp. National Academy Press, Washington DC. http://www.nap.edu/books/0309053269/html (accessed March 7, 2010).

Nebert, D. 2003. *OpenGIS® Catalog Services Specification*. OpenGIS Project Document: OGC 03–108, Open GIS Consortium Inc.

NSF CIC. 2007. Cyberinfrastructure Vision for 21st Century Discovery. National Science Foundation Cyberinfrastructure Council. http://www.nsf.gov/pubs/2007/nsf0728/index.jsp (accessed March 7, 2010).

Nyerges T.L. 2004. GIS in urban-regional transportation planning. In *The Geography of Urban Transportation*, 3rd ed., eds. S. Hanson and G. Giuliano, pp. 163–198. The Guilford Press, New York.

OASIS BPEL. http://www.oasis-open.org/committees/tc_home.php?wg_abbrev=wsbpel (accessed March 7, 2010).

OGC WFS. http://www.opengeospatial.org/standards/wfs (accessed March 7, 2010).

OGC WMS. http://www.opengeospatial.org/standards/wms (accessed March 7, 2010).

OGC WPS. http://www.opengeospatial.org/standards/wps (accessed March 7, 2010).

O'Looney, J.A. 2000. Beyond Maps: *GIS Decision Making in Local Government.* ESRI Press, Redland, CA.

Parker, D. C. 2005. Integration of geographic information systems and agent-based models. In *GIS, Spatial Analysis, and Modeling*, eds. D.J. Maguire, M. Batty and M.F. Goodchild, pp. 403–422. ESRI Press, Redland, CA.

Peng, Z. and M. Tsou. 2003. *Internet GIS: Distributed Geographic Information Services for the Internet and Wireless Networks.* John Wiley & Sons, New York.

Prakash, A., J.E. Bailey, and P. Webley. 2008. Integrating virtual globe research applications into the educational curriculum. *2008 Joint Meeting of the Geological Society of America*, Houston, Texas.

Pu, P. and Y. Wang. 2008. Preliminary research on the system architecture of the platform integrated by emergency geographic information. *Geomatics World*, 6(6):39–44.

Quigley, J. and D. Weinberg. 1977. Intraurban residential mobility: A review and synthesis. *International Regional Science Review*, 1:41–66.

RPA (Regional Plan Association). 2006. *America 2050: A Prospectus.* RPA, New York.

Scally, R. 2006. *GIS for Environmental Management.* ESRI Press, Redlands, CA.

Special Issue of Agricultural and Forest Meteorology. 2000. Papers presented at the *International Workshop on Agrometeorology in the 21st Century: Needs and Perspectives*, February 15–17, 1999, Accra, Ghana. *Agricultural and Forest Meteorology*, 103(1–2).

Special Issue on the Tropical Rainfall Measuring Mission (TRMM). 2000. Combined publication of the December 2000. *Journal of Climate.* And Part 1 of the December 2000 *Journal of Applied Meteorology.* American Meteorological Society, Boston, MA.

Sui, D. 1998. GIS-based urban modelling: Practices, problems, and prospects. *International Journal of Geographical Information Systems*, 12(7):651–71.

Tronin, A.A. 2006. Remote sensing and earthquakes: A review. *Physics and Chemistry of the Earth*, 31:138–142.

U.S. Census Bureau. 2009. TIGER. http://www.census.gov/geo/www/tiger/ (accessed December 18, 2009).

USGS. 2009. Geographic Information Systems. http://egsc.usgs.gov/isb/pubs/gis_poster/ (accessed December 18, 2009).

USGS Landsat. http://landsat.usgs.gov/ (accessed December 18, 2009).

Waddell, P. 2002. UrbanSim: Modeling urban development for land use, transportation and environmental planning. *Journal of the American Planning Association*, 68:297–314.

Wang, H. and M. Yang. 2008. Quick compilation of earthquake emergency map by using fundamental geographic information. *Geomatics World*, 6(6):20–24.

Wegener, M. 2004. Overview of land-use and transport models. In *Transport Geography and Spatial System*, eds. D.A. Henscher and K. Button. Pergamon/Elsevier Science, Kidlington.

Wegener, M. 2005. Urban land-use transportation model. In *GIS, Spatial Analysis, and Modeling*, eds. D.J. Maguire, M. Batty, and M. F. Goodchild, pp. 203–220. ESRI Press, Redland, CA.

Wright, M.J. and T.R. Sumner. 2003. DLESE Developers Workshop, 2003: Report on Workshop Outcomes, University Corporation of Atmospheric Research, 78pp.

Yeh, A.O. 2005. The integration of case-based reasoning and GIS in a planning support system. i In *GIS, Spatial Analysis, and Modeling*, eds. D.J. Maguire, M. Batty, and M.F. Goodchild, pp. 289–317. ESRI Press, Redland, CA.

Zhang, M. and T. Lomax. 2007. Estimating congesting index at the link level with TransCAD GIS. In *The American Society of Civil Engineers (ASCE) Conference Proceedings of the 7th International Conference of Chinese Transportation Professionals*, ed. R. Liu, Shanghai, China.

Zhong, Y. and X. Ding. 2008. Ultralight aviation low altitude digital remote sensing system for Wenchuan earthquake application. *Geomatics World*, 6(6):30–33.

10

Vision for Geoinformation Science

Kennith Foote, Myra Bambacus, Qianjun Miao, Huayi Wu, Wenwen Li,
Jing Li, Jiqiang Tan, Yanli Tang, Min Sun, and Hongwei Yu

CONTENTS

Geoinformation has been utilized in many different domains as we saw in Chapters 8 and 9. Integratively, more general usage of Geoinformation could fundamentally change how we view our world and subsequently help solve comprehensive problems for our home planet. Geoinformation Science education will help us prepare professionals to implement the integrative visions. This chapter introduces three visions (including Digital Earth, GEOSS, and EcoCity) and the education for preparing the workforce to achieve these visions.

10.1 Digital Earth*

10.1.1 Introduction

The vision of a *Digital Earth* was populated by U.S. Vice President Al Gore (1998). It refers to the technological innovation that allows practitioners to discover, access, and utilize an unprecedented amount of Geoinformation of our planet in a multi-resolution, multi-dimensional manner. Such a vision embraces a philosophy that anyone on the planet, linked virtually through the Internet, is able to freely access a virtual world of knowledge resources, both historical and predictive data.

The idea of building a digitalized world where the entire planet can be visually viewed has been the dream of human beings for a long time dating back to the early twentieth century, when the first electronic digital Atanasoff-Berry Computer (ABC) was developed and became the basic foundation of the beginning of a digital world (Mollenhoff, 1988). In 1946, the creation of the first general-purpose Electronic Numerical Integrator and Computer (ENIAC Goldstine, 1972) improved computing, making it one thousand times faster than electro-mechanical machines. The ENIAC was heralded as a "Giant Brain." By the next decade, geographers began to explore quantitative methods to solve spatial problems by using computer technologies (National Research Council, 1997). In 1968, Fuller's work on Spaceship Earth provided a detailed description of the functional requirements and specifications for how to model the world using modern computers. His futuristic vision was regarded as the immediate forerunner of a Digital Earth (Foresman,

* This section is contributed by Myra Bambacus and Wenwen Li.

2008). This concept gained recognition when Al Gore proposed the same concept in 1998 as a milestone for the evolution of Digital Earth. Since then, government agencies, non-profit organizations, and other industries have endeavored in the cooperative study of our planet and promoted the efficient sharing of resources to address complex application and fundamental scientific problems.

From the fall of 1998 to the fall of 2000, the U.S. Digital Earth was initiated by NASA and its federal partners, including NOAA, FGDC, USGS, USACE, EPA, and NSF (Table 10.1). The initiative tried to accelerate the technological development in standards, protocols, and tools toward the full realization of the Digital Earth vision. In 1999, NASA was selected to lead the Interagency Digital Earth Working Group (IDEW) due to its reputation for technology innovations and focused study of planetary change (Craglia, 2008). The Digital Earth initiative coordinated 23 government agencies (Table 10.1) to sustain development in Earth-oriented applications for the Digital Earth enterprise. The 3-year effort had several outcomes, including the current widely accepted WMS specification (Beaujardiere, 2004), DERM (Evans, 2004), and several other applications with a focus on climate change and environmental protection.

The Digital Earth Initiative and vision has been quickly adopted internationally and championed by China, through the Chinese Academy of Sciences. In 1999, the first International Symposium on Digital Earth was held in Beijing to provide a venue for the extensive international cooperation in implementing Al Gore's Digital Earth vision. In this symposium, the vice minister of the Ministry of Science and Technology of China presented the impetus to develop

TABLE 10.1

Government Agencies Engaged in IDEW

Department of Agriculture (DOA)	Department of Interior (DOI)
Defense Advanced Research Projects Agency (DARPA)	Department of Justice (DOJ)
Department of Commerce (DOC)	Department of State (DOS)
Department of Defense (DOD)	Department of Transportation (DOT)
Department of Energy (DOE)	Environmental Protection Agency (EPA)
Department of Health and Human Services (DOHHS)	Federal Geographic Data Committee (FGDC)
Federal Emergency Management Agency (FEMA)	National Science Foundation (NSF)
National Aeronautics and Space Administration (NASA)	Tennessee Valley Authority (TVA)
National Archives and Records Administration (NARA)	The Library of Congress (LOC)
National Imagery and Mapping Agency (NIMA)	The Office of Management and Budget (OMB)
National Oceanic and Atmospheric Administration (NOAA)	United States Army Corps of Engineers (USACE)
United States Geological Survey (USGS)	

Advanced Geoinformation Science

Digital Earth based on georeferenced information in China. Dr. Xu discussed the necessity for China to enhance the Research and Development (R&D) of Digital Earth in terms of (1) maintaining sustainable development by utilizing modern technologies to enhance land, water, and environmental protection; develop technology-intensive agriculture practices and intensify disaster prevention in order to improve people's living standards; and (2) realizing a rationalized plan to develop science and technology, especially in ES by developing high-quality sensors; improving EO satellite technology, satellite-borne data processing technology, and satellite-plane-ground data receiving technology (Xu, 1999). The establishment of the Center for Earth Observation and Digital Earth sets another milestone in Digital Earth effort by the Chinese and China's enthusiasm in such an international program. Along with cooperative efforts from worldwide scientists and scholars, the theoretical and practical study of Digital Earth has been greatly promoted.

Japan has also played a prominent international role in Digital Earth to promote regional cooperation and initiatives (Wiki, 2009). Digital Asia, supported by Keio University and JAXA, aims at providing the whole world with easy access to Geoinformation over the Internet among all countries of Asia. An important milestone was to build the Digital Asia Network (DAN) to enable connecting the available Web-based GIS from participating organizations and government agencies (Shin'ichi and Ryuzo, 2003).

Apart from government agencies, industries worldwide have also been actively involved in the research and development of Digital Earth. In 2004, Google Earth was released, which presents satellite images and digital maps at various resolutions in order to visually view the Earth's surface and the

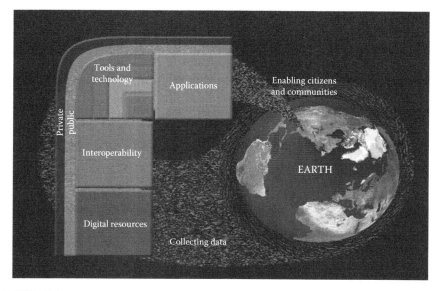

FIGURE 10.1
(See color insert following page 144.) Components of Digital Earth.

utilization of DEM, which makes it available for 3D terrain display. Google Earth allows users to manage their own 3D spatial data through KML, which also allows people to build and share 3D models of objects (such as buildings) on Google Earth using Google's SketchUp product. The emergence of Google SketchUp was considered the most successful 3D Digital Earth prototype (Grossner and Clarke, 2007). In 2005, Microsoft started Virtual Earth, focusing on massive data visualization and user-friendly interaction. It has solved many problems in dealing with massive simultaneous users by using thousands to millions of computers.

10.1.2 Components of Digital Earth

The Digital Earth community has already developed and implemented major components that contribute to the larger Digital Earth vision. These components together demonstrate evidence of significant progress necessary to address the economic, social, and environmental issues that face us today. Figure 10.1 demonstrates the components of a Digital Earth system according to the previous Digital Earth Office, including

- Digital Resources
- Interoperability
- Tools and Technologies
- Applications

At the outset, the vast availability of digital resources around our planet is the basic component in the Digital Earth paradigm, made possible through the rapid development in EO technology, led by NASA. In 2006 alone, NASA's EOSDIS produced over 3 terabytes (TB) of ES science data on a daily basis (NASA, 2007). This Geoinformation is widely utilized in different applications, such as navigation (Rae-Dupree, 2006), transportation (Peytchev et al., 2001), urban planning (Stevens et al., 2007), and emergency response (Rauschert et al., 2002). Currently, NASA's ES efforts are concentrated in six areas: (1) building and operating Earth observing satellite missions in cooperation with international and/or interagency partners; (2) making high-quality data products available to the general science community; (3) conducting and sponsoring cutting-edge research, for example for example, field campaigns to complement satellite measurement, modeling, and analysis of non-NASA mission data in six thematic focus areas (Atmospheric Composition, Carbon Cycle and Ecosystems, Climate Variability and Change, Weather, Water and Energy Cycle, and Earth Surface and Interior); (4) conducting an applied science program to improve the utilization of the data through the United States; (5) developing technologies to improve EO capabilities, providing the seed technologies for the next generation of Earth observing instruments; and (6) enhancing education and public outreach.

To integrate EO technology with applications, Li et al. (1999) proposed an architecture from the aspects of (a) contents, (b) key technologies, and (c) applications from Geoinformation and informationization perspectives. Benefiting from international efforts, global spatial data (from collections of direct satellite and airborne remote-sensing data as well as local measurements) are obtained to support applications across a wide range of scales. However, since spatial data is always in a variety of formats, it is difficult for them to exchange and interoperate with each other. Thus, an interoperability module (on top of Digital Resources), which defines the standards, terminologies, and protocols for communication between heterogeneous data sets, is of great importance. Figure 10.2 demonstrates the interoperability stack in a DERM (http://cartome.org/draft-derm.htm). A complete geo-processing workflow consists of (1) data providers storing spatial data (coverage data, feature data, and other data) within their local database and registering metadata information with a link to their local database into globally accessible catalogs; (2) data consumers conducting spatial queries to discover the relevant spatial data sets when needed; and (3) a geo-processing service automatically retrieving the real data sets from data providers' databases, hooking up various kinds of data sets into an integrated map, and then visualizing it through a multiple dimensional geobrowser on demand. The interoperability happens in each of the steps: (a) The OGC Web Registry Service (DCIS), OGC Catalog Services, and ISO 23950(ANSI Z39.50) all make sure that catalog access can be conducted in an interoperable infrastructure; (b) the OGC Simple Features access for SQL, COM, CORBA, and OGC WFS and OGC Grid Coverage Access for OLE/COM Web

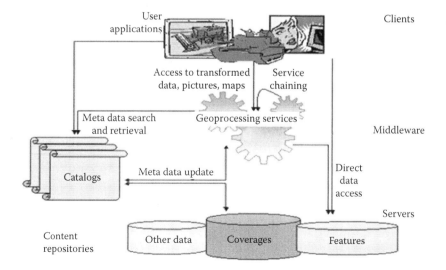

FIGURE 10.2
The Interoperability stack. (Adapted from Evans, J.D. 2004. *Digital Earth Reference Model.* http://www.cartome.org/draft-derm.htm [accessed December 1, 2009].)

Coverage Service allow data providers to exchange heterogeneous data at both feature and coverage levels; and (c) OGC WMS serves as a simple application client and enables the seamless map production and integration from various distributed data resources.

A technology stack that is used to build a Digital Earth always comprises a lightweight online or desktop geobrowser, a high-speed communication network, and a high-performance server infrastructure, which benefited from significant technology breakthroughs, such as cloud computing (Milojicic, 2008). Cloud computing is a recent trend that delivers applications such as Web services through Internet. Its emergence promises to streamline the on-demand provisioning of computing infrastructure, software, and data as a service (Dikaiakos et al., 2009). Taking advantage of cloud computing technologies enables the quality of geospatial services to be increased and reduces the associated costs.

With all the supporting modules in the previous paradigm, Digital Earth benefits people at varying scales, from individuals to communities to countries. It also provides an effective solution to bridge the scales from the global to local environment for global sustainable development. There has been a rapid development in building *Digital Communities, Digital Towns, Digital Rivers, Digital Cities, Digital Country, Digital Continents*, and, ultimately, the *Digital Globe*. For instance, we see the *Digital Olympic, Digital Beijing, Digital Asia*, and, finally, the *Digital World* (Guo and Wang, 2004) (Figure 10.3).

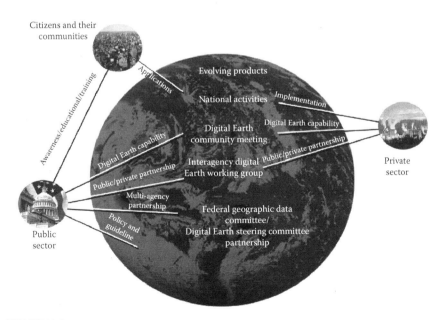

FIGURE 10.3
(**See color insert following page 144.**) Roles of Public Sector, Private Section, and General Public in a Digital Earth Environment. (Courtesy of previous NASA Digital Earth office.)

10.1.3 Empowering a Digital Earth Environment

To better meet the challenges of understanding the ES and improving the prediction of ES phenomena toward empowering the Digital Earth environment for addressing fundamental scientific questions and complex application problems, efforts and close cooperation from public and private sectors, citizens, and communities are greatly needed. The public sector is responsible for developing policies and guidelines, enhancing multi-agency partnerships, and providing appropriate training of the Digital Earth vision and related technologies to the public at large. For example, NASA promotes the full and open sharing of all NASA data within the research and applications communities, private industry, academia, and the general public. To this end, NASA provides open access to data with no periods of exclusive access. Most of the data is provided at no charge to the user, except in cases impacted by international agreements. NASA actively encourages a free and open data policy with other international organizations. NASA is also a participant and contributor to the Inter-agency Working Group on Digital Data (IWGDD). The private sector should continuously promote the implementation of Digital Earth and related products due to their superiority in technology. Both public and private sectors should extend Digital Earth capability and intensify public-private partnerships in order to satisfy citizens and their communities' various application demands.

10.1.4 Future Challenges

Although it has been only 11 years since the concept of Digital Earth was proposed, we have made significant research achievements in both theoretical mechanisms and practical implementations. Many modules of Digital Earth are not only available but also used daily by hundreds of millions of people worldwide (Craglia et al., 2008). However, there are still a number of challenges in current developments, for example (a) metadata for describing that the digital resources are difficult to synchronize and update in a timely manner (Li et al., 2010); (b) resource discovery, access, and utilization needs to be fully automated by utilizing ES and application knowledge (Li et al., 2008); (c) how to better conduct distributed geographic information processing (Yang and Raskin, 2009) for sharing data, computing, and processing capabilities within Digital Earth; and (d) how to build a Geospatial CI (Yang et al., 2010) that can provide the infrastructural support for Digital Earth. Future developments of Digital Earth include addressing the above issues and providing a user-friendly, smart, and more understanding resource discovery, integration, and visualization system by (1) transitioning from research to operations; (2) characterizing uncertainty in model forecasts for weather, climate, and natural hazards; (3) increasing computing capacity to handle volume and ranges of data from Earth observatories; (4) accessing observations and model outputs throughout the Global Spatial Data Infrastructure; and (5) establishing gateways or portals (Yang et al., 2007) to provide access to scientists, decision makers, educationists, and citizens.

10.2 Global Earth Observation System of Systems*

10.2.1 Introduction

Transitioning into the twenty-first century has not automatically solved global problems, such as extreme poverty, environmental vulnerability, and climate change. To confront these challenges, the General Assembly of the United Nations adopted the Millennium Declaration on 8 September 2000 during the Millennium Summit. The millennium declaration setup eight international development goals, designated the Millennium Development Goals (MDGs), which 192 UN member states and at least 23 international organizations have agreed to achieve by the year 2015. To attain these goals, cooperation among international governments on the acquisition and sharing of EO information is crucial.

The World Summit on Sustainable Development, Johannesburg South Africa, 2002, highlighted the urgent need for coordinated observations relating to the state of the Earth. The Summit of the Heads of States of the Group of Eight Industrialized Countries in June 2003 in France (Evian) reinforced the importance of EO as a priority activity. The first EO Summit in Washington, in July 2003, adopted a declaration, stating the political commitment to move toward the development of a comprehensive, coordinated, and sustained EO system of systems. The Summit established an *ad hoc* intergovernmental GEO, co-chaired by the European Commission (EC), Japan, South Africa, and the United States and set them with the task of developing a 10-Year Implementation Plan.

The third EO Summit established the intergovernmental GEO. The GEO Secretariat was created in the World Meteorological Organization (WMO) in Geneva in 2005. The GEO invites all the Member States of the United Nations and the Participating Organizations, which act at the intergovernmental, international, and regional level to join, on a voluntary basis, one of the most challenging coordination actions in the world. In 2009, GEO's members included 80 countries and the EC. There are 56 organizations that are participating members of the GEO. Besides these members, there are six observers of GEO, including one country and five organizations.

The GEO is the official organization to coordinate efforts toward building a Global Earth Observation System of Systems (GEOSS). The 10-Year

---

* This section is contributed by Huayi Wu, Jing Li, and Min Sun.

Implementation Plan, "Framework for GEOSS 10-Year Implementation Plan: From Observation to Action—Achieving Comprehensive, Coordinated, and Sustained Earth Observations for the Benefit of Humankind," is the formal guidance for GEOSS.

The structure of GEO can be divided into four parts, which are plenary, Executive Committee, Secretariat, and four Standing Committees. In the Executive Committee, 12 regional members are selected to serve as the regional representatives. They are Morocco and South Africa from Africa; Brazil, Honduras, and the United States from the Americas; Russia from the CIS; China, Japan, and Thailand from Asia; and the EC, Germany, and Italy from Europe. The EC, the United States, China, and South Africa are co-chairs in this committee.

The four GEO standing committees are Science and Technology, Architecture and Data, User Interface, and Capacity Building and Outreach. These four committees guide the implementation of GEOSS's 10-year plan. The GEO participants agree that GEO is based on a voluntary and legally nonbinding basis. The GEOSS provides the products, sharing them through the existing national and international mechanisms without a centralized funding approach. The estimated budget for GEO Secretariat is about 3–4 million dollars per year, as *seed money* for the project. With this infrastructure in place, GEOSS products are easy to share.

10.2.2 GEOSS Architecture

The GEOSS is a framework that is accessible, comprehensible, and operable. Figure 10.4 shows the general architecture of GEOSS. In the GEOSS, data are collected and stored in the system following certain standards. In the assimilation process, not only ES data but also other data are used. After assimilation, these data sets serve as the input to ES models in different scientific domains. The outputs from the model then help decision makers to make predictions and further analysis. The feedback of the prediction and analysis then go back to the GEOSS to make modifications.

In the GEOSS architecture, interoperability either in data or models promises compatibility between systems in GEOSS.

The architecture supports nine societal benefit areas, which are (1) Reduction and Prevention of Disasters. Through the integration of EO with other information in GEOSS, decision makers can use the information to reduce vulnerability, strengthen preparedness, early-warning measures, and future planning. (2) Human Health. Environmental changes have impacted human health either in changing the climate or in changing the landscape. The task of GEOSS is to determine environmental friendly parameters and variables, which will benefit people's health. Biological indices are observed to predict the occurrences of potential diseases. (3) Energy Management. Today, energy management is crucial. The GEOSS can help detect energy resources using remote-sensing technologies; estimate greenhouse gas emissions;

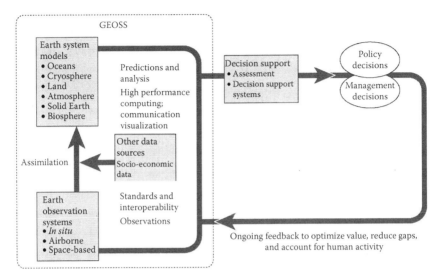

FIGURE 10.4
GEOSS architecture. (Adapted from Duchossois, G. 2006. The global Earth observation system of systems (GEOSS): An example of international cooperation. *Second ESA/ASI Workshop on International Cooperation for Sustainable Space Exploration Sarteano*, Siena, May 9–12, 2006.)

monitor changes in hydropower, solar, ocean, and wind energy sources; and predict energy consumption. The information in GEOSS is a significant resource for planners to make full use of all energy sources. (4) Climate Change. The GEOSS supports climate change studies. Data from GEOSS features high spatial and temporal resolution; so, the models within GEOSS can be used to produce reasonably accurate climate simulation data. (5) Water Management. GEOSS collects water data from multiple sources, such as satellite-based radar altimeters, ground-level, and in situ monitors. The GEO also seeks to develop global, national, and regional models for water resources. Among the models, interoperability is emphasized to exchange data among them. (6) Weather Forecasting. Weather forecasting is one of the fundamental topics in EO. Besides integration, the GEOSS develops common data formats and standards through collaborative effort by experts from different disciplines. The forecasting will lead to a broad range of benefits. (7) Ecosystem. The GEOSS monitors the changes of land, ocean, and coastal areas. The spatial information helps researchers estimate the trends in ecosystems, such as the changes of vegetation, desertification, and sustainable timber harvests. (8) Agriculture. In terms of agriculture, the GEOSS will assist in increasing productivity. Sustainable management of agriculture is the ultimate goal of GEOSS in this category. This dimension requires weather forecasting, climate monitoring, water management, and so on. (9) Biodiversity. The role of GEOSS in biodiversity is to provide high-quality data and analysis to help in the conservation of biodiversity. The popular analyses are pollution estimation, global biogeography modeling, and the migration of species.

The GEOSS are built on a series of cross cutting areas. The cross cutting areas emphasize the improvements of GEOSS itself as a system. According to the 10-year plan, the areas define the content of GEOSS. They are (1) Architecture and Data Management. A core part of GEOSS is the data and information associated with interoperability and standards. To regulate the data from multiple sources, the GEOSS has developed a series of technical specifications, which are used to define data collecting, processing, delivering, and sharing. Interoperability facilitates the communication among data, models, and components. This requires an efficient message exchanging mechanism. (2) User engagement. Since the GEOSS is a collaborative work among different countries and international organizations, it is necessary to realize the needs of global communities. The users may be decision makers, engineers, scientists, government officials, and so forth. Therefore, user engagement is essential for keeping the GEOSS updated, which will be beneficial to the larger user community, especially for developing countries. (3) Capacity-building. With respect to the World Summit on Sustainable Development, capacity building pays attention to the participation of all countries to use GEOSS and contribute to it as well. The aim is to build global capacity for sharing data, models, and related resources. (4) Outreach. The outreach of GEOSS encourages the members of GEO to promote the GEOSS to various research areas not limited to the Earth system sciences. This area includes both global and regional approaches to enlarge the influences of GEOSS.

Since the GEOSS is an integration of various systems, it is necessary to introduce a mechanism to help connect different systems, services, and data. To achieve this aim, GEOSS Common Infrastructure (GCI) was developed. The GCI is a set of common core services facilitating the integration of GEOSS. The GCI has several components: (a) the component and service registry; (b) the standards and interoperability registry; (c) the user requirements registry (in development); (d) best practices wiki; (e) Web portals; and (f) a clearinghouse.

The component and service registry allows the definition of membership through registration. This information can then be found in the GEOSS system. The registration can absorb systems, services, and data into the GEOSS, which are then arranged by the social benefit areas. The standards help maintain the interoperability of services, data, models, and relevant information. The nomination mechanism has two categories of classification, which are standards and special arrangement. The "standards" candidates are the nominations maintained by standards, whereas the "special arrangements" ones are *ad hoc* community practices. In the GEOSS system, the standards and special arrangements are used for accessing the original candidates. The wiki offers a platform to exchange and deliver knowledge or GEOSS techniques. The Web portals and clearing house offer the public access to EO products. The Web portals allow client access to the products though the user interface. The portals are supported by tools such as portlets. The content of a portal may include the general information of ES products. After obtaining the resources, the portal should be able to visualize the resources. Finally, the linkage to original resources is required. In the GEOSS clearinghouse, metadata and registered catalog services can be found. The metadata describes the registered systems, services, data, and other products.

The major role of GCI is to publish and use EO products. Before publishing, the contributor must agree with the 10-year plan and be willing to provide necessary information. The contributor can log on to the GEO portal and follow the registration process from the link http://geossregistries.info/geosspub/. In terms of using GEOSS registered products, the users can go to the Web portals and take advantage of portals as well as products.

Generally speaking, GCI facilitates the implementation of "system of systems" with this registration mechanism. To keep the system updating, regular clean up is needed in the next phase (Figure 10.5).

10.2.3 10-Year Implementation Plan

The 10-Year Implementation Plan is one of the most essential plans of GEOSS (GEO, 2005). It defines the major role of GEOSS from 2005 to 2010. This plan explains the role of all major aspects of GEOSS. The 10-year plan elaborates the input of the GEOSS system, technologies of GEOSS, research areas that GEOSS can contribute to, and how to contribute. By examining the plan, we

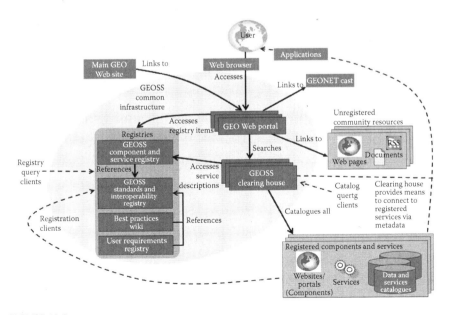

FIGURE 10.5
Structure of GCI. (Adapted from GEO. 2009. The GEOSS Common Infrastructure, http://www.earthobservations.org/gci_gci.shtml [accessed December 16, 2009].)

see how the GEOSS as a global system incorporates the joint efforts of organizations or countries to solve the EO problems. It requires the participation of government as well as international organizations. This user-driven system provides interoperable products and shared observations that are accessible, comparable, and understandable. This distributed and sustainable system encourages the input of the global community and benefits a broad range of scientific domains through the easy interoperability of data, models, and analysis. The plan also points to the future development of the next generation of GEOSS.

To be specific, the key points of the 10-Year Implementation Plan are (1) Vision of GEOSS. The vision of GEOSS puts forward the benefit of the global community by "coordinated, comprehensive and sustained Earth observations and information" (GEO, 2005). This vision is determined by the needs of GEOSS. (2) Purpose and Scope of GEOSS and the Group on EO. The purpose is "to achieve comprehensive, coordinated and sustained observations of the Earth system, in order to improve monitoring of the state of the Earth, increase understanding of Earth processes, and enhance prediction of the behavior of the Earth system." Therefore, the GEOSS was established for the study of ES in a global community. The scope of GEOSS is to provide a framework of data, models, and systems to the users from global communities. The GEOSS utilizes all available EO data and models to make predictions, analyses, and decisions. The "system of systems" determines the scope of GEOSS is broad with a concentration on ES. (3) Benefits of GEOSS. The benefits are the societal benefit areas and user involvement. These two aspects have been discussed in Section 10.2.2. (4) Technical Approach, Capacity Building, and Outreach. The GEOSS should incorporate a range of techniques to realize its purpose in the following aspects: observations and modeling; products, data management, and radio frequency protection; architecture and interoperability; and data sharing and research facilitation. The GEO will try to establish a well-developed data system to provide products. These products should be used for national, regional, and international decision making. Therefore, a timely observation system should be built to meet these needs. Besides providing data, the GEOSS is attentive to standards and representations of data. In this whole process of collecting, processing, sharing, and storing data, interoperability defines an interface to connect all the components. The SDI is used in the system to serve as a spatial framework. In terms of the cost of the data, there should be certain principles such as the full and open exchange of data, a quick response to data requests, and so on. Finally, these products should support relevant research, for example for example, model development, which then improves the performance of the system. Capacity building and outreach have been described in Section 10.2.2. (5) Governance. The plan outlines the functions of GEO, which are to implement and update the plan; fill in the gaps in data; coordinate efforts of the GEO; promote data exchanging, sharing, and delivering; improve the interoperability; attract experts from different scientific domains;

and revisit work plans, budget, and administrative work. (6) Funding and Measuring Progress. This plan defines that resources should be provided by voluntary contribution among the participants of GEO. This reduces the costs on data, models, and other recourses, freeing funds for other GEO activities. (7) The Transition Period. Based on the implementation, this plan will be adapted as circumstances and needs warrant.

10.2.4 GEOSS Pilot Project: The National Integrated Drought Information System

A drought occurs when a region has a deficiency in its water supply for an extended period of months or years. Usually, a drought is caused by a consistent shortage of precipitation, far below normal averages and insufficient to meet the needs of humans and the environment (Wilhite and Buchanan-Smith, 2005). A drought has a substantial impact on the ecosystem and agriculture of the affected region. Although droughts can persist for several years, even a short, intense drought can cause significant damage and harm the local economy. This global phenomenon has a widespread impact on agriculture.

Drought is not rare in the United States; over the last century, it has been an ever-present threat. Virtually all parts of the United States are drought prone, and drought occurs somewhere in the country each year. Recent severe droughts in the United States—beginning in 1996 and affecting nearly all parts of the country—are indicative of a growing drought hazard. Besides the climate patterns such as El Nino and La Nina, the North Atlantic Oscillation is another factor for the rise of droughts in the United States. Above-normal dominations of high pressure systems in any particular regional vicinity of the United States are known to be related to droughts.

The economic, environmental, and societal impacts of drought are severe and extremely costly in the United States as elsewhere. In 1995, the Federal Emergency Management Agency (FEMA) estimated that the average drought cost to the United States ranges from US $6 to $8 billion annually (FEMA, 1995). According to estimates by the Texas Agriculture Extension Service, the 1996 Texas drought was estimated to cost producers in Texas US $1.9 billion, reducing the overall state economy by about US $5 billion. Vulnerability to drought is increasing in all parts of the United States due to population growth and population shifts, especially in the water-short western states and in the southeast; land-use changes; global climate change; and increased water resource demands. The U.S. population has increased by about 50 percent since 1970 to more than 300 million, much of that occurring in water-scarce western regions. Land-use changes due to development and other activities reduce water storage and degrade water quality. Global climate change directly and indirectly impacts the hydrologic cycle, reducing water availability and increasing vulnerability to drought in many regions of the United States.

To deal with the constantly occurring hazard of drought, the National Drought Policy Act of 1998 established the National Drought Policy Commission to ensure collaboration between different government agencies on drought-related issues. The Commission issued a groundbreaking report, "Preparing for Drought in the 21st Century" in 2000 (National Drought Policy Commission, 2000). Following the Commission's recommendations, the National Integrated Drought Information System (NIDIS) was envisioned in a Western Governors' Association Report in 2004. The NIDIS Act was introduced in the U.S. Congress and was signed by the president in 2006.

The NIDIS has been identified as a high-priority item to be developed as a part of the U. S. Integrated Earth Observation System (IEOS), the U.S. contribution to GEOSS. The challenge of predicting drought highlights the underlying theme of GEOSS. A variety of factors go into determining where a drought will form and how long it will last, ranging from rain and snow amounts to the intensity of the solar radiation that reaches the ground and evaporates soil moisture. The NIDIS U.S. Drought Portal was officially launched on 1 November 2007 (http://www.drought.gov). The portal is registered in GEO component repository as part of GEOSS system (Figure 10.6).

The vision of NIDIS is to create a dynamic and accessible drought information system that provides users with the ability to determine the

FIGURE 10.6
NIDIS Web portal.

potential impacts of drought and the associated risks they bring, and the decision support tools needed to better prepare for and mitigate the effects of drought.

The goal of NIDIS is multifold:

- Develop the leadership and partnerships to ensure successful implementation of an integrated national drought monitoring and forecasting system;
- Foster and support a research environment that focuses on impact mitigation and improved predictive capabilities;
- Create a drought "early warning system" capable of providing accurate, timely, and integrated information on drought conditions at the relevant spatial scale to facilitate proactive decisions aimed at minimizing the economic, social, and ecosystem losses associated with drought;
- Provide interactive delivery systems, including an Internet portal, of easily comprehensible and standardized products (databases, forecasts, GIS-based products, maps, etc.); and
- Provide a framework for interacting with and educating those affected by drought on how and why droughts occur and how they impact human and natural systems.

The NIDIS is an integrated system in that it collects and transmits data using technologies like remote sensing. It also works with private sectors to facilitate multiple studies. The NIDIS brings together a variety of observations, analysis techniques, and forecasting methods in an integrated system that will support drought assessment and decision making at the lowest geopolitical level possible. The tools will allow users to access, transform, and display basic data and forecasts across a range of spatial and temporal scales most suited to their individual needs. There are four basic types of drought information tools:

1. Data access tools facilitate the retrieval of data from the different agencies that collect and archive it.
2. Analysis tools add value to the raw data through computer data transformation, modeling, and statistical analysis.
3. Data display tools enable visual display of raw and analyzed data in ways to enhance its value to users. The GIS software enables the examination of georeferenced information.
4. Forecast tools are a specialized analysis combining statistical properties of available observations and models of future developments to make forecasts.

The key components of the NIDIS are as follows (WGA, 2004):

a. Improve and expand the compilation of reliable data on the various indicators of droughts, from both the physical or hydrological data (such as a national surface observing network) to the socioeconomic and environmental impacts data (such as agriculture losses and wildfire impacts).

b. Integrate and interpret that data with easily accessible and understandable tools, which provide timely and useful information to decision makers and the general public.

Similar to any system in GEOSS, the NIDIS provides the analysis results for scientific researches, which, in turn, will then improve the performance of the system. In the next phase, the NIDIS should be officially formed to enhance its functionality. More tools and data should be added as well to broaden the influence of NIDIS. Both research needs and practical needs, such as drought preparedness, should be considered in the system. Finally, this system will move toward education for users.

The GEOSS is a comprehensive system that supports the studies in EO sciences. Besides the applications in hydrology, the GEOSS also plays an important role in various fields of study such as atmospheric sciences, which have been discussed in the previous section. With the support of GEOSS, the EO studies have been extended from EO to other disciplines, such as socioeconomic domains. It is expected that the 10-year plan will lead to a well-established system.

10.3 EcoEarth*

Our home planet is confronting unprecedented challenges including (1) rising sea levels due to melting ice in polar regions, (2) global climate changes, (3) balancing ecosystems, and (4) developing clean energy. These challenges pose a great threat to the Earth due to the emphasis on economic growth at the expense of sustainability. We have been exploring the Earth, utilizing the Earth, and, consequently, must protect the Earth. When the U.S. former Vice President Al Gore put forward the vision for Digital Earth, he foresaw the need for a platform that allows us to represent and describe the current situation and changes in the Earth due to human activity. To address these challenges, from the outset, we consider dimensions of a balanced ecosystem in different aspects of EcoEarth. From the prehistorical agriculture living to industrial living, digital living, information living, toward knowledge living (Table 10.2), technological advancements have equipped us with

---

* This section is contributed by Qianjun Miao, Jiqiang Tan, Yanli Tang, and Hongwei Yu.

TABLE 10.2

Comparison of the Developing Phases of Human Society

Phase Contents	Agricultural Living	Industrial Living	Digital Living	Information Living	Ecological Living
Main body	Physical force and labor power	Product and size	Technology and standard	Innovation and services	Sustainable and harmonious
Object	Entity resources	Products	Information	Knowledge	Resident environment
Method	Based on handwork	Based on machine	Based on computer	Based on network	Based on resources
Technology	Traditional methods	Industry Revolution	Data Warehouse	Resources sharing	Clean technology
Orientation	Material value	Value and profit	Expand profit	Virtual value	Future value
Position	Past Perfect	Present Perfect	Present	Present Progressive	Future
Dominant	Modesty utilize	Predacity utilize	Excavated utilize	Public utilize	Sustainable utilize
Base	Nature gift	Extensive consume	Digital advantage	Intensive Services	Saving resources
Objective	Autarky	Collecting wealth	Virtual world	Spatial-temporal integration	Unity of people and nature

capabilities to respond to the questions of When, Where, What Object, and What Change (4W) to expect and to deliver information to Anyone, Any information, Any time, and Any where (4A). All these advances have an ecological price. By learning from past painful experiences, we believe we must change living patterns to ensure that the Earth is ecologically balanced and sustainable.

Eco-civilization should become one of the objectives we pursue. We must give a high priority to developing an ecological infrastructure. We must promote ecological human residential environments. We must evolve an ecological human cultural and social shape and must develop ecological science and technology including advanced ecological measurement tools to realize these objectives.

The EcoEarth objective is to build an Auroville—like community, as a model for sustainable development. The key point is to build EcoEarth starting from the concept of eco-civilization and move toward constructing EcoCities.

10.3.1 Ecology and Eco-Civilization

With the continuous increase of the scale of human exploitation of natural resources and the growing intensity of ecological impacts, awareness of ecological concepts has been expanding to what might be termed broadly as *ecological knowledge*. Build on this interest, we must deepen public understanding of ecological science, strengthen ecological consciousness, and disseminate the eco-civilization concept in the cognitive, institutional, technical, and behavioral areas to achieve implementation.

Ecology includes three aspects: relationships, an area of knowledge, and a state of harmony. (1) *Ecology* refers to the interactive relations between the environment, the individuals, and unity of life and lives, including all humankind. (2) As a discipline, ecology includes several different levels as (a) a methodology and philosophy for people to understand nature and rebuild the environment, (b) an engineering technology to model the environment and simulate nature, (c) a natural esthetics, and (d) the integration of "harmonious ecological relations" and "ecological positive cycle."

Ecological philosophy is dialectical: harmonious but imbalanced, exploiting without exhaustion, adaptive rather than being destructive, and emphasizes circulation over regression. Ecology is complicated; therefore, we must recognize and comprehend its complexity in the view of ecologic philosophy, simplify and simulate the complexity by the measures of the ecological sciences, plan and manage the complexity using the ecological technologies, and transform the complexity of ecology into social sustainability.

Eco-Civilization is the concrete expression of material civilization, spiritual civilization, and political civilization in the relationship between nature and socio-ecology, involving institutional civilization, cognitive civilization, physical state civilization, and mental state civilization, having different representation forms at different social development stages.

The narrow definition of Eco-Civilization envelops the preferred activities, ideas, and consciousness, thus enabling people to rebuild and comply with nature. It is through the concrete practices of living that people become civilized and make progress. The generalized definition of Eco-Civilization integrates practices and artifacts of material production and consumption modes, social organization and managing institutions, value opinions, resources exploration, and environment impact modes so that people and nature are harmoniously developing during the practice of reconstructing nature, adapting nature, protecting nature, and enjoying nature.

10.3.2 Building EcoCities

EcoCity has already become a concept that is often used as a political or commercial buzzword and, therefore, often a superficial synonym. An EcoCity includes three different types:

- Eco-communities, where people mostly live an ecologic lifestyle. They grow their own food, often make their own clothes, and heat their homes using wood as fuel.
- Larger urban units with more energy-saving buildings, recreational parks, and balanced life styles in contrast to traditional dwellings, where development adversely impacts ecological balances.

- Some cities that do not seem ecological might have some genuine EcoCity aspects, such as where more attention may have been pad to the energy consumption of buildings or where there is a denser public transport network than cities, in general.

10.3.3 Goal of an EcoCity

An EcoCity is neither a new style of architecture nor a new way of making profits, but rather it is one of the most important solutions to the ongoing global crisis. It possesses two basic characteristics:

- Makes economical use of natural resources—materials, energy, and space.
- Does not over pollute the environment—land, water, or atmosphere.

10.3.4 Technical Support of EcoCities

Technology forms the basis for an EcoCity. Here the aim is to introduce new crosscutting technologies to realize conservation and pollution reduction aims of the EcoCity. Interaction among these solutions is vital and, therefore, these interactions should automatically take into consideration from the beginning. In the future, the EcoCities will become more common and influence the whole society. Therefore, it is imperative to redefine the principles of planning and design including educational approaches in relation to the EcoCity idea.

When the principles of planning are redefined, this also logically leads to the redefinition of architecture and urban planning. The cities will need new types of transportation management and infrastructure; this redefines the whole city operation. Further, energy production, transportation, communication, recycling, and self-sustainability for food and water are all radically different from today's cities. The recycling of materials will become an essential part of the production process. This requires new products that are produced in new kinds of factories.

10.3.5 Key Indicators of EcoCity

Resources: No matter in the initial or operational stage—resource consumption is one of the main differences between conventional cities and EcoCities.

City Function Indicators: Environmental restoration to guarantee a sustainable life cycle; a function of balance; and nurturing life. For human beings, ecosystems have three functions: ensuring food, supporting daily life, and providing production materials.

Economic Indicator: An economic indicator includes economic vitality index, regional economic activity index, and industrial circle index.

Social Development Indicator: An integrated social system with well-balanced democracy, religion, morality, legal system, management, population structure, and so on.

10.3.6 Urban Ecosystem

An urban ecosystem is a complex ecosystem of society-economy-nature, which is human behavior orientated, natural ecosystems related, and ecological process driven. It contains the natural, economic, and social subsystems of the city. It is the fundamental mission of urban sustainable development to identify the coupling relationship in levels of time, space, process, structure, and function among the three subsystems and within themselves; to plan overall the internal and external, local and global, and recent and long-term conflicts of an urban complex ecosystem (Figure 10.7); and to promote the harmonious development of the city in efficiency, fairness, and vitality.

10.3.7 Practical Solutions

An EcoCity generates the energy needed with a minimum pollution of the air, water, or soil. For this purpose, a large part of the city area will be reserved for solar panels, wind generators, geothermal heat pumps, and bioenergy

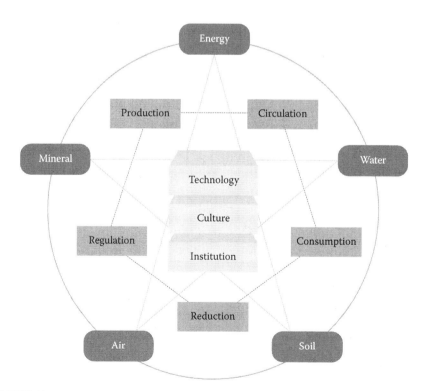

FIGURE 10.7
The urban complex ecosystem.

production. These areas are at a distance from the residential areas. Bioenergy production releases carbon dioxide into the atmosphere, but plants reabsorb the carbon from the air. Thus, a balance is maintained, and the carbon burden of the atmosphere does not increase.

An EcoCity also includes artificial pools for fish farming, and vegetables will be grown in greenhouses where ambient temperatures are too low for normal field cultivation. Artificial pools can be integrated with the biological water purification system and energy supplied to the greenhouse the same way as in other parts of the city.

In addition to food and energy, the third commodity needed daily is water. An EcoCity will have a closed water circulation system. This means that water imported from outside is minimized, and wastewater transported outside the city is also minimized. In an EcoCity, the water circulation system is controlled based on biological and natural purification, but it also utilizes physical and chemical purification methods. The idea of using two circulations beside each other—the drinkable and grey water—is also possible. Waste management and material recycling are seamlessly integrated with all other operations in an EcoCity.

However, there are two larger issues that are fundamentally reflected in land-use planning and zoning throughout the entire city: transportation and construction.

Data communications are of state-of-the-art quality. There are rentable facilities for information distribution all around the city, with large panels covering the walls. The panels can be used for virtual meetings with experts around the world, finding information on the Internet, contacting hospitals or universities, and trading and taking care of any number of daily matters. The facilities will be available for all inhabitants.

Construction and land-use planning are seamlessly linked with transportation. Buildings are constructed and maintained with an eye on various ecological principles; local materials are used wherever possible and energy is conserved at various stages of the process. Buildings are designed to be energy saving and long-lasting. These aspects are also taken into account in land use.

An EcoCity could house a small factory for the assembly and maintenance of the special vehicles used in the city. Production could later be expanded to supply similar vehicles to other cities. An EcoCity must also have a maintenance centre for energy production equipment. Thus, the production sector could support the basic function of the EcoCity and thus facilitate the design and construction of the next generation ecocities.

10.3.8 EcoCity and Geomatics

The fast development of IT is deeply changing our lifestyle, and it provides new technology for the sustainable development of EcoCity. Informationization can not only effectively decrease the consumption of resources and energy and lighten the burden on logistics and people flow but also improve the industrial structure, increase the urban operating efficiency, and reduce the environmental pollution. In addition, it can better implement the sustainable development strategy.

Geomatics is a technology and service sector focusing on the acquisition, storage, analysis, dissemination, and management of geographically referenced information for improved decision making. Geomatics technology has great advantages in environmental monitoring and management:

Environmental mapping: Compared with the traditional, long term, and low-speed manual updating mapping methods, using GIS technology to construct a geographic database can achieve the effect of multiple time production with one input. It can not only produce whole features on a topographic map for users but also provide various thematic maps according to the requirements of different users, such as pollution sources distribution map, air quality function division map, and others. The mapping method using GIS technology is more flexible than traditional mapping methods. Based on fundamental electronic maps, Geomatics can quickly produce high-quality environmental thematic maps by adding related thematic data.

Environment Monitoring: Storing and processing environmental monitoring data using Geomatics technology can directly visualize and analyze the environment's present situation, pollution sources distribution, environmental quality assessment, and track pollution sources. Combined with a digital map, it can query the historical monitoring data and various statistics data to carry out spatial analysis, assisted decision making for the scientific management, and decision making of the urban environment. By constructing the urban environmental spatial database and pollution sources monitoring attribute database, we can develop an urban environment GIS, to integrate the monitoring information of atmosphere, surface water, and acoustic environment and directly reflect the situation of urban environmental quality by distributing thematic maps and 3D models.

Environmental Alerting and Emergency Response: The construction of an alert system for urban major environmental pollution can realize the management of geographic locations and its attribute of accident risk sources and accident sensible areas, and it can provide air and river pollution diffusion simulation processes in pollution accidents. The system carries out the searching and location of pollution sources using GIS technology with activation command and navigation by the integration of GIS and GPS technologies. It solves the dynamic updating problem of GIS basic map by acquiring land information by using remote-sensing technology.

Environmental Assessment: Since GIS can integrate both the environmental data and the location, it is a powerful tool for comprehensive analysis and assessment. Environmental Impact Assessment carries out predictions and assessments for the possible environmental impacts of all extended reconstruction projects, and it puts forward the measures and countermeasures for preventing and decreasing impacts. Using the spatial analysis function of GIS, the various data of construction projects can be comprehensively analyzed. With regard to urban environmental quality assessment, it can integrate the data of atmosphere, soil, water, and noise and, objectively, evaluate the current situation of the whole urban environmental quality using spatial analysis function.

EcoCity Planning: At present, the general program and contents of EcoCity planning includes field investigation, database construction and analysis, ecological environmental assessment, plan and strategy design, objective prediction and index calculation, ecological function division, plan compilation, and plan implementation. The program of EcoCity planning should be standardized, systematic, and feasible.

According to the research theory of ecology, urbanology, and regional planning, which are the basis of existing EcoCity plans and practical requirements, the methods of EcoCity planning can be divided into analysis and assessment methods. These methods include the function division method, index system, and establishment method. Ecocity planning can be supported by image processing, remote sensing, GIS, mapping, and information sharing technology.

EcoCity planning platform includes method research platform, technology design platform, and plan compilation platform. The platforms represent the overall procedure of EcoCity planning and summarize the three phases of EcoCity planning (Figure 10.8).

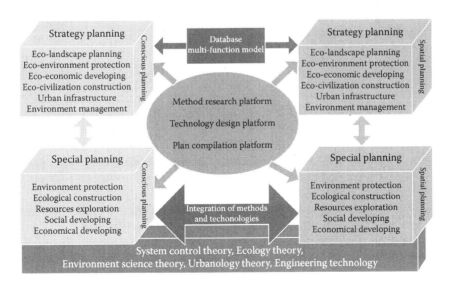

FIGURE 10.8
Framework model of EcoCity planning.

10.3.9 Future of the EcoCity

EcoCity is one of the answers to global challenges. The quicker it is adopted, the easier it will be for us to address these challenges. It is also essential to ensure that the EcoCity is not only in the interests of its own purity but also does not impact the environment of other neighborhoods, the country, or another continent. Finally, we must emphasize that the building of the EcoCities is only half of the whole story. The renovation of the traditional cities will need us to renew their infrastructure instead of tearing down buildings or changing the appearance of the cities.

10.4 Education and Advanced Geoinformation Science: Accomplishments and Opportunities*

10.4.1 Record of Innovation

The rapid pace of advancements in GIScience has been matched by a rapid innovation in GIScience education. In a span of less than two decades, GIScience education has moved from a few courses in a few departments to a major element of almost all geography and environmental studies programs, and it has become a growing presence in other disciplines as well. This expansion responds in part to the dramatic growth in demand for high-quality education and training as the GIScience industry has spread into new commercial markets, into more government agencies, and into NGOs (Gaudet et al., 2003; Phoenix, 2000). However, equally important in spurring innovation has been the diffusion of GIScience into disciplines across the social, natural, and engineering sciences.

This demand for both broadening and deepening GIScience curricula has presented opportunities for educators at all educational levels. The central challenge is that GIScience is changing so rapidly. Preparing effective courses and curricula is like aiming at a moving target and requiring, among teachers especially, a special commitment to stay abreast of constantly changing concepts, techniques, and tools. Despite these challenges, educators are responding in new and creative ways. In fact, a case could be made that over the past two decades, faculty in GIScience have spearheaded more than their share of improvements in higher education. Active pedagogy, problem-based learning, Web-based instructional materials, distance education, professional training and certification, and other innovations have all received a substantial push from GIScience educators (Carver et al., 2004; Clark et al., 2007; DiBiase, 1996). Time and again, GIScience educators have been quick to implement new

---

* This section is contributed by Kenneth E. Foote.

educational strategies (Benhart, 2000; Deadman et al., 2000; Giordano et al., 2007; Keller et al., 1996; Wentz and Trapido-Laurie, 2001; Zerger et al., 2002).

10.4.1.1 Creating Model Curricula

One of the greatest challenges faced in GIScience education was establishing its place in existing college and university curricula (Chen, 1998; Gilmartin and Cowen, 1991; Jenkins, 1991; Johnson, 1996; Lloyd, 2001; Nyerges and Chrisman, 1989; Painho et al., 2007; Poiker, 1985; Sui, 1995; Unwin, 1990, 1997; Unwin and Dale, 1990). This involved developing new courses, but for many it raised both theoretical and practical issues about how GIScience could fit into graduate and postgraduate curricula as well as the rigor of this training (Marble, 1998, 1999). This situation meant that GIScience educators tended to be open to new opportunities and strategies that would help them get started quickly and successfully, and efforts were always made to link major research initiatives such as the NCGIA and UCGIS to education.

Certainly one of the most important innovations was the development of prototype curriculum materials like the Core Curriculum in GIScience published by NCGIA in 1990 to define the knowledge base of the field (Goodchild and Kemp, 1992). These materials were instrumental in helping educators develop courses in GIScience all over the world (Coulson and Waters, 1991). Other projects like the GISAccess project, the iGETT project, and NCGIA's Core Curriculum in GIS for Technical Programs have aimed at helping community colleges' faculty get started in offering 2-year degrees and certificates in areas of rapid expansion in GIS (Allen et al., 2006).

The most recent effort in this direction was the publication of the Geographic Information Science and Technology Body of Knowledge (DiBiase et al., 2006). Although not written as a replacement for the earlier Core Curriculum, the Body of Knowledge substantially updates and expands the inventory of topics included and provides a scaffolding for planning, revision, and assessment of GIScience curricula, accreditation, professional certification, the articulation of curricula among two- and 4-year institutions, and for employee screening (DiBiase et al., 2006).

However, the Body of Knowledge does not necessarily address all at once all the issues surrounding the integration of GIScience into college and university curricula, in at least three areas recognized by the authors themselves (Dibiase et al., 2006). First, only a relatively few departments can address the full scope of the Body of Knowledge in its full depth and breadth. Its complete implementation would suggest more courses and more teaching staff than most programs can afford to invest in GIScence education. As a consequence, choices must be made about how which core concepts will be addressed in a given curricula. Though the Body of Knowledge alludes to the need to conceptualize "multiple pathways to diverse outcomes," clear suggestions are still needed and are one of the next steps intended by the editors. Second, the issue of "adaptability to varied institutions" remains an issue even today.

The dramatically different educational missions of institutions of higher education in the United States and abroad mean that justifications for GIScience in the curriculum vary greatly. In small, private liberal arts BA programs, GIScience may be stressed as a means of cultivating critical thinking and reasoning about environmental issues (Sinton and Lund, 2007). In 2-year colleges and masters programs, the employability of GIScience graduates may be the key rationale. In research universities, far different rationales are needed.

Third, more attention is needed as to how GIScience can be integrated into curricula outside geography and the environmental sciences. Sinton and Lund (2007) highlight a range of the many examples of how GIScience has been used to improve teaching in the social and natural sciences. However, more needs to be done to help educators in these disciplines get started with such innovations and to aid pedagogic transfer across disciplinary boundaries. The Center for Spatially Integrated Social Science is one of the type of initiative that may help in these efforts, but many other disciplines would benefit from parallel efforts.

10.4.1.2 *Expanding e-Learning Opportunities in Online and Distance Education*

The GIS educators have been in the forefront of efforts to put courses, curricula, and certificate programs online (Breetzke, 2007; Harris, 2003; Onsrud, 2005; Rees et al., 2009; Wright and DiBiase, 2005). The intention was to expand the potential audience for such educational opportunities by making it easier to enroll in high-quality classes, but educators also recognized that the multimedia features of the Web could be effectively used to enhance the learning experience. The visual and interactive natures of some of the elements of GIScience are particularly well suited to learning in a hypermedia environment. Though there has been speculation that online programs will supplant traditional classroom and laboratory instruction, the transition has been gradual. More work than was expected needed to be invested in finding effective models for online instruction and in reaching appropriate audiences. Early experiments in online instruction offered little more than online text and graphics. The more advanced models are now usually "asynchronous" (or self-paced) and use "blended," or a mix of instructional technologies including online text, discussion boards, blogs, chat rooms, help desks, virtual seminars and tutorials, and even Facebook and Second Life sites to promote interactions between teachers and learners.

The appeal of these online offerings is evident in the success they have had in drawing students. The ESRI in particular, but other vendors like Intergraph as well, have seen enrollment in their online seminars and courses increase rapidly (Johnson and Boyd, 2005). These online opportunities are particularly appealing to professionals and adult learners who take classes when they need them, at a pace that fits their schedules, without traveling to a conventional classroom. However, these courses are also appealing to students in traditional academic programs who may be able to choose a greater range of courses than is available at their own institution or prefer the more flexible, asynchronous schedule of online courses. The UniGIS program, an international collaboration of universities offering an MS in GIScience, has been in operation for almost two decades. Penn State and the University of Denver's online certificates in GIS have also been very successful and are two of many available, nationally and internationally.

The rapid evolution of e-learning suggests that other innovations may be just around the corner. Some suggest that the trend is toward open, flexible paths to learning and teaching, which is a greater range of blended educational resources, available from both non-profit and commercial educational institutions and business and suitable for learners of a wide range of career stages. Despite promoting interoperability in GIScience courses, curricula, and learning and teaching materials around the world, few programs involve effective collaborations. However, the rise of collaborations, such as the Worldwide Universities Network (WUN), means that frameworks may be emerging for new innovations.

10.4.1.3 Establishing Certificate Programs

Both the American Society for Photogrammetric Engineering and Remote Sensing (ASPRS) and the Geographic Information Systems Certification Institute (GISCI) now offer successful certification programs for GIScience professionals. However, academic certificate programs are also growing rapidly in both undergraduate and graduate curricula (USGIS, 2008). As of spring 2009, ESRI's (2009) online database lists 316 such programs internationally. However, the exact meaning of such certification is sometimes unclear (Obermeyer, 1993). Wikle (1999) notes that these certificate programs are "different from degree programs mostly in terms of their focus and duration. In contrast to degree programs that include general education courses, certificates are narrowly focused and require less time to complete." However, in some cases, certificate requirements differ little from what majors or minors would earn in a traditional degree program by concentrating some of their electives in GIScience. Yet such certificates can affirm or emphasize a person's proficiency and depth of training and can be helpful as students enter the workforce or advance their careers. Still, more discussion at the national or international levels is needed to reach an agreement on what a certificate in GIScience should include. Such certification may prove of more value in some areas of GIScience such as surveying, land-record and cadastral mapping, and photogrammetry than in others.

10.4.1.4 Promoting Effective Pedagogy

The GIScience educators have also been at the forefront of classroom innovation in other areas. Perhaps the most notable is their embrace of active-learning strategies. Active pedagogy, based on constructivist learning theory, seeks to shift the focus of the learning experience from teacher to student so as to engage students as active and not passive participants in the learning process. The term *active pedagogy* actually serves as an umbrella for a range of related techniques including problem-based learning, inquiry learning, discovery learning, and experiential learning, all of which have been used in GIScience education (Carlson, 2007; Drennon, 2005, Lo et al., 2002; Summerby-Murray, 2001).

The development of open, Web-based resources is also an area in which GIScience educators have taken a lead. This started with projects like The Geographer's Craft in the early 1990s (Foote, 1997) and has continued with a long list of other notable projects including online versions of the core curricula in GIScience and remote sensing; the reference materials created for the U.S. and Canadian national atlases, DiBiase's (2009) online Nature of Geographic Information open source textbook at Penn State, as well as a large number of other high-quality wiki entries, reference sites, and educational materials. This body of online, open materials means that, in GIScience, it is possible to teach many introductory and intermediate courses entirely with online materials. Finally, GIScience educators have taken the lead in exploring, at least tentatively, the use of virtual worlds and other new Internet and virtual reality techniques (Hudson-Smith and Crooks, 2008).

10.4.2 Future Opportunities and Challenges

The record of innovation in GIScience education is very strong, but a number of opportunities and challenges remain in the near horizon.

10.4.2.1 Addressing Spatial Learning and Literacy

The U.S. National Research Council report *Learning to Think Spatially* (2006) raised the important issue of how GIScience can also involving in supporting not just environmental and geographical reasoning but also spatial learning and literacy. Until recently, spatial learning has not received as much attention as verbal or mathematical reasoning. Spatial thinking is not confined to geography or environmental sciences but is instead a fundamental way in which humans understand and think about many different processes and patterns in many disciplines. However, as the NRC (2006, ix) report notes:

> "Spatial thinking—one form of thinking—is based on a constructive amalgam of three elements: concepts of space, tools of representation, and processes of reasoning. It is the concept of space that makes spatial thinking a distinctive form of thinking. By understanding the meaning of space, we can use its properties as a vehicle for structuring problems, for finding answers, and for expressing solutions. By expressing relationships within spatial structures, we can perceive, remember, and analyze the static and, via transformations, the dynamic properties of

objects and the relationships between objects. We can use representations in a variety of modes and media to describe, explain, and communicate about the structure, operation, and function of objects and their relationships. Spatial thinking is not restricted to any domain of knowledge, although it may be more characteristic of architecture, medicine, physics, and biology, for example, than of philosophy, business administration, linguistics, and comparative literature."

GIScientists are just beginning to explore the implications of this report. It implies both a range of new research questions about how GISystems can support the development of spatial thinking as well as a range of questions about how spatial thinking is involved in geographical and environmental analysis (Sinton and Schultz, 2009; Spatial Literacy in Teaching, 2008; West, 2003).

10.4.2.2 Expanding International Education

Despite efforts to internationalize GIScience education, programs still remain concentrated in relatively few countries, mostly in the developed nations in Europe, North America, and some areas of Australasia, and English tends to be the dominant language of GIScience. However, in surveying the many global issues and needs relating to spatial technologies, expanding international cooperation in GIScience education seems an important issue (Kemp and Frank, 1996; Phoenix, 2004). Previous efforts like the UniGIS program have certainly had some success, but barriers to greater sharing remain high. Differences in the ways college and post-graduate education is organized in different nations present one of the greatest challenges to sharing educational know-how, but wide variations in access to software, hardware, and staff is also an obstacle, as different standards for professional certification.

10.4.2.3 Exploring the Capabilities of Virtual Globes, Mashups, and Neogeography

The rise of virtual globes like Google Earth and NASA's World Wind are suggesting new methods for bringing some elements of GIScience into a broader range of disciplines and courses. Although the rapid pace of mapserver technologies has been generating excitement since the mid-1990s, the open API of systems like Google Earth and Google Maps make it far easier for users to combine data from many sources and customize their maps than have earlier systems. This has opened a world of "mashups" in which users can quickly and flexibly overlay data on existing maps to create new applications and tools. The major innovation of these systems is that they provide excellent visualization tools to a far larger public than ever before. These mashups do not offer all of the capabilities of many GIS, but they have instead helped spur the rise of a neogeography movement—the use of geographic and spatial data by non-expert users, the rise of user-generated spatial content, and efforts to use "crowd sourced" information effectively. All of these developments suggest new areas which GIScience education can move so that mashups and virtual globes can support spatial geography both inside and outside.

10.4.2.4 Getting Started as a GIScience Educator

Getting started in GIScience education remains a hurdle, because it often requires a greater commitment than teaching in other fields. Getting started involves mastering the literature, developing expertise with software and hardware, preparing and constantly revising laboratory materials, building and maintaining a lab, being available regularly to help students, and many other unsung tasks. Since GIS curricula are often covered by only one or two faculty members in a given department, there is often little guidance available on learning and teaching materials and curriculum plans. Although existing Web-based materials and the Body of Knowledge can help new instructors articulate some learning goals, far more can be done to help educators get started teaching in GIScience, for example, by providing a range of curriculum models that can be readily adapted to different institutional settings, an easier means of sharing and customizing learning materials, or even the use of cloud computing to eliminate the need to create special purpose GIS computer labs.

10.4.2.5 Addressing Ethics

Considerable recent interest has focused on issues of ethics in GIScience education, as highlighted in a recent NSF grant on Ethics Education for Future Geospatial Technology Professionals led by Wright, DiBiase, and Harvey (2009). The GIS technologies are raising a number of important ethical issues related to their growing power and prevalence. These include issues such as privacy when GIS are used for surveillance or when data collated by location in GIS are used to create profile or propensity files. The widespread use of GIS in decision making also means that data may be misused or misrepresented or that the inappropriate data or erroneous data may lead to harm. Equally important is how GIS technologies, due to their expense and complexity, can limit access to important data to the detriment of particular nations, organizations, or individuals who lack the resources to acquire them. It is likely that these issues will gradually enter more GIS curricula in coming years.

10.4.3 Conclusion

As educators address these many opportunities and challenges, it is likely that GIScience will remain at the forefront of innovation in higher education. It is likely to continue to help set the pace for developing new programs, deploying new technologies, serving new and different audiences, and raising important theoretical and practical issues about learning and teaching in the twenty-first century. Although the focus in this section has been on higher education and on programs for adult and professional learners, almost as much effort is being invested in introducing GIScience into K-12 education as a means of promoting spatial thinking, improving geographical problem-solving skills, and introducing techniques of environmental analysis and reasoning to a new generation of students (Audet and Ludwig, 2000).

References

Allen, J., O. Brand, R. Beck, and A. Johnson. 2006. Integrating geographic information systems and remote sensing for technical workforce training at two-year colleges. http://gistech.delmar.edu/geospatial_ws1.htm (accessed April 16, 2009).

Audet, R.H. and G.S. Ludwig. 2000. *GIS in schools*. ESRI Press, Redlands, CA.

Beaujardiere, J.D.L. 2004. Web Map Service Implementation Specifications, Version 1.3. OGC Document Number: 04-024. Open Geospatial Consortium, U.S.: 85.

Benhart, J. Jr. 2000. An approach to leaching applied GIS: Implementation for local organizations. *Journal of Geography*, 99(6):245–252.

Breetzke, G. 2007. A critique of distance learning as an educational tool for GIS in South Africa. *Journal of Geography in Higher Education*, 31(1):197–209.

Carlson, T. 2007. A field-based learning experience for introductory level GIS students. *Journal of Geography*, 106(5):193–198.

Carver, S., A. Evans, and R. Kingston. 2004. Developing and testing an online tool for teaching GIS concepts applied to spatial decision-making. *Journal of Geography in Higher Education*, 28(3):425–438.

Chen, X.M. 1998. Integrating GIS education with training: A project-oriented approach. *Journal of Geography*, 97(6):261–268.

Clark, A.M., J. Monk, and S.R. Yool. 2007. GIS pedagogy, web-based learning and student achievement. *Journal of Geography in Higher Education*, 31(2):225–239.

Coulson, M.R.C. and N.M. Waters. 1991. Teaching the NCGIA curriculum in practice: Assessment and evaluation. *Cartographica*, 28(3):94–102.

Craglia, M., M.F. Goodchild, A. Annoni, G. Camara, M. Gould, W. Kuhn, D. Mark, D. Maguire, S. Liang, and E. Parsons. 2008. Next-generation digital earth. *International Journal of Spatial Data Infrastructures Research*, 3:146–167.

Deadman, P., G.B. Hall, T. Bain, L. Elliot, and D. Dudycha. 2000. Interactive GIS instruction using a multimedia classroom. *Journal of Geography in Higher Education*, 24(3):365–380.

DiBiase, D. 1996. Rethinking laboratory education for an introductory course on geographic information. *Cartographica*, 33(4):61–72.

DiBiase, D. 2009. *Nature of Geographic Information: An Open Geospatial Textbook*. http://natureofgeoinfo.org (accessed April 16, 2009).

DiBiase, D., M. DeMers, A. Johnson, K. Kemp, A. Luck, B. Plewe, and E. Wentz. 2006. *The Geographic Information Science and Technology Body of Knowledge*. Association of American Geographers and University Consortium for Geographic Information Science, Washington, DC.

Dikaiakos, M.D., G. Pallis, D. Katsaros, P. Mehra, and A. Vakali. 2009. Cloud computing: distributed Internet computing for IT and scientific research. *IEEE Internet Comput* 13(5):10–13.

Drennon, C. 2005. Teaching geographic information systems in a problem-based learning environment. *Journal of Geography in Higher Education*, 29(3):385–402.

Duchossois, G. 2006. The global Earth observation system of systems(GEOSS): An example of international cooperation. *Second ESA/ASI Workshop on International Cooperation for Sustainable Space Exploration Sarteano*, Siena, May 9–12, 2006.

Environmental Systems Research Institute (ESRI), 2009. *ESRI's online database of academic GIS programs.* http://edcommunity.esri.com/universityprograms/results.cfm (accessed April 16, 2009).

Evans, J.D. 2004. *Digital Earth Reference Model.* http://www.cartome.org/draft-derm.htm (accessed December 1, 2009).

Foote, K.E. 1997. The geographer's craft: Teaching GIS in the web. *Transactions in Geographic Information Systems*, 2:137–150.

Foresman, T.W. 2008. Evolution and implementation of the Digital Earth vision, technology and society. *International Journal of Digital Earth*, 1(1):4–16.

Foresman, T.W., H. Guo, and F. Hiromichi. 2004. Progress with the Digital Earth global infrastructure. In *Proceedings of the Seventh Conference on Global Spatial Data Infrastructure*, Bangalore, India.

Fuller, R.B. 1968. *Operating Manual for Spaceship Earth.* Fiffer & Simons, London.

Gaudet, C., H. Annulis, and J. Carr. 2003. Building the geospatial workforce. *Journal of the Urban and Regional Information Systems Association*, 15(1):21–30.

GEO. 2005. The global Earth observation system of systems (GEOSS) 10-Year Implementation Plan. http://www.earthobservations.org/documents/10-Year%20Implementation%20Plan.pdf (accessed December 16, 2009).

GEO. 2009. The GEOSS Common Infrastructure. http://www.earthobservations.org/gci_gci.shtml (accessed December 16, 2009).

Gilmartin, P. and D. Cowen. 1991. Educational essentials for today's and tomorrow's jobs in cartography and geographic information systems. *Cartography and Geographic Information Systems*, 18(4):262–267.

Giordano, A., Y. Lu, S. Anderson, and M. Fonstad. 2007. Wireless mapping, GIS, and learning about the digital divide: A classroom experience. *Journal of Geography*, 106(6):285–295.

Goldstine, H.H. 1972. *The Computer: From Pascal to von Neumann.* Princeton University Press, Princeton, NJ.

Goodchild, M.F. and K.K. Kemp. 1992. NCGIA education activities: The core curriculum and beyond. *International Journal of Geographical Information Systems*, 6(4):309–320.

Gore, A. 1999. The Digital Earth: Understanding our planet in the 21st Century. *PE&RS*, 65(5):528.

Grossner, K. and K. Clarke. 2007. Is Google Earth, "Digital Earth?": Defining a vision. In *Proceedings of the Fifth International Symposium on Digital Earth*, Berkeley, California.

Grossner, K., M.F. Goodchild, and K. Clarke. 2008. Defining a digital earth system. *Transactions in GIS*, 12(1):145–160.

Guo, H. and C. Wang. 2004. Digital Earth: Bridging the scales from global to local for sustainable development. In *Proceedings of Millennium Ecosystem Assessment*, Alexandria, Egypt.

Harris, R. 2003. Building a GIScience community in cyberspace: Reflections on GIScOnline. *Journal of Geography in Higher Education*, 27(3):279–295.

Hudson-Smith A. and A. Crooks. 2008. The renaissance of geographic information: Neogeography, gaming and second life. Centre for Advanced Spatial Analysis,

University College London, London. http://www.casa.ucl.ac.uk/working_papers/paper142.pdf (accessed April 16, 2009).

Integrated Earth Observation System (IEOS). 2006. Air quality assessment and forecast system: Near-term opportunity plan. A Report of the National Science and Technology Council September 2006.

Jenkins A. 1991. Through a model darkly: An educational postscript. *Cartographica*, 28(3):103–108.

Johnson, A.B. and J.M. Boyd. 2005. Content, community, and collaboration at ESRI virtual campus: A GIS company's perspective on creating an online learning resource. *Journal of Geography in Higher Education*, 29(1):115–121.

Johnson, M.L. 1996. GIS in business: Issues to consider in curriculum decision-making. *Journal of Geography*, 95(3):98–105.

Keller, C.P., T.J. Davis, and R.R. Canessa. 1996. A hypertext tutor for teaching principles and techniques of GIS. *Journal of Geography in Higher Education*, 20(2): 193–207.

Kemp, K.K. and A.U. Frank. 1996. Toward consensus on a European GIS curriculum: The International post-graduate course on GIS. *International Journal of Geographical Information Systems*, 10(4):477–497.

Li, Q., C. Yang, and S. Yi. 1999. Architecture of Digital Earth. *Journal of Remote Sensing*, 3(4):281–287 (in Chinese).

Li, W. and C. Yang. 2010. An active crawler to discover geospatial web services and their distribution pattern. *International Journal of Geographic Information Science*, 24(8):1127–1147.

Li, W., C. Yang, and R. Raskin. 2008. A semantic enhanced search for spatial web portals. In *Technical report of Semantic Scientific Knowledge Integration of AAAI Spring Symposium*, pp. 47–50. AAAI Press, Menlo Park, CA.

Lloyd, W.J. 2001. Integrating GIS into the undergraduate learning environment. *Journal of Geography*, 100(5):158–163.

Lo. C.P., J.M. Affolter, and T.C. Reeves. 2002. Building environmental literacy through participation in GIS and multimedia assisted field research. *Journal of Geography*, 101(1):10–19.

Lu, Y. 1999. Building up the Digital Earth together, sharing global data resources each other. In *Proceedings of the International Symposium on Digital Earth*. Science Press, Beijing.

Marble, D.F. 1998. Rebuilding the top of the pyramid. *ArcNews*, 20(1):28–29.

Marble, D.F. 1999. Developing a model, multipath curriculum for GIScience. *ArcNews*, 21(2):31.

Milojicic, D. 2008. Cloud computing: Interview with russ daniels and franco travostino. *IEEE Internet Computing*, (5):7–9.

Mollenhoff, C.R. 1988. *Atanasoff: Forgotten Father of the Computer*, pp. 47–48. Iowa State University Press, Ames.

NASA. 2007. Earth system science data resources: Tapping into a wealth of data, information, and services. National Aeronautics and Space Administration, 65.

National Drought Policy Commission. 2000. Preparing for drought in the 21st century, Washington, D.C. http://drought.unl.edu/pubs/pfd21main.htm (accessed February 19, 2010).

National Research Council. 1997. Environmental Epidemiology, Volume 2—Use of the Grey Literature and Other Data in Environmental Epidemiology: Committee on Environmental Epidemiology and The Commission on life Sciences, p. 189. National Academy Press, Washington, DC.

National Research Council. 2006. *Learning to Think Spatially: GIS as a Support System in the K-12 Curriculum*. The National Academies Press, Washington, DC.

Nyerges, T. and N.R. Chrisman. 1989. A framework for model curricula development in cartography and geographic information systems. *Professional Geographer,* 41(3):283–293.

Obermeyer, N.O. 1993. Certifying GIS professionals: Challenges and alternatives. *Journal of the Urban and Regional Information Systems Association,* 5(1):67–75.

Onsrud, H.J. 2005. Web-casting of geographic information science graduate courses. *Journal of Geography in Higher Education,* 29(1):123–137.

Painho, M., P. Curvelo, and I. Jovani. 2007. An ontological-based approach to geographic information science curricula design. In *The European Information Society: Leading the Way with Geo-Information*, eds S.I. Fabrikant and M. Wachowicz, pp. 15–34. Springer, Berlin.

Phoenix, M. 2000. Geography and the demand for GIS education. *Association of American Geographers Newsletter*, June:13.

Phoenix, M. 2004. Global GIS Education. *GIM International*, February:35–37.

Poiker, T.K. 1985. Geographic information systems in the geographic curriculum. *The Operational Geographer,* 8:38–41.

Rae-dupree, J. 2006. Dash navigation—Real-time GPS driving. *Business Journal,* http://sanjose.bizjournals.com/sanjose/stories/2006/11/20/focus15.html (accessed January 23, 2008).

Rauschert, I., P. Agrawal, R. Sharma, S. Fuhrmann, I. Brewer, and A. MacEachren. 2002. User interfaces: Designing a human-centered, multimodal GIS interface to support emergency management. In *Proceedings of the 10th ACM international symposium on Advances in geographic information systems GIS '02*, pp. 119–124. VA, USA.

Rees, P., L. MacKay, D. Martin, and H. Durham. 2009. *E-Learning for Geographers: Online Materials, Resources, and Repositories*. IGI Information Science Reference, Hershey, PA.

Schultz, R.B., J.J. Kerski, and T.C. Patterson. 2008. The use of virtual globes as a spatial teaching tool with suggestions for metadata standards. *Journal of Geography,* 107(1):27–34.

Shin'ichi, S. and Y. Ryuzo. 2003. Introduction to digital Asia network, *Journal of Remote Sensing Society of Japan,* 23(3):257–260.

Sinton, D.S. and J.J. Lund. 2007. *Understanding Place: GIS and Mapping across the Curriculum*. ESRI Press, Redlands, CA.

Sinton, D.S. and R.B. Schultz. 2009. GIS and mapping technologies: Applications for reasoning and critical thinking. In *Teaching college geography*, eds M. Solem and K. Foote, pp. 69–80. Prentice-Hall, Upper Saddle, NJ.

Spatial Literacy in Teaching (SPLINT). 2008. http://www.le.ac.uk/geography/splint/ (accessed April 16, 2009).

Stevens, D., S. Dragicevic, and K. Rothley. 2007. iCity: A GIS-CA modeling tool for urban planning and decision making. *Environmental Modeling & Software,* 22(6):761–773.

Sui, D. Z. 1995. A pedagogic framework to link GIS to the intellectual core of geography. *Journal of Geography,* 94(6):578–591.

Summerby, M. R. 2001. Analysing heritage landscapes with historical GIS: Contributions from problem-based inquiry and constructivist pedagogy. *Journal of Geography in Higher Education,* 25(1):37–52.

University Consortium for GIS (UCGIS). 2008. GIS certificate and masters programs. http://www.ucgis.org/priorities/education/GIS_Cert+Masters_Prog/certificates.htm (accessed April 16, 2009).

Unwin, D.J. 1990. A Syllabus for teaching geographical information systems. *International Journal of Geographical Information Systems* 4(4):457–465.

Unwin, D.J. 1997. Curriculum design for GIS. *NCGIA Core Curriculum in GIScience*. http://www.ncgia.ucsb.edu/giscc/units/u159/u159.html (accessed April 16, 2009).

Unwin, D.J. and P. Dale. 1990. An educationalist's view of GIS: Some educational and sociological concerns. *Journal of Geography in Higher Education*, 14(2):166–169.

Wentz, E.A. and Trapido, L. B. 2001. Structured internships in geographic information science education. *Journal of Geography*, 100(4):140–144.

West, B.A. 2003. Student attitudes and the impact of GIS on thinking skills and motivation. *Journal of Geography*, 102(6):267–274.

Western Governors Association (WGA). 2004. *Creating a Drought Early Warning System for the 21st Century: The National Integrated Drought Information System*. Western Governors Association, Denver, CO.

Wikle, T.A. 1999. GIS education through certification programs. *Journal of the Urban and Regional Information Systems Association*, 11(2):53–60.

Wright, D. and D. DiBiase. 2005. Distance education in geographic information science: Arena symposium and an informal survey. *Journal of Geography in Higher Education*, 29(1):91–100.

Wright, D., D. DiBiase, and F. Harvey. 2009. Ethics education for future geospatial technology professionals. http://gisprofessionalethics.org (accessed April 16, 2009).

Xu, G. 1999. Building the "Digital Earth," promoting China's and global sustainable development. In *Proceedings of the International Symposium on Digital Earth*, Science Press, Beijing, China.

Yang, P., Evans J., M. Cole, N. Alameh, S. Marley, and M. Bambacus. 2007. The Emerging Concepts and Applications of the Spatial Web Portal, *PE&RS*, 73(6):691–698.

Yang, C. and R. Raskin. 2009. Introduction to Distributed geographic information processing research, *International Journal of Geographic Information Science*, 23(5):553–560.

Yang, C., R. Raskin, M. Goodchild, and M. Gahegan. 2010. Geospatial Cybinfrastructure: History, status, and future. *Computers, Environment, and Urban Systems*, 34(4):264–277.

Zerger, A., I.D. Bishop, F. Escobar, and G.J. Hunter. 2002. A self-learning multimedia approach for enriching GIS education. *Journal of Geography in Higher Education*, 26(1):67–80.

Index